The Concepts of Science

From Newton to Einstein

The Concepts of Science

From Newton to Einstein

Lloyd Motz
and
Jefferson Hane Weaver

Plenum Press • New York and London

Library of Congress Cataloging in Publication Data

Motz, Lloyd, 1910–
The concepts of science from Newton to Einstein / Lloyd Motz and Jefferson Hane Weaver.
p. cm.
Includes bibliographical references and index.
ISBN 0-306-42872-5
1. Science. I. Weaver, Jefferson Hane. II. Title.
Q125.M59 1988
500–dc 19 87-38493
CIP

Plenum Press is a Division of
Plenum Publishing Corporation
233 Spring Street, New York, N.Y. 10013

Printed in the United States of America

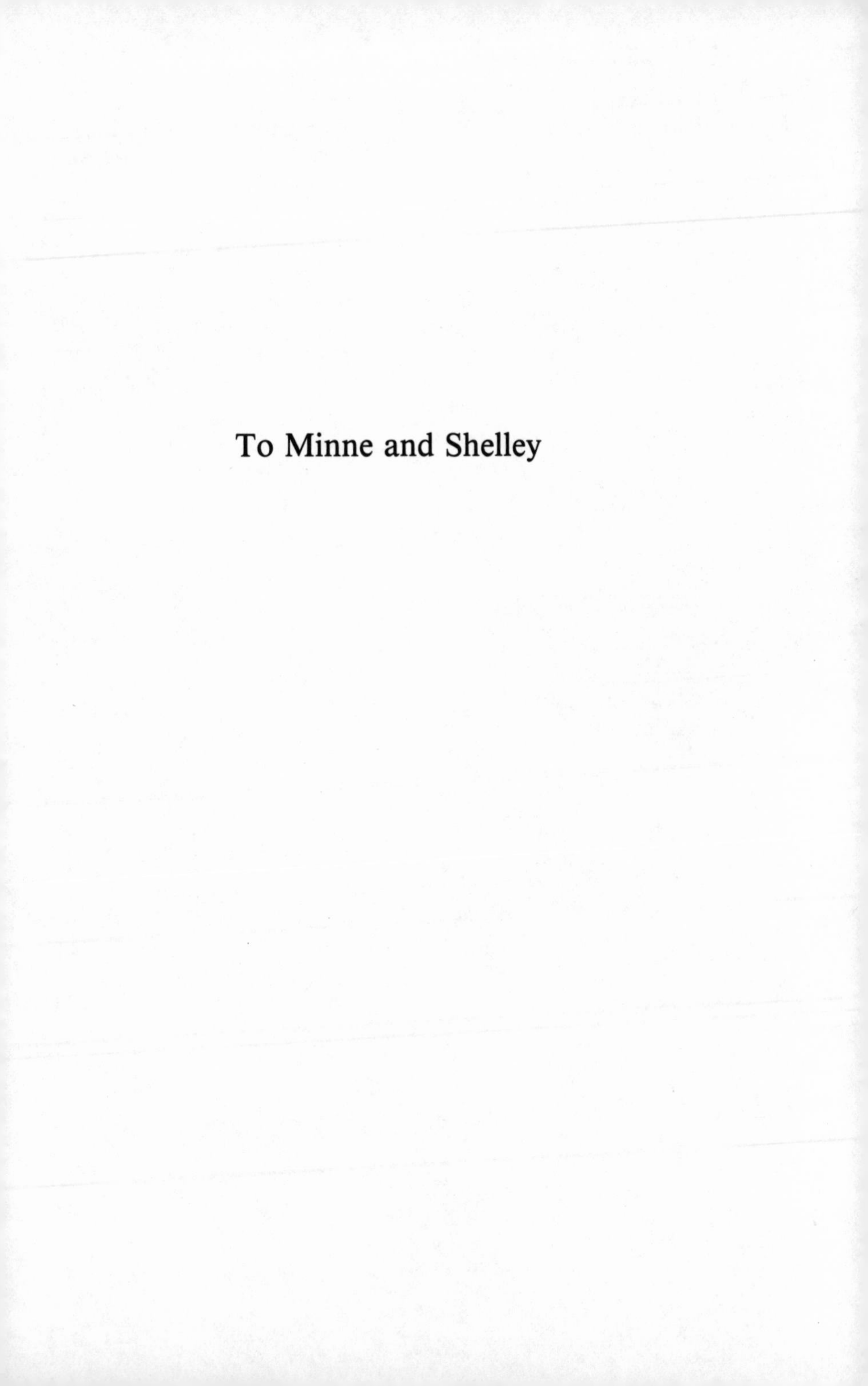

To Minne and Shelley

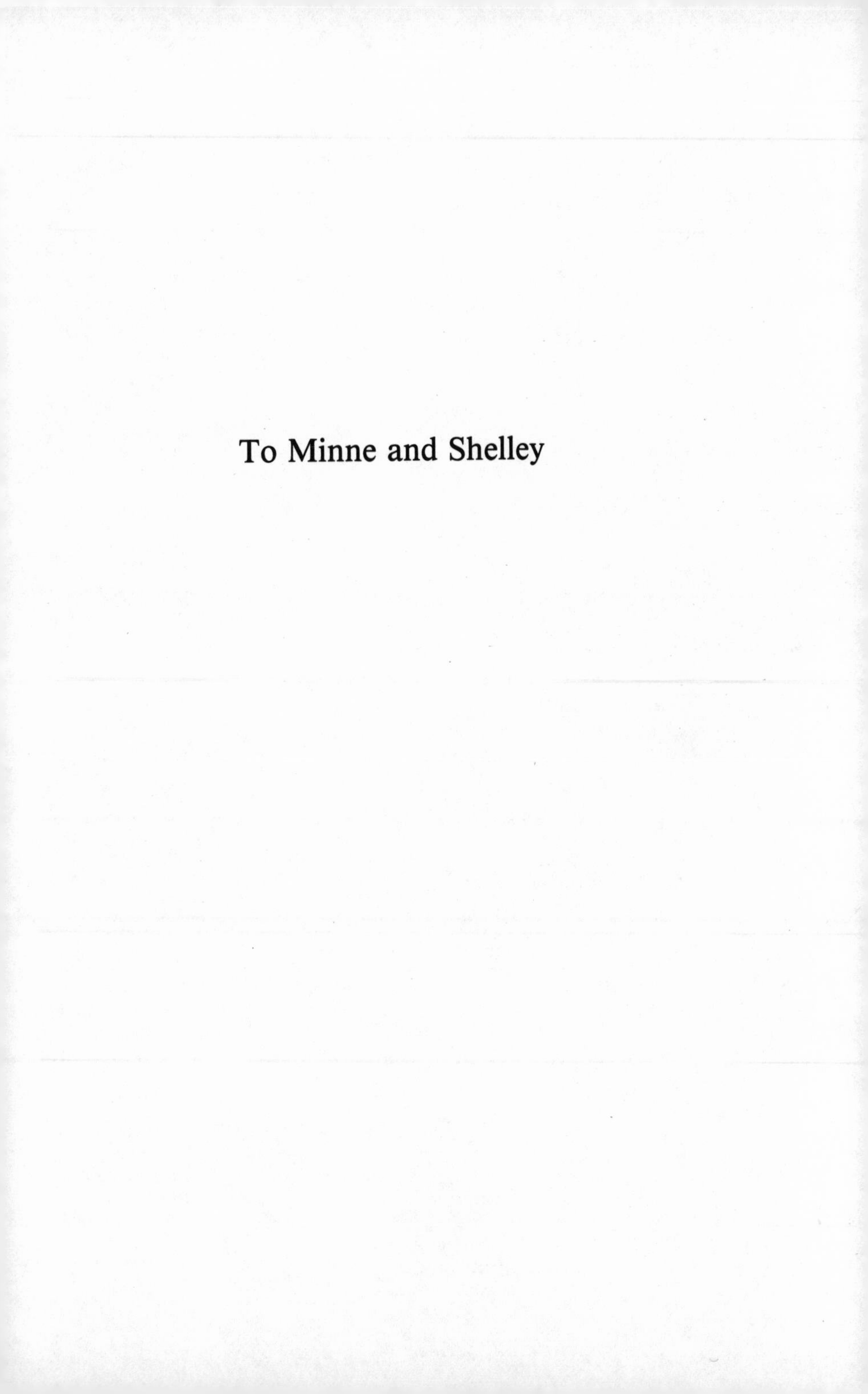

To Minne and Shelley

Preface

To the nonscientist, science, particularly in its modern dress and as pursued today, is a glittering intellectual jewel, mysterious, forbidding, and even threatening except to a few chosen ones, the scientists, who appear to be superior beings, endowed with an ability to probe and understand nature far beyond that of the average layman. This perception has led to a deplorable situation in which people who want to make science as much a part of their intellectual background as music, art, literature, and the social sciences are repelled by the technical and often obscure language in which it is presented; the word of the scientist (about his science, of course) is therefore accepted uncritically and without question even though it often raises many questions.

We have written this book to show that this state of affairs need not be so and that any intelligent readers who are willing to devote some time and thought to basic scientific concepts, as presented in this book, can acquire a deep enough understanding of scientific principles to respond to science as they do to any other of their interesting and stimulating pursuits. By emphasizing concepts we show that nature performs all its magic by working with a few forces and a few laws in a way almost anyone can understand. Thus, all Newtonian wisdom is contained in Isaac Newton's laws of motion and his law of gravity, which are simple statements about how a force acting on a body determines its dynamical history and how the force of gravity depends on the masses of two gravitationally interacting bodies and on the distance between them. We point out that these two statements alone permit us to predict the orbits of the planets around the sun, and, in fact, the motions in gravitational fields of any bodies such as man-made space probes.

By introducing such concepts as momentum, energy, and angular momentum (rotational motion), Newtonian physics can be cast into a form which is most easily understood in terms of the three basic conservation principles: conservation of momentum, conservation of energy, and conservation of angular momentum. But these principles extend far beyond Newtonian physics to pervade all of science and govern all aspects of the universe.

By developing the concepts of the electric and magnetic charges and their interactions, we give the reader an insight into electromagnetic phenomena and, in particular, into the properties of the electromagnetic field as revealed by the experimental work of Michael Faraday and the brilliant theoretical work of James Clerk Maxwell. This survey completes our treatment of what is generally called "classical physics." Since modern physics begins where classical physics ends, we show why and where classical physics is deficient and describe, in some detail, the basic concepts on which rest the two great modern theories, the theory of relativity and the quantum theory, that guide us in our study of the universe. We have not gone beyond relativity and quantum theory in this book, however, because we believe that one who understands these theories as presented here can grasp the essential features of current scientific discoveries without being burdened by many inessential details that the professional scientist must have at his fingertips to pursue his work.

Lloyd Motz
Jefferson Hane Weaver

Contents

CHAPTER 1

Structures and Forces

Here and elsewhere we shall not obtain the best insight into things until we actually see them growing from the beginning.

—ARISTOTLE, *Politics*

When, as thinking creatures, we begin to explore the world around us, we are probably most impressed by the great and wonderful variety of objects it contains, and our first response to this variety is to arrange these objects into classes, each of which consists of all objects that appear to have similar characteristics. Thus, we speak of stars, of clouds, of trees, of stones, of living creatures, and so on, without at first refining our classifications to take into account the differences that exist among members of each class.

We may say that science began when humanity became more observant and first subdivided the individual members in any one class into subclasses, based on certain differences that exist among the individuals of the class. The science of astronomy began when the ancient Greeks (and probably the Hebrews, the Chinese, the Egyptians, and others) noted that some of the objects they had first called stars did not remain in the same relative configurations (the constellations), but appeared to move and alter their positions among the other stars. The Greeks called these moving celestial objects ''planets'' (the Greek word for wanderers) to differentiate between them and the ''fixed'' stars, and thus introduced the important distinction between the objects that constitute our solar system and the rest of the sidereal universe.

The distinction between the solar system and the rest of the universe remained a purely phenomenological one until Nicolaus Copernicus introduced the revolutionary concept that the planets, the sun, the moon, and the earth are all part of a single structure (the solar system) which somehow retains its own configurational existence as a separate entity among the stars. This concept of a structure consisting of moving parts was extremely useful because it led to the idea that the solar system can be studied as a structure quite independently of the stars. The remarkable consequences of this idea are represented in Johannes Kepler's discovery of the laws of planetary motion which show that the solar system as a dynamical structure can be described by precise mathematical formulas. Although this in itself was an enormous achievement, it did not tell the most important part of the story of the solar system because Kepler's laws of planetary motion are still only phenomenological and descriptive. These laws tell us only that the planets move in certain regular orbits that are related to their distances from the sun, but these laws do not give us an insight into the dynamics of the solar system or why such a structure should exist at all. The important point was not clarified until an entirely new concept—that of a central force—was introduced by Isaac Newton. Kepler himself had surmised that the orderly motions of the planets around the sun are controlled by some kind of influence exerted by the sun on the planets, but physics had not yet presented him with the conceptual tools to solve, or even to formulate, this problem. Newton, however, was more fortunate because Galileo Galilei had already studied the phenomena of freely falling bodies and the manner in which the motion of a body depends on how it is pushed or pulled. Galileo's discoveries so greatly interested Newton that he expressed them in a set of precise laws—the three laws of motion (not to be confused with Kepler's laws of planetary motion)—which show exactly how the force acting on a body determines the motion of that body. In formulating these laws of motion, Newton established the science of mechanics and laid the groundwork for what is now called "classical physics," an essential feature of which is the concept of force. Having given a precise meaning to this concept, Newton did not find it too difficult to see that each planet is moving in accordance with Kepler's laws because the sun is exerting a force on each planet.

The introduction of such a force (which Newton called the force of gravity) to explain the motions of the planets required a great deal of imagination and boldness since, until that time, the concept of a force had been associated with the idea of contact between bodies—people exerted forces on bodies by putting their hands (or some other object) on the bodies and then pushing or pulling these bodies. The notion of pushing or pulling an object without touching it (at a distance) was unheard of, and yet this is precisely what Newton suggested that the sun does to the planets.

But Newton went much further than just postulating that the sun exerts a gravitational force on the planets. He introduced a precise mathematical formulation of this law of force and then went on to show that Kepler's laws can be derived from this law by a purely mathematical procedure. The solar system as a dynamical structure (consisting of a central body—the sun—and the planets, satellites, and comets revolving around it) could now be understood. Without a gravitational force, there could be no solar system because the gravitational force not only dictates the motions of the planets but also determines the nature of solar system structure. It prevents the solar system from disintegrating into separate bodies, each moving through space independently of the others.

Newton was not content with this single great achievement because he felt that the gravitational force might explain all the structures that are seen in nature. He therefore generalized his hypothesis and proposed the idea that the gravitational force is universal: that every particle of matter in the universe attracts every other particle. He further postulated that all these gravitational interactions produce the structures all around us: the earth, the earth–moon system, the solar system, the stars, the galaxies (the existence of which Newton surmised, but which were discovered much later), and the universe itself.

We know that Newton was much too optimistic about his gravitational force and that other forces must be assumed if we are to explain structures in the small (e.g., atoms, molecules) as well as in the large (e.g., stars, solar systems). But Newton had taken the first bold step on the path that led from a primitive understanding of nature to our modern science.

Newton's dream of explaining all the structures in nature in terms of his law of gravitation did not come true because matter is much more complex than he imagined it to be. As long as we limit ourselves to the study of large systems such as stars and solar systems, we can explain most of the observable features in terms of the gravitational force alone, and for more than 200 years after Newton's great discovery, scientists dealt primarily with this force. They made enormous advances in the refinement of the mathematical tools needed to apply the gravitational force to a system of gravitationally interacting bodies such as those that constitute our solar system. Thus, the science of celestial mechanics was born. One of the noblest developments of the human mind, this beautiful intellectual structure is associated with the names of such giants of 18th and 19th century mathematics and science as Pierre Laplace, Joseph-Louis Lagrange, Leonhard Euler, William Hamilton, Carl Friedrich Gauss, and, later, Jules-Henri Poincaré. The foundation they laid is now the theoretical basis of our space program, which would be impossible without a sound understanding of celestial mechanics.

The gravitational force is, indeed, the dominant overall force in the universe in the sense that it governs the structure of the universe itself, as well as the motions of its constituent stars and galaxies. But it plays a negligible role in the domain of the small. Other forces, which are vastly stronger than the gravitational force, must be taken into account if we are to explain the structure of matter. We shall see that there is a hierarchy of forces, increasing in strength as we go from galaxies and stars to molecules, atoms, atomic nuclei, and the basic particles themselves, such as electrons, protons, and so on. Our concern, therefore, is to study the forces in nature and to try to understand how they determine the structures of both the large and the small systems in the universe.

As we have already noted, a few hundred years elapsed before scientists recognized the importance of forces other than the gravitational force in the structural schemes of nature. Near the end of the 19th century, enough experimental evidence had been collected by physicists and chemists to indicate quite clearly that electrical and magnetic forces play the dominant role in atomic and molecular structures. The investigation of these forces finally led to our present picture of the atom and the molecule. The first three decades of this century were devoted to the theory of atomic structure, but, once mastered, this theory was followed by the discovery of a new group of forces much more powerful than either the gravitational or electrical forces, which account for the structure of the nucleus of the atom.

There is no reason to suppose that with nuclear forces we have exhausted the array of forces in our hierarchical scheme. The investigations that are now being conducted into the nature of elementary particles such as electrons and protons indicate that forces may exist in nature that are even more powerful than the nuclear forces; these must be responsible for the structure of the elementary particles themselves. Until such forces are discovered and their properties understood, the fundamental particles themselves will not be fully understood.

If the only forces existing in nature were those responsible for the very large and the very small systems, one of nature's most interesting structural forms, life itself, could not exist. Nor could we have the vast variety of molecular structures now present in the universe. We may refer to these structures as the intermediate systems which are caused by still other types of forces but which are related to the electromagnetic forces. These forces, which are responsible for the existence of molecules, are of an unusually interesting nature, and can be understood only in terms of one of the most recent theories in physics (the quantum theory).

The forces between any atom in a molecule and its nearest atomic neighbors are of utmost importance in determining the structure of the molecule. When we go from simple molecules to more and more complex ones, however, we are no longer able to relate these complex structures to specific forces between neighboring atoms in the molecules. Other long-range forces must also play an important role, since without them the behavior of such large molecular structures as RNA and DNA could not be understood. It is highly probable that such forces come into play when the atoms of a molecule are brought into the proper configuration. Indeed, the life processes themselves may thus arise, so that ultimately we may be able to explain life in terms of a long-range molecular force.

We have indicated that our primary purpose here is to study the relationship between the forces that exist in nature and the structures which these forces determine. We shall carry out this program by considering a single force at a time, analyzing the kind of structure with which it is compatible, and which, in fact, is necessitated by such a force. To do this, we must know something about the nature of force.

If we were interested only in static structures (e.g., a bridge), we would only have to know how to measure the force that one element of the structure (in this case, a girder, for example) exerts on another element of the structure, or the weight of each element, to understand such a structure. (We are all intimately acquainted with this kind of force—the force of gravity—since our muscles are constantly responding to the gravitational force of the earth as we move about and carry objects from one place to another. In fact, not only do our muscles adjust to the various forces we must contend with in our daily activities, but they give us a very good estimate of the magnitude of these forces.) If, then, we knew the weight of each girder in the bridge, we could calculate how to arrange the girders and what kind of support would be needed to keep the bridge from collapsing under heavy loads. To build such a static structure, then, we need only know how big the forces involved are because nothing else about them is pertinent.

Our concern here, however, is not with such static structures, but with dynamical structures such as the atom or the solar system. The component parts of the atom are its nucleus and electrons, and the sun and planets are the components of the solar system. In such structures, the component parts are in constant motion. Hence, to understand these structures we must know how the forces involved are related to the motions of the various components. We shall see, in fact, that very important physical laws relate the force acting on a body to the manner in which that body moves. To understand these laws, however, we must introduce the basic concepts of space and time. The reason for this is fairly obvious: the only way we have of studying the motion of a body is by noting its position at each instant of time and recording all observed positions and times. These observations can then be used to construct what we call an orbit of the body. If we also know the force acting on the body at each instant of our observation, we can relate the orbit to the force and from this relationship we can deduce how the force determines the dynamical structure to which the moving body belongs. By knowing the orbit of a planet and how this orbit is determined by the gravitational force, we can see how this force maintains the structure of the solar system.

In the next chapter we introduce the basic concepts of space and time which we need to pursue our study of the forces in nature and their relationship to its structures.

CHAPTER 2

The Concepts of Space and Time

Nothing puzzles me more than time and space, and yet nothing troubles me less, as I never think about them.

—CHARLES LAMB, *letter to T. Manning*

In the previous chapter we outlined in a general way the relationship between the structures in our universe and the forces that control these structures. We saw that we cannot understand these structures without first understanding the nature of the forces. There are, however, two different levels in which we can speak of forces. On the one hand we can introduce a procedure for measuring forces without concerning ourselves about how a force is related to other basic physical entities, or, on the other hand, we can start with other basic physical concepts and then see how the force concept is related to them. As we have already noted, if we were interested only in static structures such as a bridge, we would only have to know how to measure forces to analyze the structural relationships in the bridge. But we shall be dealing with systems and structures consisting of moving parts, so we shall have to know how the forces that prevent the individual components from moving away from each other are related to the motions of these components. This means we must first acquire a greater understanding of the nature of motion itself, which, in turn, requires an understanding of space and time concepts. But such concepts cannot come to us on the basis of pure reasoning; they must stem from a proper synthesis of experimental data and theory.

If science were merely a matter of collecting and classifying data, it would be of little interest to the creative mind. Fortunately, science is much more than that; it consists of two complementary facets, each of which in itself is barren, but which together constitute the most remarkable of all human activities. One of these facets, the collection and careful classification of data, is the domain of the experimental or observational scientist. The other facet, the formulation of the laws of nature, which are broad principles that correlate the data of the experimentalist and permit the scientist to predict the behavior of a physical system, is the domain of the theoretician. Until quite recently a scientist with a broad enough background (e.g., a man like Enrico Fermi) could be both an experimentalist and a theoretical physicist and could contribute greatly in both domains; but experimentation has become so complex and the equipment required so incredibly technical and involved that doing experimental work is a full-time task. The serious theoretician is, then, generally excluded from experimental work, although he must remain in constant contact with the experimental results and know a great deal about experimental techniques since these results are the source of his theoretical ideas. The experimentalist is, in general, excluded from theoretical work by the constant demands made on him by his experimental work and by the complex formalism of theoretical work. But again, the experimentalist must know what the theoretician is doing since the theories proposed by the theoretician are the source of ideas for the experimentalist. Although both physicists work in their own domains, they influence each other in a very intimate and important manner.

To the layman, whose knowledge of the physical sciences is generally limited to the conclusions he can draw from superficial newspaper and popular magazine articles on science, it appears that a theory in physics or astronomy is the least stable of all intellectual concepts because it seems to him that theories come and go as rapidly as the fashions in clothes. The layman generally distrusts theories and accepts only what he considers to be "hardheaded facts," the data obtained through experiment and observation. This preference is based on a mistaken concept of the nature of science and the relationship of theory to experiment and of the relationship of both of these to what we call the "truth" about the world. Generally when a person does not accept a theory, it is not because he does not like theories but actually because that particular theory does not agree with his own prejudices (his own "theories"), which he likes to describe as "common sense." If there is one thing that the developments of 20th century physics have taught us, it is that we cannot trust "common sense" because we know that what is usually called "common sense" is a person's accumulated prejudices or his own "theories," based on his very limited experience. If we were to use such "common sense" as a criterion for the truth or falsity of a physical theory, we would have to reject the two most basic theories of modern science, the theory of relativity and the quantum theory.

However much we may distrust theory in general, there is hardly a decision we make or a "fact" that we accept which is not based on a theory which we have already incorporated into our scheme of the universe or of our immediate world. Whether we know it or not, what we call a "fact" as distinguished from a "theory" is in most instances based on a theory. If we accept the observed "fact" that Jupiter or some other planet is in some particular position at this moment, we are really accepting the theory of planetary motion. To learn the position of a planet, we do not observe the planet directly nor do we ask someone who has observed it to tell us where the planet is located. Instead we go to the *Nautical Almanac* and take its position as given there. This position, of course, has been calculated from planetary theory, the fundamental ingredients of which are Newton's law of gravity and his basic laws of motion. But suppose we were of a very doubtful nature and mistrusted all theories. To find the position of Jupiter, we then observe it either directly with our eyes or with a telescope. It is impossible to say that we know the position of Jupiter at the moment we see it without calling upon any theory because we can deduce very little about the "true" position of Jupiter at the moment we "see" it without accepting some theory about the propagation of light and the formulation of images on the retinas of our eyes. Does light travel in straight lines or not? How is the position of the image of Jupiter on the retinas of our eyes related to the "true" position of the planet? Does light take a finite time to reach us from Jupiter? In deducing the "true" position of Jupiter from our observations, is it permissible to assume the theories of plane (Euclidean) geometry? Do the motions of the earth and of Jupiter have any effect on our deductions? And so on and on. What appears to be a simple observed fact carries with it an array of physical and geometrical theories. We are forced to conclude that there is no way of separating fact from theory. Just as we insist that theory be confirmed by fact (by experimentation), so we must insist that observational results can have no meaning until they base themselves on some particular theory. Of course, we hope that all our actions, at least our scientific ones, are guided by the correct theory.

We appear to have arrived at something of an impasse: on the one hand theories are to be accepted only if they can be checked by facts, but on the other hand we can interpret observed facts only via some theory, so that we appear to be in a vicious logical circle. But this is only apparent for we shall see that the test of the validity of a theory is not merely a matter of how well it agrees with observation. There are certain internal consistencies, quite independent of observed facts, which a theory must have before it can be accepted. The mathematical structure of the theory itself may be a guide to whether the theory is "true" or not. It is clear that in constructing a physical theory we must formulate it without violating the basic mathematical laws or mathematical axioms we adopt. Having done this we then find that the inner dynamics of the mathematics itself reveals to us new facts of our theory and leads to what we may call new theories. An example of this is the relationship between Newton's law of gravity and Kepler's laws of planetary motion. Kepler deduced the laws of planetary motion empirically from the observed data, but we can just as well start with Newton's law of gravity as a basis. Then Kepler's laws are a mathematical consequence of Newton's law of gravity. If Newton's law of gravity is correct, then we know that Kepler's laws must be correct also and we can accept them without checking them against the observed data.

Most of our knowledge (at least the data that we generally have in mind when we speak of facts) is not obtained by observation but rather by computations based on accepted theories. Examples of this are the positions of the planets, the internal properties of stars, the overall shape of the earth, the structure of nuclei of atoms, the structure of atoms, and so on. The question as to whether facts or theories guide our actions (at least as scientists) thus becomes meaningless because of the intimate relationship between these two scientific realms.

We have used the words "true," "correct," "valid," and others quite often in the previous paragraphs as though these terms were well defined and easily understood, but if we attempt to define them precisely, we soon discover that this is a difficult task indeed. Before discussing the scientist's meaning of "truth," we note that "truth" occupies a very special position in our hierarchy of concepts. That "truth" is one of the great virtues in humanity's scheme of things seems obvious to everyone, although just why we accept it as a virtue is difficult to say. We are not talking here about the pragmatic virtues of truth, for these are obvious, nor are we concerned with its moral aspects. Life would be intolerable if we were in constant uncertainty about routine events, but the truths that guide us and that we must know for an orderly daily life are rather trivial and lead to no profound questions. It is really of little interest to us, outside of properly regulating our daily routine, that an accurate clock tells us the correct time or that we know the truth about plane schedules. Each event that we become aware of as we pursue our daily activities is, of course, a truth in the sense that it has happened, but these events do not represent the kind of truth that we are talking about here. The "truth" we are concerned with here is a much grander one. It has to do with the structure and behavior of things in our universe and, indeed, of the universe itself. When we ask about the truth of a physical theory, we are really asking whether the theory reveals to us some universal property of the universe (or some part of it) which cannot be discerned in unrelated observed events or in isolated facts.

The curious thing about this aspect of "truth" is the remarkable appeal that it has for the human mind. Two of the most distinguishing characteristics of humanity which set it apart so definitely from the rest of the animal kingdom is the human compassion for living things in general and humanity's pursuit of truth. We consider this pursuit as one of the most glorious and exalting experiences we are capable of and ascribe to truth elements of beauty as well. Emily Dickinson, in one of her short poems, speaks of "one who died for beauty" as being a brother to "one who died for truth," since the "two are one," and Edna St. Vincent Millay opens her sonnet on Euclid with the line

''Euclid alone has looked on beauty bare,'' thus equating the ''truth'' of Euclidean geometry with the purest form of beauty. Shakespeare, in one of his sonnets, notes that truth enhances beauty, as expressed in the lines:

> O, how much more doth beauty beauteous seem
> By that sweet ornament which truth doth give.

We also assign a quality of truth to the beautiful, and tend to equate any intellectual or emotional experience (like listening to Bach or looking at a great painting) that exalts us to a revelation of some great truth. We thus speak of a ''valid'' experience or a ''moment of truth,'' or the ''validity of a religious revelation.'' But this exaltation of our emotions and sensations to the level of great truths obscures the nature of scientific truth and opens the door to mysticism and metaphysics, which have no place in science.

The scientist, and we, in this discussion, dismiss all of these concepts of ''truth'' as having nothing to do with ''scientific truth,'' which we must accept as the only discipline that can give us a picture of the world. The confusion that arises in the mind of the layman when he speaks of ''truth'' in the sense discussed above stems from his misunderstanding of how the scientist arrives at his truths and how these truths are related to the real world. Only a systematic scientific analysis of carefully collected data can lead to a great scientific truth, that is, to a universally valid principle. But the observed facts themselves, which constitute the data that the scientist uses in his synthesis of a truth, are, in general, not the simple kind of facts, like seeing a bird in flight or hearing a musical tone, that one associates with our usual sense perception. Today scientists obtain their facts as readings on exceedingly complex electronic detectors and recorders. Sense perceptions play hardly any role in all of this fact gathering and the coordinating and linking together of these data into a rational structure require mathematical skills of the highest order. This sort of procedure is quite mystifying and often disturbing to the layman because he is aware that great changes have occurred in the thinking and attitudes of the physicist during the last half-century, and he is never quite sure when he will be called upon to discard all that he has been taught and accept some new idea or strange theory.

This attitude is based on a misconception of the way physical science progresses. As we shall see, the introduction of a "new theory" does not completely invalidate the previous theory, but rather incorporates it into a more general and larger theoretical structure. The old theory still remains valid in a restricted domain but must be replaced by the extended theory to explain observations and experimental data that lie outside the restricted domain. The old theory thus becomes a special case of the extended theory. As an example of this process, we may consider the description of a moving body. As long as the speed of the body is small compared to the speed of light, the laws of Newtonian mechanics give an adequate description of the motion of the body, but for speeds that are not small compared to the speed of light, we must use the theory of relativity.

The layman is subject to another misconception because he looks upon experimental physics as a more or less random collection of data and he cannot understand what prompts the physicist to seek one set of data rather than another set of data. But the procedure of the experimentalist is quite different because his search for facts is based on existing theories which lead him to certain questions about nature. His experiment, therefore, is in the form of a question which his apparatus presents to nature. The data which he collects are then the answers presented by nature. He then compares these answers with the answers he deduces by applying the known theory to the physical situation represented by the experiment. If these two sets of answers agree to within the accuracy of the measuring instruments, this agreement is taken as one point in favor of the theory but this single agreement does not prove the theory. We may assume the theory to be proved only if this question and answer procedure has been applied to all possible aspects of nature and if the answers given by the apparatus agree with those deduced from the theory in each instance.

Before leaving this point, we must say a few words about what we mean by agreement between theory and experiment in science. The experimental physicist prepares his experiment with a definite question in mind, the theoretical answer to which he has already obtained from the applicable theory. The experimental answer is given by his apparatus as a series of readings on meters, or in some other form, which may ultimately be represented by curves on a graph. But in analyzing these data to obtain the answer given by his apparatus and to compare it with the theory, the experimental physicist must assign to each number (reading on the meter or other recording device) an uncertainty or error which stems from the inaccuracy of the instruments. These errors may be extremely small but they must not be neglected in comparing the experimental answer with the answer given by the theory. For only if we know the range of error in the experimental data can we assign any significance to the comparison between theory and experiment. The best we can do at any stage in the progress of science is to say that a particular theory agrees with the measured data only to within the accuracy of the apparatus used. It may well be that when more accurate measurements with better instruments can be made, agreement between theory and experiment will no longer hold and the theory will have to be revised or even discarded. This is often the hope of the experimentalist and the theoretician who see such disagreements as the source of new and deeper insights into nature and its laws.

A case in point is the observation of the motion of the planet Mercury before and during the 19th century. Since Newton's law of gravity had achieved great success in predicting the motions of the planets (in accordance with Kepler's laws of planetary motion), there was no reason to believe that Mercury's motion would not also be governed by the Newtonian law, and this was found to be so when Mercury's motion was studied with the best available telescopes and instruments before the 19th century. As far as the observable data were concerned at that time, Newton's law was correct. But with the improved observational techniques developed in the 19th century, a very small discrepancy between the observed motion of Mercury and the predictions of Newtonian theory was detected which ultimately, in the hands of Einstein, led to a new and more beautiful theory of gravitation.

It is interesting to note, in connection with this example of a discrepancy between theory and observation, that the discovery of such a discrepancy may come too early in the history of a science to lead to a useful development. If, during Newton's time, it had been found that Mercury's motion did not agree with Kepler's laws, Newton might have been discouraged in his pursuit of a gravitational theory (as, indeed, he was for a time, owing to the incorrect measurement of the moon's distance) and the remarkable development of Newtonian physics in the 18th and 19th centuries would not have occurred. Since Newton knew nothing about non-Euclidean geometries and four-dimensional space-time concepts, without which relativity theory could not have been developed, he could not have used the behavior of Mercury to lead him to Einstein's more advanced theory of gravitation. This had to await the discovery of the non-Euclidean geometries in the 19th century and of physical space-time in the early years of the 20th century.

The laws of nature are incompletely perceived at any stage in the history of science because each law apparently consists of a series of terms. As we progress in our study of nature and as our experiments and observations reveal more and more accurate data, the new terms in the theory are also revealed. These new terms do not invalidate those terms of the theory that we already know but rather must be added on to these terms. Newton's law of gravity is the first term (the first approximation) in a much larger theory—the Einstein theory. For most purposes, Newton's law gives excellent results, but to describe the motion of a body in a gravitational field with greater accuracy than that afforded by Newton's law, we must use Einstein's theory to which Newton's law is a first, and very good, approximation.

The previous paragraph may leave us with a feeling of futility and frustration insofar as arriving at ultimate truth is concerned. If a physical theory can never be fully perceived because it is a sum of many (possibly an infinitude of) terms, the discovery of each of which in turn requires more and more accurate data, how can we ever discover the full truth? This question did not bother classical physicists because they had no reason to suppose that, as their measuring technology improved, they could not discover more and more about nature with ever-increasing accuracy. They knew that there would always be an error in their measurements, but they felt that such errors and disturbances could be reduced without limit.

Today this belief has been challenged by the two dominant theories that govern the physical world: the theory of relativity and the quantum theory. Each of these theories, in its own way, has shown us that the relationship of the observer to the world that he is observing and the laws of nature themselves cannot be neglected in formulating these laws. In a sense, the observer himself becomes part of these laws so that objective reality, as we see it today, is quite different from what the classical physicist deemed it to be.

Relativity theory has shown us that the "objective reality" of

Newtonian absolute space and Newtonian absolute time are really subjective entities that have different meanings for different observers and, in fact, depend on the motion of the observer. Quantum theory, on the other hand, has taught us that the information we can obtain about any system in nature is limited by an unavoidable disturbance we introduce into the system by our measurement. This disturbance is part of the very structure of nature itself and cannot be eliminated no matter how refined our instruments and measurements may be. This important fact casts doubt on the principle of causality itself.

Before leaving the question of the relationship of truth to science we must say a few words about the role of mathematics in the formulation of the laws of physics. Since physics deals with precisely measurable quantities that can be expressed numerically, it is clear that a correspondence must exist between the laws that govern numerical quantities (arithmetic) and the laws that govern the measurable quantities of physics. Since the measurable quantities of physics can be added, subtracted, multiplied, and divided just like numbers, these measurable quantities themselves constitute an arithmetic. Starting from this basic arithmetic we can then develop the other branches of mathematics as they apply to measurable quantities. If some universal relationship is found to exist among a group of such physical quantities, one can then use the algebra or other branch of mathematics to express this relationship.

Suppose that whenever we measure a certain physical quantity of a system (e.g., the height of a column of mercury of a given cross section) and then measure some other quantity associated with this system (e.g., the weight of the column of mercury) we always find that the second measurable quantity is the same multiple of the first one. We then express this relationship by saying that the second quantity is equal to a definite number times the first quantity. In the case of the column of mercury, suppose that we always find that the weight of the column equals 13 times its height which we can write as $w = 13h$ where w stands for the weight and h for the height of the column.

This relationship, written in mathematical form (an algebraic equation), may be taken as a law of nature since it expresses something definite about mercury. This is only a very simple example of the kind of relationships among measurable quantities that the scientist seeks, and one can see that such relationships may, at times, be so complex that very sophisticated mathematical procedures must be used to express them. Indeed, it may happen, and, in fact, does happen, that the precise physical meaning of the relationship cannot be completely described in terms of our everyday experience and our ordinary concepts of space, time, and causality. The mathematical equation that expresses the relationship becomes the only "reality" that we can understand. But we must not suppose that the mathematics is the reality and that mathematics and truth are one.

Mathematics is still to be considered as only a tool, but an extremely powerful one, which enables us to subject the physical world to profound questioning. Once we have a general mathematical relationship which we have reason to believe is a law of nature, we can apply this relationship to systems of all kinds and under all kinds of conditions. The relationship then permits us to predict (with certain reservations which we shall soon discuss) what the system will do under the given conditions or to deduce the past history of the system.

We may again consider the application of Newton's law of gravity to a planetary system such as our own solar system. We can derive Kepler's three laws of planetary motion and obtain an ellipse for the orbit of each planet, but we soon see that these ellipses are only the first approximations (but very good ones) to the correct orbits, and are valid only if the gravitational pull of the sun alone on each planet is taken into account and the gravitational actions of the planets on each other are left out of the account.

If we were skillful enough mathematical programmers, with high-speed electronic computers at our command, we could feed Newton's law of gravity into the computer (plus the program for computing an orbit taking all gravitational interactions into account) and the computer would then trace out the past and the future of the solar system. This process, in a sense, enables us to predict events, but the very nature of the mathematical relationships that permit us to deduce the orbit of the planets from the law of gravity limits the accuracy of our predictions. Our computer can give us a picture of our solar system in the future or in the past only if we feed into it not only the law of gravity but also the position and the velocity of each planet at this moment. These data are derived from observations and, as we have seen, all such observations are burdened with observational errors. Each such error is magnified more and more as we carry out our calculations to more and more distant times (past or future). Owing to such errors, the calculated configuration of the solar system departs more and more from the actual configuration as we go forward or backward in time. This may not appear to present a serious obstacle to pursuing such calculations to as high a degree of accuracy as we may desire, since all that appears to be required for this is to determine the position and the velocity of each planet at this moment with ever-increasing accuracy. Until the advent of the quantum theory there was no reason to think that, with instruments of increasing accuracy, the precision with which measurements of position and velocity of an object could be made could not be increased indefinitely.

The quantum theory has shown, however, that this conclusion is incorrect and that nature itself is so constructed as to prevent the simultaneous determination of the position and the velocity (more precisely, a quantity called the momentum) of a particle. The more accurately we determine one, the less accurately we know the other. In a sense, this inherent indeterminacy destroys the basis of causality as it was understood in classical physics, since the laws of nature do not permit us to predict the future from the present. It is now incorrect to say (with absolute certainty) that if A occurs at this moment, B will follow later, where A may be pictured as the cause of B.

Although the most frequent role of mathematics in its relationship to physics (in the day-to-day work of the physicist) is that of a tool, this is not its most important or most dramatic role. Since mathematics is a logical structure with its own internal consistency and laws, these laws become part of the laws of physics. Owing to this feature of mathematics, new aspects of nature are revealed by purely mathematical manipulations of the known physical laws. The very nature of the mathematical formulation of these laws leads the brilliant theoretical physicist to other mathematical relationships that reveal new and often strange, and, at first, incomprehensible (in terms of our previous experience) physical phenomena.

Of all the mathematical disciplines that the physical scientist uses, geometry seems to occupy a special place vis-à-vis the physical universe, since this universe is itself a geometrical structure. We intuitively assume that the geometry which describes the universe in its totality is the same as that which our daily experience in our small region of space seems to reveal to us. Because of this intimate relationship between geometry and the way the world looks to us, we tend to go one step further: we ascribe a kind of absolute truth to the geometry of our limited daily experience. This view has led to serious errors about the nature of space and time in the past since our casual experiences (with light rays and the material bodies all around us) seemed to indicate that our physical world is governed by Euclidean geometry. There appeared to be a kind of universal truth associated with this geometry and this geometry itself was accepted as part of the very laws of nature, therefore, never to be questioned. The great German transcendental philosopher Immanuel Kant considered the axioms and propositions of Euclidean geometry to possess a validity beyond experience and reason as though they had been divinely imposed upon the world as part of the act of creation.

But we now know that the geometry that governs the world must stand the test of experimental and observational investigation like any other physical law. We must keep in mind that geometry is a logical system of theorems constructed from a complete, consistent set of axioms. As students, we were told to accept the Euclidean axioms because they are "self-evident truths." But the word "truth" in this statement does not mean that these axioms have been checked against the world and found to be physically true. The expression "self-evident truths" simply means that the axioms constitute a logically consistent set of statements and that they apply to a very limited region of our physical world under the very restricted conditions of our daily experiences. The famous parallel postulate of Euclidean geometry illustrates this point very clearly. This postulate states that given a straight line and a point outside that line, only one line can be drawn through the point parallel to the given line. This axiom does indeed appear to be a self-evident truth for, try as we may, we seem unable to draw more than one line through the point without cutting the original line. But a little thought will show us that this is "true" only because we have drawn the lines on a flat piece of paper. Actually, we do not even know that the paper is flat but we assume it to be so and then, of course, this axiom seems obvious.

If we accept the parallel postulate as a "self-evident truth," we must accept this statement as a truth about the world. But this raises a serious question, for the parallel postulate is equivalent to saying that the sum of the angles of a triangle equals 180 degrees. Hence, if we construct a triangle using three points in space and measure the three angles of the triangle, we should get 180°, but it is clear that we will not get 180° because of the unavoidable errors in our measurements. How then do we really know that the entire discrepancy between our measurements and the assumed value of 180° is due to error? Certainly part of it is, but part may arise because the geometry of our space is non-Euclidean so that the parallel postulate does not apply. When we say that the sum of the angles of a triangle is 180°, we really mean that if this triangle lay on a perfectly flat plane and the lines of the triangle were perfectly straight, the sum of its angles would then equal 180°.

Another consequence of the parallel postulate, which we merely mention in passing, is that the circumference of a circle is equal to 2π times its radius, r, or $2\pi r$. Again, this statement can never be verified experimentally by actually comparing the length of the diameter of a circle with the length of its circumference, as the ancient Greeks tried to do.

Until the middle of the 19th century, any questioning of Euclidean geometry would have been only of academic interest and would have meant very little since Euclidean geometry was the only kind of geometry known at that time. But the work of Carl Friedrich Gauss, Nikolay Lobachevsky, Wolfgang Bolyai, and Georg Riemann demonstrated that Euclidean geometry is but a very special kind of geometry that separates two much broader geometries somewhat in the way that a point separates a line into two parts or a line separates a plane into two parts.

Lobachevsky was one of the first geometers of the period to question the parallel postulate. He first tried to derive this postulate as a theorem from the other axioms of Euclid but failed. If the parallel postulate is left out of this group of axioms, Euclidean geometry is not complete and certain of its theorems cannot be deduced. Lobachevsky therefore decided to replace this axiom by the following one: Given a straight line and a point outside this line, an infinite number of straight lines can be passed through the point without cutting (that is, being parallel to) the given line.

Of course we cannot visualize a construction of this sort on a sheet of paper because paper has a flat surface. But on certain curved surfaces such "straight" lines can be drawn. To understand this concept, we must, of course, change our definition of a "straight line."

With his new parallel postulate replacing that of Euclid, Lobachevsky showed that a different kind of geometry (non-Euclidean), complete in every respect, can be constructed. This geometry is called Lobachevskian or hyperbolic geometry.

Somewhat later, the great German geometer Riemann replaced Euclid's parallel postulate by still another kind of parallel postulate which may be stated as follows: Given a straight line and a point outside this line, no straight line can be drawn through the given point parallel to the given line. Riemann showed that a complete geometry can be constructed with this axiom. This geometry is called Riemannian or elliptical geometry; it is a geometry in which no parallel lines exist. Again it is clear that we cannot have such a geometry on a flat piece of paper, but it does apply to the surface of a sphere where "straight lines" are arcs of great circles.

From the discussion above we see that, logically, a whole range of geometries are permitted and that Euclidean geometry exists only if a very special condition (the parallel postulate or the sum of the angles of a triangle equals 180°) is fulfilled—a condition that can never be accurately verified in practice. It may be argued that other geometries are so foreign to our experience that common sense dictates that we accept Euclidean geometry as correct for our three-dimensional space, but this kind of reasoning can lead us into grievous errors. In any case, hyperbolic and elliptical geometries are more than mere mathematical curiosities when we are dealing with surfaces—two-dimensional manifolds—so that these geometries are as "real" and as "physical" as one imagines Euclidean geometry to be.

The reader may argue that our world is three-dimensional and in three dimensions only Euclidean geometry makes sense. But this attitude denies the very essence of scientific inquiry. Like all other scientific results, our acceptance of one kind of geometry or another for our universe or for any region of space in it must be determined by precise measurement and experimentation. Such observations must depend on the propagation of light from point to point in our space and therefore the question as to the nature of our geometry is intimately related to the properties or behavior of light. Finally, it appears from what we have said that the validity of Euclidean geometry can never be rigorously demonstrated.

As we have already noted, our purpose in this book is to give the reader an understanding of the relationship between the structures and forces in nature and just how these relationships are governed by the laws of nature. The laws themselves are precise statements (generally formulated mathematically) that relate various sets of measurable quantities to one another. However, the physical quantities that may appear in a law are not all of the same basic character because such quantities can be divided into one set which we call fundamental and another set which we may call derived. As in all other logical disciplines, the concepts of physics (the basic physical science) are constructed from certain undefinable but measurable quantities, which may be taken as the fundamental elements of the physical sciences. Just as the mathematican does in geometry, the physicist introduces as few of these undefinable quantities as may be necessary to construct all the derived quantities which are needed to formulate the laws of physics. We may compare the structure of science (physics) with that of a language in which certain basic words must be accepted without a formal definition and in which all the other words are then defined in terms of these basic words.

Since the fundamental physical quantities cannot be defined in terms of simpler quantities, we must somehow introduce them into our scientific structure by outlining a procedure for measuring these quantities and thus obtaining a set of numbers. The measuring procedure itself, of course, is not a definition of the quantity that is being measured, but the definition is then of no importance to us. As long as we can assign to each fundamental quantity a number which is obtained from a precise and carefully defined set of operations, we can use these basic measured quantities to construct all the other quantities that may be of interest to us or may be required to express a law.

Suppose, for example, that to describe the behavior of a physical structure such as the earth–moon system, we must know the value at each moment of some quantity Q which is defined as the product of two fundamental quantities A and B. If we now have a definite procedure for measuring A and B at each moment, we can calculate Q and thereby analyze the structure, even though we have not defined A and B as concepts, but merely have given a procedure for measuring them.

With these preliminary remarks in mind, we can now consider which physical quantities we should take as fundamental and which as derived. This choice is determined by the simplicity of the operations that are required to measure such quantities, and here our daily experience serves us as a guide. As we look about us and study events of various kinds, we are aware that each such event occurs at a particular time and at a particular position in space. A particle moving from point to point is an example of a series of events; the particle at each moment is an event in our sense of the term. Various physical characteristics are associated with this series of events (such as the speed of the particle and other features of its motion) but the simplest description we can give is to specify the position of the particle at each moment of time. The reason for this is that it is much easier for us to determine the position of a particle by measuring its distance from some fixed reference point at some moment and to look at a clock at that moment than it is for us to measure directly such things as its motion (its velocity, acceleration, and so on). From the particle's position at each moment we can obtain its path (orbit) and then derive (from this orbit) the various characteristics of its motion (velocity, acceleration, and so on) by standard arithmetic procedures. In other words we take distance and time as our fundamental physical quantities and consider as derived such quantities as velocity and acceleration. We may, of course, proceed differently by taking velocity and acceleration as basic quantities, and derive distance and time from them, but we then run into unnecessary complications.

It may seem that choosing distance as one of our basic physical quantities is an arbitrary choice, and in a sense it is; but this choice suggests itself by our daily experience. We live in a spatial world and the spatial relationships among the physical objects that we meet in our daily physical activities shape these activities in very definite ways. From the moment we get up in the morning until we get back into bed we are constantly adjusting our steps, our posture, the motion of our hands, the orientation of our bodies, and so on, to the physical objects that surround us and to the geometrical patterns and arrangements of various physical objects. But these continuously changing physical adjustments that we make as we move about, to keep us from physical collisions or to permit us to operate efficiently, are carried out so automatically that we are hardly aware that we are doing so.

We are constantly altering the positions of various parts of our bodies with remarkable precision to meet our momentary needs. Consider, for example, the very act of reaching out to pick something up.

We do so with hardly any thought as to the distance of the object; we place our hand at just the right distance almost at the moment that we decide to reach for the object. Again, as we write and form words our hands move across our pages and our fingers oscillate up and down by just the right amount to form each letter correctly (as to size and shape) to give each of us his unique handwriting.

The reader can think of even more remarkable examples of this kind of precision (e.g., the behavior of the tongue in forming words) which may be developed to an amazing degree in certain individuals. The great violin virtuoso arranges his fingers from moment to moment on the strings of his violin with almost unbelievable accuracy—a minute deviation from the position of one of his fingers ruins the concert. The great athletes can gauge spatial relationships with remarkable precision and can make their bodies do whatever is required to triumph. Obviously, then, we are born with an inherent sense of distance (shown even in the infant's appropriate motion toward the source of its food) without which we would constantly be colliding with objects all around us. This distance sense, however, is not sufficiently accurate, even when most highly developed, to serve as a basis for introducing the precise procedure for measuring distances as required in science. Moreover, it is difficult to translate the subjective estimate of a given distance as described above, to an objective numerical value that all observers can agree upon. Even though a skillful person "knows" quite accurately how far he must shift a hand or a finger to achieve his goal, he cannot express this numerically with the same accuracy. Since scientific laws are essentially statements of relationships among numerical quantities, we must present a precise procedure for measuring these quantities and expressing them numerically. We emphasize, however, that the procedure we describe is not a definition of the concept of distance but rather a precise method for obtaining a number that can be used when comparing theory with observational data.

Since the measurement of any quantity is essentially a procedure in which we compare the measured quantity with an accepted unit of that quantity, we begin our discussion of the measurement of distance by introducing a unit of distance. This unit is simply a carefully machined "straight edge" (let us say a platinum rod) of arbitrary but definite length which is adopted by the general acceptance of the scientific community. Before outlining how this unit is used to measure a distance, we must say something about the expression "straight edge." Since the distance between the two points is to be measured along a straight line, the unit itself must be "straight" in the same sense that the line is straight. Since the concept of straightness has meaning only in the frame of reference of some particular geometry, we assume that the geometry of our space is Euclidean so that our "straight edge" (unit length) is straight in the commonly accepted meaning of the term.

To check whether our unit is straight, we need only construct three such units and bring each one into coincidence with the other two. If we obtain perfect coincidence in all three cases, we may conclude that each rod is straight—in the usual sense (provided, of course, that our geometry is Euclidean). Another way to check the straightness of our rod is to compare it with a beam of light from a laser which we accept as defining a straight line (for Euclidean space). If the geometry of the physical space itself is non-Euclidean, all physical bodies must be constructed in accordance with such a geomtery and the test for straightness that we have just described merely shows that our unit is "straight" in the non-Euclidean sense.

With our unit length at hand, we can now describe how we use this unit (that is, the set of physical operations we must perform) to obtain a number that we may call the distance between two points A and B which are located at opposite ends of a long rod. We first bring our unit length (a ruler, for example) in coincidence with the "straight line" (the rod) connecting the two points, with one end of the unit (ruler) coincident with point A and the other end coincident with some point on the rod (point C, for example) which is located between A and B. We then repeat this set of operations, starting from point C and so on until we reach point B. The number of times we must do this operation to reach B is called the distance from A to B along the straight line (rod).

We may, of course, measure the distance from A to B along any other path which does not lie parallel to the rod itself but goes out a certain distance from the rod and then bends inward at the other end. The number we obtain along such a path is larger than the one obtained along the straight line ACB. Indeed, if we draw many different paths from A to B, we obtain a different number for each path. We find, by using this procedure, that there is one path (ACB) that is smaller than any other path. This path is called a straight line (in the frame of reference of the particular geometry that applies).

In describing the operations we must perform to measure the distance from A to B we considered the simplest case in which our unit length fits an integral number of times into this distance (on completing the last step of operations the end of our unit falls exactly at the endpoint B) but this will happen only by sheerest accident and, in general, point B will coincide with some point between the two ends of our unit. We can still make a precise measurement in this case if we carefully divide our unit into equal divisions, e.g., by dividing our ruler into inches. The more such subdivisions we introduce, the more accurate will be our measurement of the distance. But we can see at once that there is a physical limitation to how far we can continue subdividing by putting scratches on our ruler. Since each scratch has its own width, we find, after a certain number of subdivisions, that the scratches touch each other and we can go no further. In other words, the limitation of the accuracy of this procedure is set by how fine a scratch can be ruled on the measuring rod or ruler. Today, with modern techniques, as many as 15,000 lines can be ruled in the space of an inch.

Although various units of length (e.g., the inch, the foot) have come down to us from antiquity, we shall deal almost entirely with the metric system which is based on a standard platina–iridium bar (deposited in Paris) whose carefully machined length is called a meter. This bar was originally intended to equal one ten-millionths of the distance on the earth's surface between the north pole and the equator. Legally, the meter is defined as 39.370113 inches, although an exact measurement shows it to be 39.370147 inches. The centimeter (cm) is one tenth of a meter, and 1000 meters is called a kilometer (km).

Although we have indicated that the accuracy with which we can measure a distance using the procedure described above is limited by how finely we can subdivide the unit length by means of scratches, we can still imagine going beyond this limiting point and using the structure of the rod itself. If the bar has a crystalline structure consisting of atoms arranged in definite patterns and spaced at definite intervals from each other, we can, in principle, use this pattern and these spacings in place of the scratches on the bar to measure distances. This sort of indigenous spacing would permit us to measure distances with a much greater precision than by using a bar with physical scratches on it, but this procedure meets with practical difficulties. Moreover, even this technique is limited in accuracy by the vibrational motions of the atoms in the crystal.

We can get around the difficulties associated with using a bar by using a beam of light instead. The basis for this is that light consists of waves so that, in principle, we can compare the distance we wish to measure with a train of waves. To do this we must know the wavelength of the light we are using (the distance between two successive crests of the wave) and we must also know how to determine the number of such wavelengths that are contained in the distance AB. There are very definite procedures for doing this which we shall not discuss now. The important point is that optical techniques can be used to obtain a much higher precision than can be obtained with rods.

In our discussion of the measurement of a distance, we spoke of the distance between two points. It is clear, however, that we do not mean mathematical points in the operation described above. The points A and B, if drawn on a piece of paper, are two ink spots which exhibit a rather complex physical structure when viewed with a microscope. When we lay down our unit length we assume that the "point" we are talking about is in the ink spot because the latter is assumed to be so small compared to the total distance from A to B that we may treat the spot and the "point" as identical. Here an analogy may clarify the situation. New York and San Francisco are represented on a map of the United States as two points, and we can lay off a ruler along the line connecting them to measure the distance between them (using the scale to which the map is constructed) even though New York and San Francisco are certainly not mathematical points. When we say that San Francisco is a certain number of miles from New York we are treating these two cities as points and not worrying about the exact distance between a particular place in one city and a particular place in the other city.

With this difference between a physical and mathematical point in mind, we can still consider the possibility of measuring the distance between two mathematical points. We may begin by trying to choose these points within the two ink spots A and B, and the centers of these spots (if we picture them as small disks) might suggest themselves as the points to choose. But how do we find the center (C) of a disk? We again find ourselves involved in measuring the distance between two spots, A and B. In other words, we are no nearer our goal of using mathematical points by this procedure than we were before. By using the center (C) of the spot (instead of the whole spot) as the starting point of our procedure, we increase the accuracy of our measurement but we are still measuring distances between physical structures and not between mathematical points. Once we recognize this distinction and accept it as an unavoidable restriction on the measurement of distances, we can try to increase the accuracy of our measurement by making the physical spots A and B as small as possible. We might, for example, pick out a single atom in each spot and measure the distance between these two atoms. However, we immediately encounter two problems using this procedure. To begin with we must introduce some way of locating (or seeing) each atom. This involves special types of optical devices since atoms are so small that we must use very short-wave radiation (X rays) to observe them. When the X-ray beams reveal the atoms to us, we can measure the way the incident X-ray beams are reflected from the two atoms and then, using the known laws of optics, we can deduce the distance between them.

In deducing the distance, however, we note a very important feature inherent in the measuring process: The two atoms are not at rest but are moving about constantly so that the distance between them changes continuously. This means that we must look at the two atoms simultaneously to obtain the distance between them at a particular moment. At some moment later, however, the distance will be slightly different and our previous measurement will no longer apply. We can overcome this problem to some extent by performing many sets of measurements of the sort just described. We then obtain a set of distances which differ from each other but which cluster around some mean of distances which can be computed by the usual statistical procedures. We may then call this mean value the mean distance between the two atoms and hence define this number as the distance between the two points A and B about which the atoms are oscillating. If we carry out a procedure of this sort, we can obtain much greater accuracy, but again we reach a point where our accuracy is limited by the physical structures we are dealing with. The individual distances that we must measure to obtain the mean value are inaccurate for two reasons. First, the atom itself is not a point but a complex structure consisting of a nucleus and a cloud of electrons, so that each such measurement is inaccurate to the spatial extent of the atom. We may try to overcome this inaccuracy by looking at the nucleus of each atom or at some particular electron in each atom as the markers for our two points A and B, but this would lead to more serious difficulties associated with the second of the two reasons for inaccuracy mentioned above.

The second source of inaccuracy stems from the process we must use to look at the two atoms. To do this we use either a beam of radiation or a beam of particles which can be bounced off each atom. This beam has a structure which must be taken into account in estimating how accurately we can locate (or see) each atom. The structure of the beam of light is given by the wavelength of the light. The smaller this wavelength, the more accurately we can locate an atom with it. If we want to locate an electron inside the atom, or the atom's nucleus, we must use very short-wavelength light such as X rays or gamma rays. But in doing this we run into the following remarkable property of light which makes it very difficult, if not impossible, to carry out repeated measurements between the two atoms to obtain the mean distance: The shorter the wavelength of light, the more energetic it is; that is, the more violently it disturbs a particle from which it is reflected. This is a consequence of Planck's quantum theory of radiation and means that the more accurately we try to locate each of the atoms, or some part of the atom, the more violently we disturb it. In fact, by trying to locate either atom with extreme accuracy, we tear it apart or knock it completely out of its position so that the distance between the two atoms is irreparably disturbed. As a result, the idea of using the average value of a set of distance measurements is invalidated.

This discussion shows that the very idea of the distance between two mathematical points is meaningless from a physical point of view. Moreover, we find that the measurement of very small distances (e.g., the diameter of the nucleus of an atom) is limited by the interaction of the physical entities; that is, the points that define the distance and the means used (e.g., radiation, electrons) to locate these points.

Before considering the measurement of large distances, we must introduce the concept of an angle. This can best be done by noting that a body can experience two general kinds of motion. On the one hand, the body in its entirety can be moved from one locality to another without changing its orientation. This kind of motion is a displacement or a translation of the body. When this occurs, all the points of the body are shifted parallel to each other by exactly the same amount. During the translation of the body a line connecting any two points of the body remains parallel to itself. The magnitude of the translation or displacement is related to translational motion.

Another kind of motion which does not result in a change of the body's locality (position in space) but only its orientation (the side that faces the sun, for example) is rotation. A rotating body has a line of points in it which do not move. If we draw such a line through the earth, it enters at the north pole and exits at the south pole. This line is called the axis of rotation of the body. All other points of the body move in circles with their centers on the axis of rotation. Any line that connects two points of the body, but which is not parallel to the axis of rotation, changes its orientation during this motion in that it does not remain parallel to itself. Another important characteristic of rotational motion is that every point of the rotating body returns to the same position repeatedly.

It is obvious that the amount of translation a body undergoes is measured by the distance between its initial position and its final position. But how do we measure the rotation of a body? To see how to do this, we consider a door which is attached to two hinges; the line connecting these hinges is the axis of rotation of the door. When the door is shut, it lies within the plane formed by the doorframe. When viewed from overhead, the doorframe appears as a line and the open door as a second line which is joined to the first line at the top hinge. If we rotate the door about its hinges (about its axis of rotation), the amount by which the door has been rotated to bring its from its original orientation (in the doorframe) to its new orientation (outside the doorframe) is given by the angle which the two lines above make with each other. Thus, just as distance is a measure of the translation of a body, angle is a measure of its rotation. When we measure an angle, we measure the amount of rotation a body experiences.

To introduce the unit of angle we again consider the door in the above example and suppose that it can be rotated all the way around until it again coincides with the doorframe and is back to its original orientation. We then say that the door has been rotated through 360°.

In other words, one complete turn or rotation is defined as 360° so that the degree (a unit of angular measure) is defined as one 360th of a complete rotation. One sixtieth of a degree is called a minute and one sixtieth of a minute is a second.

We can get an idea of how large a degree is by noting that the eye of an observer on the earth turns through one half a degree when it shifts its focus from one point on the rim of the full moon to another point on the opposite side. To get some idea of the magnitude of a minute, we consider two points whose separation is such that they can just be seen as distinct by an observer with normal eyesight. The angle through which this observer's eye must turn to shift focus from one point to the other is almost one minute; this is defined as the normal eye's resolving power. If we picture this amount of turning divided into 60 equal parts and take one of these parts, we have the second.

Although the degree is used extensively in expressing the size of an angle, a more useful unit is the radian which is related to the radius of a circle; that is, the line connecting the center of the circle to a point on its circumference. We now consider an arc of the circle (a piece of the circumference) equal in length to the radius. The angle formed at the center of the circle by the two radii drawn to the two ends of the arc equals one radian. We can determine the number of degrees in a radian from the number of radians in a complete turn. But one radian (by the definition above) is the same part of a complete turn as an arc with a length of one radius is of the complete circumference. Since the arc length associated with a radian just equals the radius r of the circle and the circumference of a circle equals $2\pi r$, where π approximately equals 3.14159, the number of degrees in one radian equals 360° divided by 2π; with the value of π given above, we find that one radian equals about 57.3°. Put more succinctly, the number of times the radius of a circle can be laid off along the circumference is approximately 6.28318; hence the number of degrees in a radian is approximately 57.3°.

The advantage of using the radian instead of the degree as the unit of angular measurement is that the radian leads to a simple formula in which we can express the length of the arc of a circle in terms of its radius. If the length of an arc of a circle is s and the radius of the circle equals r, then s equals the radius multiplied by the angle θ (expressed in radians) subtended at the center of the circle by the arc; that is, $s =$

$r\theta$. As an example, suppose that the radius of the circle is 10 inches and the central angle θ (the angle subtended by the arc) is 36°. How long is the arc of this circle? Since 36° equals one tenth of 360°, it equals π divided by 5 radians. The arc length s therefore equals 10 times $\pi/5$ inches or 2π inches.

An angle is measured with an instrument called a protractor. This instrument is essentially a circle or the arc of a circle (most often a semicircle) which has been subdivided into a number (e.g., 360) of equal parts. To measure the angle formed by two lines, one places the center of the protractor at the point of intersection of the two lines and reads off the number of subdivisions on the protractor contained between the two lines. If each subdivision represents one 360th of the circumference of the protractor, the number of subdivisions contained between the two lines is the magnitude of the angle in degrees.

Although we can always measure an angle with a protractor and express it in degrees or radians, there is another very useful and convenient way of representing an angle which is based on the measurement of a distance. To return to our overhead view of the door, if the door is opened partly we see that an angle θ is formed at the hinge by the intersection of the top of the door and the top of the doorframe. If at a unit distance from the hinge a piece of string is attached to the top of the door and extended perpendicularly to the top of the doorframe (it makes an angle of 90° with the door frame), the length of the string from the door to the doorframe is related in a very definite way to the angle θ formed at the hinge by the doorframe and the door. The wider the door is opened, the longer the string from door to doorframe becomes. Thus, there is a direct relationship between the size of the angle θ at the hinge and the length of the string. The wider the door is opened, the larger the angle θ and the longer the string is. For each value of the angle θ, there is just one length of string. We may therefore use the length of the string instead of degrees or radians to specify the size of the angle θ. Mathematicians have prepared special tables in which the length of the string (the perpendicular length from door to doorframe) is given for many different values of the central angle θ. We call the length of the string the sine of the angle θ. However, the sine of an angle is not a length but appears to be a length only because we attached the string at a unit distance from the hinge. If this distance is r, the sine of θ is the length of the string divided by r. The sine of an angle is thus a pure number; it is dimensionless because it is the ratio of two lengths.

We notice that the sine of an angle increases from 0 to 1 as the angle goes from 0° to 90°. This follows since the length of the string increases from 0, when the door is closed, to 1 as the door is opened until it is perpendicular to the doorframe; the string then lies exactly along the top of the door and just reaches the doorframe. As the angle θ increases from 90° to 180°, the sine drops from 1 to 0. At this point, the door is parallel to the doorframe but on the other side of it. From 180° to 270° the sine goes from 0 to -1 and then from -1 to 0 as the angle changes from 270° to 360° and the door completes one turn about its hinge and returns to the doorframe.

Instead of using the length of the string (extending from the top of the door to the top of the doorframe) to define the angle θ, we can also use the horizontal length on the doorway between the hinge and the point where the string intersects the doorframe. Corresponding to each angle there is just one such horizontal length, so that it represents the angle θ just as well as the length of the string does. This length is called the cosine of the angle θ; it decreases from 1 to 0 as the angle θ goes from 0° to 90°, then becomes negative and equals -1 when θ equals 180°. The cosine goes from -1 to 0 as θ goes from 180° to 270°. It then varies from 0 to 1 as θ changes from 270° to 360°. Mathematicians have also prepared cosine tables for many different values of θ.

Another important and useful angle specification is the tangent of θ which is the sine of θ divided by the cosine of θ or the length of the string divided by the horizontal length along the doorframe. Now that we have introduced the concept of angle we can use it to measure distances which cannot be measured by the usual procedure outlined in the previous sections. That the measurement of an angle permits us to obtain a distance is not surprising since, as we have just seen in the definition of the sine of an angle, each value of an angle corresponds to a definite distance in the unit circle (a circle of unit radius).

Very definite and insurmountable difficulties are associated with measuring very small distances or with obtaining unlimited accuracy in the measurement of a distance. The inaccuracy in such a measurement becomes relatively less important as the distance being measured becomes larger and larger. A star may be treated as a point and the earth (or sun) as a point when we measure the distance between them. The error we make in doing this is completely negligible but other problems arise in measuring such large distances. It is clear that we cannot apply the procedure of laying off a unit of length to measure large distances such as that from the earth to the moon. We must therefore devise other methods which depend on the known properties of triangles. This procedure can be illustrated by the following example. We wish to determine the distance between where we are standing (which we shall designate as point A) and another point B that is physically inaccessible to us (for example, on the opposite side of a river). We first choose a third point C on our side of the river and carefully measure the distance from our position (point A) to point C using the standard procedure described above. We now look at point C through a telescope and then rotate the telescope until we are looking at point B which gives us the angle between the two imaginary lines connecting our position with points C and B. Then we pick up our telescope and walk over to point C and sight point A (our original position) before rotating it to focus on point B. These two sightings give us the angle between the two imaginary lines connecting our new position with points A and B. Because we know the distance between points A and C, and the two angles BAC and BCA, the triangle itself is completely determined and the distance between point A and point B can be calculated by applying the known laws of geometry. This procedure—known as triangulation—assumes that Euclidean geometry is valid and relies on the laws of optics, especially the law that light travels in straight lines. It also requires that we be able to measure the base of the triangle by laying off a unit length. The method of triangulation, with some modifications, is used to measure astronomical distances. This is the method of parallax.

The method of triangulation depends on setting up a triangle, one side of which (the baseline) is measurable directly and another side of which is the distance to be determined. Obviously, the baseline must be as large as possible when measuring vast stellar distances. As long as we do not go beyond the solar system, however, the diameter of the earth suffices as a baseline.

The measurement of the earth's diameter (or rather, its circumference) has a history which goes back as far as 250 B.C. when the Alexandrian librarian Eratosthenes measured the earth's circumference by essentially the same method that is used in modern geodetic work. His method can be understood if we imagine parallel rays of sunlight striking the cities of Syene and Alexandria when the sun was directly overhead at Syene. By measuring the length of a shadow cast by a vertical stick at noon at Alexandria, Eratosthenes was able to calculate the angle between the sun's rays and the stick which equals the angle subtended at the earth's center by the arc (on the earth's circumference) from Syene to Alexandria. He found this angle to be about 7° which gave him the fraction 7/360 which is the ratio of the arc from Syene to Alexandria to the earth's circumference. Eratosthenes found that the earth's circumference is as many times greater than the distance from Syene to Alexandria as 360 is greater than 7. He knew the distance between the two cities to be about 490 miles and thus obtained nearly 25,000 miles for the earth's circumference. This ingenious method is essentially the same as that used today.

If we measure the distance between two points on the surface of the earth that differ in latitude by 1° but are on the same meridian of longitude (the same great circle running from the north pole to the south pole), we obtain the circumference of the earth by multiplying this distance by 360. This distance between the two points, called the "length of a degree," varies as we go from the equator to the north pole. It is smallest at the equator and largest at the north pole, although the difference at these two extremes is quite small. That the length of a degree is not the same everywhere on the earth means that the earth is not a perfect sphere. Since the length of a degree increases with the flatness of a circle, the earth must be flatter at the poles than it is at the equator. The earth is thus an oblate spheroid.

The length of the degree at the equator is 68.7 miles so that the equatorial circumference of the earth is 68.7 miles × 360° or 24,732 miles. The length of the degree at the poles is 69.4 miles. If we divide the equatorial circumference by the number π we obtain the equatorial diameter of the earth. The equatorial diameter is 12,756 km and the polar diameter is 12,714 km so that the distance through the earth from pole to pole is about 27 miles shorter than the distance through the earth at the equator. The mean diameter of the earth is 12,742 km.

The earth's diameter can be used as a baseline for measuring astronomical distances trigonometrically within the solar system. If two observers sight the sun at the same moment from opposite sides of the earth, they can determine the angle p which the earth's diameter subtends at the sun; this is called the geocentric parallax of the sun. The distance of the sun from the earth d can then be calculated from the relationship $dp = D$ or $d = D/p$ where D is the diameter of the earth, p is expressed in radians. This relationship is not exact but it is accurate enough for our purposes since p is a small angle. The arc connecting our two observers may be replaced by an imaginary line D that passes through the center of the earth. Using this line D as the base of our triangle, we find that the average value of p for the sun is 8.794 seconds. If we now use 8000 miles for D in the above equation, we obtain 93,000,000 miles for the mean distance of the sun. This distance is called the astronomical unit and is one of the most basic units in astronomy because it enables us to measure the distances of celestial objects far beyond the solar system.

Although the earth's diameter is a satisfactory base for measuring distances within the solar system, it cannot be used for stars because their distances are enormously larger than the earth's diameter. The most distant known planet in the solar system, Pluto, is about $4\frac{1}{2}$ billion miles from the earth while the nearest star, Alpha Centauri, is almost 30 trillion miles from us, or about 2000 times farther than Pluto. If this star were viewed from opposite sides of the earth, the angle p would be far too small to be measured. We must therefore use a larger baseline to apply the parallax method (triangulation) to measuring stellar distances; the astronomical unit provides just such a baseline. If the star is not too far away, it appears to shift against the distant background stars when viewed from the opposite extremes of the earth's orbit around the sun. If we measure the angle formed by extending lines from the earth to the star when the earth is at opposite ends of its solar orbit, we can find the distance of the star because the diameter of the earth's orbit is known (two astronomical units or roughly 186,000,000 miles). The parallax p'' of the star (expressed in seconds of arc) is equal to half of this angle; it is the angle subtended at the star by the astronomical unit A. Since p'' is extremely small even for the nearest star (about three-fourths of a second of arc), A may be treated as the arc of a circle of radius d (the distance of the star from the sun). Using the formula $d = (A/p'')(206{,}265)$ where the numerical factor 206,265 is just the number of seconds in one radian (which is introduced into the distance formula since p'' is expressed in seconds and not radians), we can calculate the

distance of any star if its parallax can be measured. The distance is then expressed in terms of astronomical units. A convenient astronomical unit of length is the light-year, which is the distance light travels in one year—about 6 trillion miles. Since we know that light reaches us from the sun in 8 minutes, we can express A in light-years and obtain the distance formula in light-years. The above formula tells us that if the parallax of a star were one second of arc, its distance would be 3.26 light-years.

The trigonometric parallaxes of stars can be measured as described above as long as they are not so far away that the semiannual change in the earth's position (a displacement of two astronomical units) has no observable effect on their positions in the sky. If stellar positions could be observed with infinite accuracy, the parallax of a star could be measured however far away the star might be; but the measurement of parallax is, like all other measurements, burdened with errors which ultimately limit the method of trigonometric parallaxes. This method is reliable out to about 500 light-years, but beyond that the error in the measured parallax is about equal to the parallax itself, making the results unreliable. In spite of this difficulty, astronomers can probe deep into space and determine the distances of objects out to billions of light-years. The methods used are indirect and based, to a great extent, on comparing some known intrinsic feature of the celestial object (e.g., its intrinsic luminosity or diameter) with this same feature as it appears to us. If we know how this feature as observed changes with distance, we can find the distance by comparing the observed with the known intrinsic feature. For example, since the brightness of a star diminishes with increasing distance in a known way, the distance can be determined if the intrinsic luminosity of the star is known. All such procedures ultimately depend on accurate measurements of the distances of the nearby stars by the trigonometric method described above. The reason for this dependence is that the intrinsic features of the nearby stars can be found only if their distances are known.

Up to this point we have considered only distances between two points that are fixed with respect to us, but it often happens that we must deal with moving distances, as in the case of a rod that is moving past us at a given speed. How do we measure the distance between the two endpoints of this rod without running along with it? We must use some kind of sighting procedure which permits us to view the two endpoints simultaneously and to find the marks they coincide with on a measuring rod that is fixed with respect to us. This procedure introduces the concept of simultaneity into our measurements, which is not present in the measurement of fixed distances. When dealing with moving lengths, time must also be considered. Owing to this intrusion of time into our distance-measuring operations, we defer a detailed discussion of this aspect of measurement until after the concept of time has been discussed.

CHAPTER 3

Vectors and Coordinate Systems

'Tis strange—but true; for truth is always strange;
Stranger than fiction

—LORD BYRON, *Don Juan*

The concept of distance (or length) is the first of the basic physical entities which constitute the foundation on which we build the body of knowledge (consisting of laws and derived measurable quantities) called science. In the previous chapter we did not attempt to define the distance concept but, instead, gave a precise physical procedure for obtaining a number which we call the distance between two points. We also introduced the angle concept to permit us to measure rotation. But the angle concept permits us to introduce another important idea—that of direction—which leads to an extension of the distance concept.

When we described the procedure for obtaining the distance between two points, we said nothing about the spatial relationship between the two points. Since all points of space are identical because space is homogeneous and the same in all directions, it does not matter, as far as the distance between the two points is concerned, in what direction point B lies relative to point A. The distance is the same because the set of operations we perform to obtain the distance is the same regardless of the direction in which we proceed from A. In spite of this isotropic quality of space, a thorough study of physical concepts and laws requires that we introduce the concept of direction when we consider distances. It is important, however, not to limit ourselves to distances in introducing direction, but rather to do this in as general a way as possible, since quantities other than distance are associated with direction. All such quantities, taken as a class, are called vectors; they will be shown in boldface type. A vector may be defined as a physical entity that has two aspects—magnitude and direction. Physical quantities (magnitudes) that are not associated with a direction are called scalars and include such things as temperature, volume, the number of objects in a group, and so on. The magnitude of a vector is also a scalar because it represents the distance between two points regardless of direction. The vector concept is hardly strange to us if we use a map, as most of us must, when we take a trip. Knowing the direction we must travel (on the roads) is far more important than knowing the precise distance of our trip. Because the map shows both distance and direction, however, it is a vector diagram.

The importance of vectors in science stems from the directional aspect of the laws of nature; these laws are statements that correlate groups of various vectors to each other. Vectors simplify the statement of the law because a vector specifies both the magnitude and direction of a physical entity so that these two important quantities do not have to be treated separately.

The motion of a planet around the sun illustrates the power of vectors to describe the planet's orbit in its totality. One of Newton's greatest achievements was his deduction of this orbit (an ellipse) from his law of gravity. As the planet moves around the sun, its distance and direction from the sun (given by a vector) change; the orbit is thus given by a set of continuously changing vectors which can be deduced by simple algebra from Newton's laws.

All vectors have the same mathematical properties regardless of the physical entities they represent. When we deal with the vector that is associated with distance, we refer to it as the displacement from point A to point B, rather than as the distance from A to B, which is just a number. The easiest way to represent a vector graphically in showing the displacement r between two points A and B is to draw a directed line (a line with an arrow at one end), starting from point A and ending at B. Associated with the displacement is its magnitude, which is simply the distance from A to B, and its direction. If we draw a line from point A to another point O, and use this line as a reference line, the direction of the vector from A to B is given by the angle the line from A to B makes with the reference line. The displacement (as a vector) is thus completely specified when the distance from A to B and this angle are given. A vector is completely known only if both its magnitude and direction are given. A map gives us exactly this kind of information.

Instead of giving the magnitude of the vector and the angle numerically, we can specify the vector by the graphical representation described above. This specification requires that we introduce a magnitude scale similar to the type of distance scale that is used on a map. A definite distance is associated with each unit length of the line that is used to represent the vector. If each half inch of the directed line described above represents 10 feet, for example, the magnitude of the vector is 30 feet if the directed line is one and a half inches long. Although this scaling process is done here explicitly for displacement, the same procedure is used for all other types of vectors. In doing so, however, we must remember that the number we get by measuring the length of the directed line is not a distance but the magnitude of some other vector quantity.

Although a vector in space is completely specified by three numbers or coordinates, we use only two numbers in the above example because we limit the vectors we deal with here to those lying in the plane of the paper. If the vector projected out of the plane of the paper, the angle defined above would not suffice to give its direction, since the vector could then coincide with any line lying on the surface of the right circular cone whose axis extends from point A to point O. If we picture a cone placed base down on a table and a wire that passes through the center of the base of the cone from the surface of the table to the top of the cone, the point where the wire meets the table represents the point O while the other end of the wire at the top of the cone represents point A. Every point on the circumference of the base can be connected to A by a vector (a line on the surface of the cone) which makes the same angle with the wire at point A. Since each of these points can be the point B, the vector A to B is not completely specified by a single angle. Alternatively, an infinite number of nonparallel two-dimensional planes can pass through the wire and the vector A to B may lie on any one of these.

Since vectors are measurable physical entities that can be expressed numerically, they can be manipulated by a variety of mathematical operations. The addition of vectors, for example, can be shown using a map on which the distances and directions of cities are given from any point on the map. If we follow a road which takes us from a point on the map through five different towns to our destination, a sixth town, we may represent each step in our journey from town to town as a vector if each town is connected to the next town by a straight road whose distance on a simple scale (10 miles to the inch, for example) is given by the length of the line connecting the two points on the map representing the two towns. Any such map is a graphical representation of vectors. The sum of these vectors on the map is then the vector that takes us directly from our starting point to the sixth town, our final destination. If we take a trip which originates in New York City and includes visits to Chicago, Seattle, San Diego, and Dallas before ending in Atlanta, the sum of the vectors traced out on this trip is the vector between New York City and Atlanta. The order in which we add the vectors does not affect the final result. Had we started our trip in Atlanta and traveled to Dallas, San Diego, Seattle, Chicago, and, finally, New York City, the sum of the vectors would be the same.

The numerical relationship between vectors may be shown if we walk 3 miles from point A east to point B and another 4 miles north from point B to point C. Although we will have walked a total of 7 miles, it is clear that the magnitude of the resultant displacement (the imaginary line between point C and point A) is smaller than the sum of the lengths of the lines from A to B and B to C. In other words, the distance that we would have to walk from point C to point A is less than 7 miles.

The length of the resultant vector between points C and A (the directed line connecting A and C) depends on the angle formed by the intersection of the component vectors (the directed lines between A and B and B and C). If the angle is zero so that the lines AB and BC lie along the same line, then the lengths of the resultant vector (displacement) equals the sum of the lengths of the component vectors. In that case, vectors may be added, just as we add numbers. If we walk east for 4 miles, then walk west for 3 miles, retracing our original steps, the resultant displacement will be 1 mile because we will end up 1 mile east of where we began. If we walk east for 3 miles, than walk north for 4 miles so that the angle between the two vectors is 90° so that the vectors are perpendicular to each other, we apply the theorem of Pythagoras (which states that the square of the hypotenuse of a right triangle equals the sum of the squares of the other two sides) to find the resultant vector. In other words, we add the squares of 3 (9) and 4 (16) which equals 25. The square root of 25 is 5 so the resultant displacement is 5. In short, we would have to walk 5 miles southwest to get back to our starting point. These examples show that the magnitude of the sum of two vectors ranges from the difference of their magnitudes to the sum of their magnitudes (in the example given, from 1 to 7) depending on the angle of intersection of the component vectors.

We apply these simple vector ideas constantly in our movements. If we go from a point on one side of a street to a point farther down on the other side, we walk diagonally across to the distant point rather than first along the street and then across it to the point on the other side. In taking this shortcut we are performing vector addition.

To be even more specific, imagine that you are going to a city which is northeast of your house by first traveling due east and then due north along two roads that intersect at right angles east of your starting point. The eastward part of your trip is then given by an east-pointing vector whose length on some scale (10 miles to the inch, for example) gives the distance to the north-pointing road. If this point is 30 miles east of you, a 3-inch line is the graphical representation of the vector. In the same way, a 4-inch line along the north road is the vector giving the second leg of your trip if the city is 40 miles north of the intersection point of the two roads. Your complete trip is the sum of the two vectors, one pointing east and 30 miles long (its magnitude) and one pointing north and 40 miles long. A road that leads directly from your starting point to the city brings you there along a single direction (northeast) and is therefore equivalent to the two legs of your original trip. The vector (of magnitude 50 miles) that represents this road is given by a 5-inch line and is the sum of the eastern and northern vectors.

Although displacements may be used to illustrate the properties of vectors, many other vector quantities occur in nature. An important example is a rotation, or angular displacement. If we have a wheel mounted on an axle and we designate two separate points along the rim of the wheel as A and B, we can represent a rotation as a vector by rotating this wheel about the axle so that point B is now in the position formerly occupied by point A. The amount of displacement of point A caused by the turning of the wheel is represented by the angle which is measured in degrees of arc. In other words, if points A and B were on opposite sides of the wheel so that one half of a complete turn of the wheel would be required for point B to reach the position formerly occupied by point A (and vice versa for point A), the angular displacement would be equal to 180°. Angular displacement is a vector quantity, since both direction (the direction of the axle) and magnitude (the size of the angle) are associated with this operation. A rotation is therefore defined as a vector whose direction coincides with the axis of rotation (180° in our example) and whose magnitude equals the numerical value of the angle or rotation. Since the axis of rotation (the wheel axle) can point in any direction, we have the same directional situation in rotations as in displacements.

There is one important difference between rotations and displacements: In the example we gave for the addition of two displacements, the order in which the two component vectors representing the displacements are added does not matter because the resultant displacement is always the same. But this invariance does not hold for rotations—if we rotate a body about two different axes, the final result depends on the order in which the rotations are carried out. If you rotate a box horizontally (about a vertical axis) one quarter of a revolution and then rotate it vertically (about a horizontal axis) one quarter of a revolution, the final orientation of the box is not the same as it would have been if you had reversed the order of the horizontal and vertical rotations.

When we introduced the vector concept, we specified a direction in space by first choosing an arbitrary line and then giving the angle that the vector makes with this line. However, this does not really define the direction of the vector completely. If we have a line OA (where O is the origin of the line), we still do not know which vector it might be because its position is not sufficiently specified. Because we do not have enough information, we are unable to say which of the vectors lying on the surface of the cone described above whose altitude is OA (the distance from the base to the height of the cone) is being considered.

To choose the particular vector we are talking about from all the other vectors, we must introduce at least one other line, e.g., the line OB which begins from the same point as OA but is not parallel to OA. If OB is perpendicular to OA (the intersection of the two lines forms a right angle), we can now specify the position of the vector in the plane determined by the lines OA and OB because all lines parallel to these lines provide us with a grid on which we can plot the position of any point or line in the plane. So long as the vector lies completely within the plane, we thus have a satisfactory system for defining its position. But we cannot specify a vector completely in space by means of the two perpendicular lines OA and OB because the plane determined by these two lines can be tilted in an infinitude of ways in space; a plane has only two dimensions—length and width—while space has three dimensions—length, width, and depth. This means that our vector is not determined until we fix this plane by choosing a third line OC which shares a common origin with the other two lines but which is perpendicular to both of them. This arrangement can be visualized by looking at a corner of a room. Where three lines meet, two of which intersect along the floorboards and one of which intersects the other two lines and rises up to the ceiling. The three lines OA, OB, and OC radiate outward from the corner of the floor and together constitute a coordinate system in space. The point O is called the origin of the coordinate system because it is the common point from which all three lines begin and the lines OA, OB, and OC are called its axes. Such a system enables us to specify a vector completely because we can lay off yardsticks along all three spatial dimensions—length, width, and depth.

The concept of the coordinate system is essential for understanding all dynamical structures in nature and the forces that maintain these structures. Since the complexity of the coordinate system we introduce depends on the nature of the space we are dealing with, we start with the simplest space—the points on a straight line. Such a space is a linear of one-dimensional space. We can set up a coordinate system in this linear space by choosing some point O, on the line as the origin. The position of any point, such as A, is then specified by giving its distance from O. If A is to the right of O on the line, the distance is taken as positive; if A is to the left of O, however, it is taken as negative. The position of any point on the line can be given as a single number (positive or negative) which is the corrdinate of the point relative to the origin O. This linear space is a one-dimensional space in the sense that only one number is required to locate a point in it. Because a one-dimensional space has little value to us, we next consider a two-dimensional space. The simplest two-dimensional space that can be constructed is the continuum or manifold of all the points on a plane, such as the points on a page. As with the one-dimensional space, we begin by choosing some point as the origin of our coordinate system. It is clear, however, that we cannot specify the position of a point such as A on the plane by just giving its distance OA from O because to say, for example, that point A is 5 cm from O is quite ambiguous; such a point A could be any one of the infinity of points lying on the circumference of a circle whose center is at O and whose radius is 5 cm. Hence, we need two numbers to locate a point A relative to the origin O in a plane because a plane is a two-dimensional manifold or continuum. There are various ways of introducing these two numbers, and each of these corresponds to a particular type of coordinate system. The simplest procedure is to start with two straight lines passing through O which we label OX and OY. These two lines are the axes of our coordinate system. The position of point A in this coordinate system can be specified by giving two distances, the distance of A from the y axis as measured along a line parallel to the x axis (the X coordinate of point A) and the distance of A from the x axis as measured along a line parallel to the y axis (the Y coordinate of point A). The position of every point on the plane can be specified in this way since we can cover the plane with two sets of parallel lines, one set parallel to the x axis and the other set parallel to the y axis. Every point on the plane is then at the intersection of two such lines. This set of lines defines a Cartesian mesh, and the lines OX and OY themselves define a Cartesian coordinate system, named after the great 17th century French philosopher and geometer, René Descartes.

The most important practical application of a Cartesian coordinate system is laying off streets and avenues in a planned city. The address of each house in such a city is given as two numbers (the coordinates) which locate it in the city. One number gives the street the house is on and the other (the house number) gives the position of the house with respect to the two avenues between which it lies.

This two-dimensional Cartesian coordinate system is only one among many such coordinate systems that can be obtained by changing the angle between the two lines OX and OY. If this angle is less than 90°, we call such a coordinate system an oblique Cartesian coordinate system. If the angle between OX and OY is increased until the two lines are perpendicular to each other, we then have a rectangular (right angle) Cartesian coordinate system. In general, when the two sets of intersecting lines of a coordinate system are perpendicular to each other, we speak of an orthogonal coordinate system.

The larger the angle between the two intersecting baselines in a coordinate system, the more the direction of the vertical axis OY (which extends upward from the origin O) is independent of the horizontal axis OX (which extends rightward from the origin O). If the vertical axis OY is perpendicular to the horizontal axis OX in the same way that the line running along the floorboards is perpendicular to the line which goes upward from the floor of the corner to the ceiling, the direction OY is completely independent of OX in the sense that no matter how far we move along OY, we advance neither to the right nor to the left of the point O, that is, along OX. We can think of an elevator moving vertically up a glass shaft as moving along the OY direction and a line on the lobby floor of the building housing the elevator as the OX direction. If the sun is shining directly over the elevator, the only shadow cast by the elevator is the square at the bottom of the elevator shaft, if the elevator shaft is perpendicular to the floor of the lobby; as the elevator moves straight up and down, the position of the shadow in the shaft does not change, but is confined to the base of the elevator shaft. If the shaft is tilted, however, the shadow moves along the lobby floor as the elevator rises. The amount of the shadow's displacement along the lobby floor depends on the extent to which the shaft is tilted. What becomes apparent is that once we no longer have a completely vertical elevator shaft, there is some horizontal displacement of the shadow along the lobby floor as the elevator rises. Because there is now some movement along the horizontal direction as well as the vertical direction, the two directions are no longer completely independent. Since only one line can be drawn through point O perpendicular to the line OX, there are only two independent directions in the plane. Once the horizontal direction OX, for example, is chosen, only one other horizontal direction on the surface is independent of (perpendicular to) OX. For this reason the plane (or any other surface) is a two-dimensional manifold, and the position of every point on it is specified by just two numbers (the coordinates of the point). Any direction from O to a point B on the plane can be represented as a mixture of the two independent directions along the axes, OX and OY, that define the orthogonal coordinate system. We might imagine the direction OB as a line drawn by a child from the corner of the floor along the wall between the horizontal direction OX and the vertical line OY. If the child marks a point B′ on this line OB and then draws a vertical line from B′ downward until it intersects the OX axis at a point x and then draws a horizontal line from B′ that intersects OY at y, the child has created a rectangular Cartesian coordinate system. The key to understanding this coordinate system is that any point on any line other than the two independent axes OX and OY can be specified in terms of those two axes by drawing straight lines from the point in question perpendicular to each of the independent axes. This is similar to specifying the position of any piece on a chessboard by referring to the particular letter and number along the two borders to which perpendicular lines from the piece itself can be drawn.

A coordinate system describes relationships between events in nature in terms of the coordinates (positions) of these events. One of the most important relationships between two events is the distance between them. Depending on the type of coordinate system used, the formula for this distance (expressed in terms of the coordinates of the two points) is more or less complicated. Obviously, it is advantageous to use a system in which the distance formula has its simplest form, for only then can the purely geometrical relationships between events be most easily separated from the physical relationships—if this can be done at all. This consideration in determining the choice of coordinate system extends beyond the mere geometrical relationships for, as we shall see, the proper choice of a coordinate system can simplify the expressions for the very laws of physics themselves.

The distance between two points P_1 and P_2 is a physical reality (here we deal with the Newtonian, or the pre-Einsteinian, concepts of space) and must therefore be the same regardless of how we express that distance in terms of the coordinates of the two points. If we use the oblique coordinate system, the formula for the distance contains not only the coordinates along both OX and OY for each point, but also the cosine of the angle that is formed by the intersection of OX and OY. In a rectangular system, however, the angle (90°) between the vertical and horizontal axes does not enter into the formula for the distance between the two points. If we square the distance between the x (horizontal) coordinates of the two points and add that to the square of the difference between the y (vertical) coordinates of the two points, we obtain the square of the distance between the two points. This operation is written algebraically as $d^2 = (x_2 - x_1)^2 + (y_2 - y_1)^2$. This is the simplest expression we can obtain for the square of the distance between two points; it is just the ordinary theorem of Pythagoras in a slightly more complicated notation. This expression for the square of the distance between two points (or between two physical events) gives us a deep insight into the geometry of the space in which the points are embedded. For this reason it is important to use the simplest coordinate system possible to reduce the formula for d^2 to its simplest form. To illustrate the practical application of this distance formula, we picture a city laid out with parallel streets equally spaced, serially numbered, and perpendicular to equally spaced avenues. If the houses on all streets are equally spaced and numbered accordingly, the distance between any two houses is obtained by squaring the difference between their street numbers, adding to that the square of the difference between their house numbers, and then taking the square root of this sum.

The rectangular Cartesian system with its two sets of perpendicular lines divides the surface into a square mesh, whereas the oblique system divides the surface into a diamond mesh. The type of mesh used merely alters the way the surface looks to us when the mesh is superimposed on it, but does not alter the intrinsic geometrical relationships on the surface.

Having considered straight line coordinate systems, we next consider a more general type of system consisting of sets of lines that need not be straight. This might appear to be a needless complication which can teach us nothing that we cannot learn using a Cartesian coordinate system, and this is true as long as we are dealing with a plane (flat) surface. But if we are studying events or relationships on a curved surface such as a sphere, we must introduce a coordinate system consisting of sets of curved lines—a curvilinear coordinate system—since straight lines cannot be laid off on a curved surface.

This question was treated in its most general form more than a century ago by the great German mathematician Carl Friedrich Gauss, who pioneered a radical geometry consisting of sets of curved lines. A coordinate system consisting of sets of such general curves is called a Gaussian coordinate system. It divides the surface into an irregular mesh consisting of boxes of varying sizes. The two sets of curves that constitute the Gaussian coordinate system may be chosen in an infinity of ways, but must be so chosen that no two curves in the same set cut each other at any point except, possibly, the origin of the coordinate system.

In general a Gaussian coordinate system in which complex curved lines are used complicates things because the distance formula discussed above becomes more mathematically unwieldy. The simplest Gaussian coordinate system on a plane is the polar coordinate system which consists of a series of circles concentric to the point O (the origin of the coordinate system) and straight lines radiating outward from this point. We might visualize such a coordinate system by imagining an archery target on which a series of lines radiate from the center of the target outward in all directions. Every point on the target plane lies on just one circle and on one line. These lines and equally spaced circles divide the plane into meshes of irregular size; the boxes of the mesh become smaller as one moves closer to the origin O.

The polar coordinates of a point are specified by indicating on which circle and on which line the point lies. To do this we choose some line—such as OX—radiating from the origin, as a reference line. Any other line relative to this reference line is then specified by giving the angle it makes with OX at the origin O. The circle the point lies on is designated by giving the circle's radius r (its distance from O). In a polar coordinate system, only one of the coordinates of a point is a distance; the other coordinate is an angle. The two Cartesian coordinates of a point (its position relative to the vertical and horizontal axes) are simply related to its polar coordinates the radius and angle. By expressing the relationship between the Cartesian and the polar coordinate system, we can transform the specification of the position of a point in a Cartesian coordinate system to its specification in a polar coordinate system and vice versa.

The expression d^2 for the square of the distance between two neighboring points (expressed in terms of the polar coordinates—radii and angles—of two points) is more complex than it is in terms of rectangular coordinates. If the first point is designated by a specified radius and angle coordinate, the polar coordinates of the second point are obtained by adding the change in the radius and the angle to the coordinates of the first point. The square of the distance d^2 between the two points is the sum of the change in the radius squared and the square of the radius multiplied by the square of the change of the angle.

Any time we depart from rectangular Cartesian coordinates we complicate the expression for d^2. But as long as we are on a plane (a flat surface, so that Euclidean geometry applies) we can always introduce a rectangular Cartesian coordinate system and thus get back to the simplest expression for d^2. In other words, the complexity in the expression for d^2 is due entirely to our choice of coordinate system, if we are dealing with points on a flat surface.

The simplest kind of coordinate system on a plane is the rectangular Cartesian coordinate system; the formula for the square of the distance between two points is then the simple law of Pythagoras—any departure from this kind of coordinate system leads to a more complex expression for the square of the distance. In fact, if we can introduce a coordinate system on a surface such that the formula for the square of the distance between any two neighboring points on the surface is the Pythagorean formula, the surface must be flat. If a surface is not flat, we cannot do this and the square of the distance formula cannot be reduced to its simplest Pythagorean form. The expression for d^2 not only gives us an insight into the kind of coordinate system being used, but also the geometrical properties of the surface itself. The importance of these insights for our understanding of the laws of physics becomes more evident when we discuss the theory of relativity which relates gravity to space-time curvature stemming from a non-Euclidean geometry.

A sphere is the simplest example of a surface that is not flat and therefore does not admit a single (the same) Cartesian coordinate system everywhere upon it. A coordinate system on the surface of a sphere is important in geography and navigation, since the earth is a sphere and we must know how to locate objects on its surface. Although the coordinate system we introduce on the earth's surface is not Cartesian, it is in many respects similar to the polar coordinate system on the plane which consists of straight lines radiating from a point and circles concentric to this point. To construct a coordinate system on the surface of the earth we choose the northern pole as the origin O because it is on the earth's axis of rotation and thus remains fixed as the earth spins.

We now draw through the point O the curves that correspond to the radiating straight lines of the polar coordinate system on a plane. Curves must be used because we cannot draw straight lines on a sphere. A straight line connecting two points on a plane is the shortest distance between them, but the shortest distance between two points on a sphere is the arc of a great circle; we must therefore use great circles on the sphere to correspond to the radiating straight lines of the polar coordinate system on the plane. The radius of a great circle equals the radius of the sphere itself—it is the biggest circle that can be drawn on the sphere; a plane drawn through this circle cuts the sphere in half. The great circles on the earth which originate at the north pole (they also pass through the south pole) are called circles of longitude. Just as we started with the reference line OX in our plane polar coordinate system, so we choose the great circle passing from the north pole through Greenwich in England as our reference circle on the earth. This particular circle of longitude is called the prime meridian. The longitude of any point on the earth is the angle between the plane that contains the prime meridian and the plane that contains the circle of longitude passing through the point. It is measured westwardly from Greenwich; this angle corresponds to the angle of the polar coordinate system on the plane.

We now consider the curves on the surface of the sphere that correspond to the concentric circles of our polar coordinate system. These curves are circles but they are not concentric. They are all the circles that can be drawn with their centers on the earth's axis of rotation. These circles, which we call circles of latitude, get larger and larger as we move away from the north pole. The center of the largest such circle coincides with the center of the earth and is called the equator.

To locate a point on the surface of the earth using this longitude–latitude coordinate system, we proceed as we did in the polar coordinate system on the plane and give its angle (its longitude) and also the distance we must move from the north pole to reach it along the circle of longitude that passes through it. Although this corresponds exactly to the polar coordinates of a point on the plane and we still keep the angle of longitude, we use in place of the distance from the pole the angle which the arc of the circle of longitude from the equator to the point subtends at the center of the earth. This angle is called the latitude of the point. The coordinates of the point are thus these two angles which are not distances. These coordinates can be used to express the distance between two neighboring points on the surface of the earth but the process is more complex than the Pythagoras formula because it contains the cosine of the latitude.

The celestial coordinate system, which is extremely important in astronomy, is similar to the latitude–longitude coordinate system on the earth. As astronomers collect more and more data about the stars and other celestial objects, they must keep accurate records, and this means, in part, specifying as accurately as possible the positions of the objects they are studying. For this purpose, a coordinate system is superimposed on the sky. To specify the position of a celestial body on the sky (where it is in relation to other bodies, *not* how far away it is) the astronomer treats the sky as a spherical surface and introduces a coordinate system on it that corresponds to the system of longitude and latitude on the earth.

If the earth's axis of rotation is extended infinitely far it pierces the sky at a point called the north celestial pole (NCP); this point remains fixed in the sky and all the stars rotate around it during each 24-hour period. We use this point as the origin for our celestial coordinate system and, just as on the earth, we lay off a system of great circles through this point. These great circles, called hour circles, meet at the south celestial pole (SCP) and are the counterpart on the sky of the terrestrial circles of longitude. Each point on the sky lies on just one hour circle. We now draw a series of small circles, all of whose centers lie on the line connecting the two poles; these small circles are equivalent to the circles of latitude on the earth. They are called circles of declination, the largest of which is called the celestial equator, which is concentric with (has the same center as) the earth's equator. Every point on the sky lies on a circle of declination.

We can then locate any celestial object by specifying the hour circle and the circle of declination on which it lies. For measuring purposes, we take as our reference hour circle (corresponding to the prime meridian through Greenwich) the hour circle which passes through the vernal equinox which is the point on the celestial equator which coincides with the position of the sun when spring begins. The sun passes this point on about March 21st of each year when it is moving (apparent motion) northwardly on the sky.

One of the coordinates of the celestial point, its right ascension, is the angle between the plane that contains the hour circle which passes through the vernal equinox, and the plane of the hour circle which passes through the point. This angle corresponds to longitude on the earth. The other coordinate is the declination of the point. This is the angular separation between the point and the celestial equator measured along the hour circle from the equator to the point. It is equivalent to latitude on the earth. The position of a star listed in a stellar catalog is given in terms of its right ascension and declination.

Two-dimensional coordinate systems are all we need as long as we are dealing with points or objects on a surface. To follow the course of events as they unfold in the real world, however, we need a three-dimensional coordinate system since the space of the real world is three-dimensional. This means we need three sets of intersecting lines or curves, the simplest of which is obtained by extending the rectangular Cartesian coordinate system to three dimensions. This can be done by introducing a third set of parallel lines which are perpendicular to the plane defined by the mutually perpendicular lines OX and OY (the X and Y lines). This new set of lines may be referred to as the Z lines and can be combined with the X and Y lines to form a three-dimensional rectangular Cartesian coordinate system. These three sets of mutually perpendicular lines divide space into cubes. We can now locate any point in three-dimensional space by specifying the three Cartesian coordinates, x, y, z, of this point relative to the origin O. This rectangular system has the same advantage in space that a two-dimensional rectangular system has in a plane: the formula for the distance between two points is simpler than in any other coordinate system. If we have one point with coordinates x_1, y_1, z_1 separated by distance s from another point with coordinates x_2, y_2, z_2, then the square of s is the sum of the squares of the difference $(x_2 - x_1)$, $(y_2 - y_1)$, $(z_2 - z_1)$ between the coordinates of the two points.

Just as we can (and sometimes must) introduce Gaussian coordinates on a surface, we can do the same in space. Instead of discussing such coordinate systems in detail, however, we give instead an example of such a coordinate system, one which is used extensively in physics and the other sciences. The three-dimensional rectangular Cartesian coordinate system X–Y–Z is our base, but this time we use three spherical coordinates to represent a point in space. The first coordinate is the distance of the point from the origin O; the second coordinate is the angle that a line drawn from the origin to the point makes with the Z axis; the third coordinate is the angle that the Z–Y plane makes with the plane that contains both the point and the Z axis. Only the first of these three coordinates is a distance; the other two are angles. This tells us that the expression for d^2 (the square of the distance between two neighboring points) cannot be as simple in terms of the three spherical coordinates as in the Cartesian system. The reason is that the two angular coordinates are not squares of distances, so these quantities have to be multiplied by the squares of distances to relate them to d^2. This makes the formula for d^2 more complicated than in the rectangular Cartesian coordinate system.

Although we introduced vectors quite independently of coordinate systems, we can understand their properties most easily by referring them to coordinate systems. The important point here is that a vector is a physical entity whose magnitude and direction (the physical attributes of the vector) are entirely independent of the particular coordinate system we use. We can therefore choose any coordinate system we please to describe relationships between vectors without altering the physical significance of these relationships. This is very useful since we can often simplify such relationships greatly by the choice of an appropriate coordinate system. Another important advantage of using a coordinate system to study vectors is that we can then replace vector operations by ordinary algebraic operations. Any vector in space can be broken up into three mutually perpendicular and independent components—one component along each of the axes of the coordinate system. All of these components are as much vectors as the resultant vector because they all have direction and magnitude. The resultant vector is equal to the sum of the three component vectors. Since the components of a vector specify it completely, the resultant vector $\mathbf{A}$ consists of a vector $\mathbf{A}_x$ along the x axis, a vector $\mathbf{A}_y$ along the y axis, and a vector $\mathbf{A}_z$ along the z axis.

If we introduce another coordinate system with the same origin O but with three mutually perpendicular axes pointing in different directions from the axes in the original coordinate system, the resultant vector is not altered by this change even though the three component vectors in the new system are not the same as the three component vectors in the original coordinate system. The reason for this is that the angles that the resultant vector makes with the axes of the second coordinate system are different from those it makes with the axes of the original coordinate system.

Before leaving vectors we define the product of two vectors, **A** and **B**. Since a vector has both magnitude and direction, it is not immediately clear how we multiply the two vectors which intersect at some angle. It is easy enough to see how the magnitudes of the two vectors are to be treated in multiplication, but the treatment of their directions is not so simple. To obtain the magnitude of the product we first multiply the magnitudes of two vectors **A** and **B** to obtain the quantity AB. However, this is only the first step in obtaining the magnitude of the product. We must see how we are to treat the angle, and here we have two choices, depending on whether we want the product of the two vectors to be a scalar or a vector, for both of these results are possible. For the scalar product of **A** and **B** we multiply the product AB of the magnitudes by the cosine of the angle between the vectors which gives what is called either the scalar product of the two vectors or the dot product, because it is written with a dot between **A** and **B**; that is, **A·B**.

The significance of the scalar product can be seen by dropping a perpendicular line from the head of vector **A** to the vector **B** which gives vector **C**. Vector **C** is simply the projection of vector **A** on vector **B** or the component of **A** parallel to **B**; its length equals A multiplied by the cosine of the angle between **A** and **B**. The scalar product of **A** and **B** is just the ordinary product of the magnitude of **B** and the magnitude of the projection of **A** on **B**. It also equals the product of the magnitude of **A** and the magnitude of the projection of **B** on **A**. From this definition of the scalar product we see that if **A** and **B** are perpendicular to each other (the angle equals 90° and the cosine of the angle is 0), then **A** multiplied by **B** is 0. This is an example of a product which can be 0 without either of the factors being 0. If the two vectors **A** and **B** are parallel so that the angle between them is 0 and the cosine of the angle is 1, the product is AB. The scalar product of a vector by itself gives the square of the magnitude of the vector.

An important property of the product of two vectors is that the value of this product does not change when we go from one coordinate system to another. Another kind of important product in the study of vectors is the vector product or cross-product of two vectors. To define this product which itself is a vector, we begin with two vectors **A** and **B** (which radiate from a common origin at right angles to each other like the two lines which diverge from the corner along the floorboards in the example of a coordinate system discussed earlier) and construct a vector **S** perpendicular (like the line running from the corner on the floor up to the ceiling in the earlier example) to the area (a rectangle) bounded by vectors **A** and **B**. We choose for the magnitude (or length on an appropriate scale) of **S** the numerical value of this area. This vector **S** is the vector product of **A** and **B**. Since we could have chosen **S** to be in the opposite direction just as well, we must differentiate between these two cases by choosing the direction of **S** in such a way that the directions of **S**, **A**, and **B** are arranged like the thumb, the forefinger, and the second finger on the right hand, respectively. This means that the vector product $\mathbf{A} \times \mathbf{B}$ is not equal to $\mathbf{B} \times \mathbf{A}$, since the direction of **S** in the latter case is just opposite to that of the former. As a result, $\mathbf{A} \times \mathbf{B}$ equals to $-\mathbf{B} \times \mathbf{A}$.

It is clear that coordinate systems are extremely useful in keeping a record of events or in analyzing geometrical relationships and plotting the orbits of moving bodies, but the importance and usefulness of coordinate systems far transcend these functions when we want to formulate the laws of nature. Since these laws are derived by correlating events, we see that such laws are most readily represented in the coordinate systems in which the events themselves are recorded. We emphasize that we are not limited to any special choice of coordinate system when we study such events and formulate laws; we may choose a coordinate system in which the laws and relationships are most easily understood or assume particularly simple forms.

It may not be immediately obvious to us which coordinate system is most suitable to represent a particular set of events, so that it is, in general, necessary to shift from one coordinate system to another before the most felicitous choice is made. This procedure is called transforming from one coordinate system to another. As we have already seen, the formula for the distance between two points or events can change considerably as we go from one type of coordinate system to another but the distance itself is unaltered.

The question that follows from this is whether the laws of nature depend on the choice of coordinate system or whether they are independent of the type of coordinate system we use. This question did not assume particular importance for a long time, since there appeared to be no reason to question the validity of the classical laws of physics, whether formulated in one coordinate system or another. The basis for this idea is that a law of physics describes something intrinsic about nature and this intrinsic property does not change with different coordinate systems. However, another aspect of coordinate systems is related to this which is more important and which was most fully exploited by Einstein in his theory of relativity. This aspect has to do with the mathematical form of the law, for it is clear that although we may demand that the content of a law must not be altered when we transform coordinate systems, there appears to be no reason to insist that the mathematical form remain the same. As we shall see later, however, this invariance is precisely what must be demanded of a law, for it is impossible to separate the mathematical expression of the law from its physical content.

This principle of invariance (that the mathematical formulation of a law must remain unaltered when we transform from one coordinate system to another) is an extremely powerful tool for finding and testing the correct laws of physics. All we need to do is introduce a transformation of coordinates when we have such a law and then see if this changes the expression for this "law." If the transformation does change the expression, the expression cannot be accepted as a law and it must be altered appropriately. If the expression does not change for some particular transformation of coordinates, this is not in itself conclusive evidence that a law has been discovered, for this is true only if the expression is invariant (does not change) for all types of transformations of coordinates. This requirement may appear to limit the usefulness of this principle but there are very general ways in which coordinate transformations can be represented which permit us to check physical laws almost at a glance.

Up to now we have considered only transformations which involve changes in the geometrical aspects of the coordinate system. However, other kinds of coordinate transformations that involve changes in the motion of a coordinate system are very important. Before considering these transformations, we must describe how we are to measure the motion of a body; for this we must introduce the derived quantities velocity and acceleration.

CHAPTER 4

Time and Velocity

The universe is not hostile, nor yet is it friendly. It is simply indifferent.

—JOHN HAYNES HOLMES, *A Sensible Man's View of Religion*

In the previous chapter we introduced distance as the first of our fundamental physical quantities from which we can construct only two other quantities: area and volume. We must introduce at least one other basic physical quantity which we can combine with distance to derive the basic quantities in terms of which the laws of nature are to be expressed. This new basic entity is time. That we need time in addition to the three spatial dimensions to describe nature shows us that the laws of nature cannot be derived from the three-dimensional geometry of space. But even the addition of time to the spatial quantities is not enough for a complete description of nature and other basic quantities must still be introduced. For the time being, however, we limit ourselves to the consideration of time.

Just as in our discussion of space we did not attempt to define distance but, instead, showed how it can be measured and expressed in terms of a unit, so, too, we proceed with time. Indeed, we cannot do otherwise because we can no more define time than we can define space. Our only recourse is to outline a set of operations which give us a number that we can call the interval or time span between two events.

In choosing these operations, we run into some difficulties that we did not encounter in introducing the distance-measuring operation. When we dealt with space we began by introducing a unit of distance which, itself, is a palpable physical body (a rod) that everyone can see. But what do we mean by a unit of time and how is such a unit to be introduced? We can begin by noting that we are born with a time perception which, though not as obvious as space perception, is still acute. It is fairly easy for us to estimate a distance with fair accuracy, but our estimate of a time interval may be completely erroneous. The reason for this discrepancy is that we estimate the distance visually—comparing it with some other visual object whose length we know. The estimation of an interval of time, however, is a mental exercise and therefore a highly subjective matter. Our consciousness of the flow of time stems from two different kinds of experience: (1) the continual changes that occur in our environment and in ourselves and (2) periodic phenomena—events which continually repeat themselves. Both of these groups of phenomena are essential for our perception of time and our measurement or estimate of time intervals. It may appear, at first, that we need only periodic processes for the measurement of time intervals but for the other aspect of time perception the first category of events is essential—this is the direction of the flow of time. We speak of the past, the present, and the future, and associate these terms with the manner and temporal order in which events unfold. If there were no inhomogeneities in our surroundings and no changes from moment to moment, we would have no way of differentiating the past or the future from the present. We could use periodic phenomena to measure the time that has elapsed in such a universe, but the time span would be meaningless, since it would have no relationship to what we now mean by "before," "now," and "after."

From our daily experiences, as they are related to the changes that go on all about us, we learn how to arrange events in an orderly time sequence. If we are shown two pictures of the same individual taken with a long enough time interval between them, we can generally tell which picture was taken first by the aging of the individual. This example is but one of the many kinds of changes in nature that indicate the direction of the flow of time.

Our estimate of time intervals can be very precise in spite of the limitation of subjectivity when we are dealing with a particular time span and seek to assign a numerical value to it. We can train ourselves to perform acts very quickly that require incredible precision in arranging a temporal sequence. When we discussed distance, we saw that the violinist (or pianist) has the remarkable ability to place his fingers just where he wants to, at precisely the right moment and with the right frequency. The same skills are required of the athlete. Moreover, we all perform, almost unconsciously, many acts in a precise sequence which we could not do without an innate sense of time. In addition to deliberate and semideliberate acts, our bodily functions, such as the heartbeat, for example, are performed involuntarily with a rhythm (frequency) over which we have no control. All these phenomena can be used to measure time spans, since they divide the flow of time into a series of small, equal intervals. A time span can be labeled by the number of such intervals that it contains. But we must now be more precise in our specification of time measurement, since we can never be sure that the periodic bodily functions, described above, occurring in different persons, represent the same basic intervals (the same units of time).

Owing to the variations from person to person, we must use some natural, inanimate periodic phenomenon, which is the same for everyone to measure time. Many such phenomena are all around us and we can use any one of them to introduce the basic period for time measurement. The beat of a particular pendulum or the oscillations of some special spring can be used, but the fundamental period that is used as a unit of time today is based on the rotation of the earth.

As seen from a body outside the earth—e.g., the moon—the earth rotates at a remarkably steady (but not exactly constant) rate. This period of rotation is called the sidereal day. It is the interval between two successive risings (or settings) of any given star. This sidereal day is not quite the same length as our ordinary, or solar, day, but is slightly shorter. The solar day is the interval between two successive risings (or settings) of the sun. The ordinary hour is defined as one 24th of the solar day and one 60th of the hour is called the minute. One 60th of the minute is called the second. We use the second as our basic time unit and express time intervals in terms of this unit.

The sidereal day—the time it takes the earth to rotate once on its axis—is about 4 minutes shorter than the solar day because the earth is revolving around the sun while rotating around its own axis. This combination causes an apparent change in position of the sun relative to the other stars, so that it rises and sets about 4 minutes later than the stars do.

The second of time is based on the rising and setting of the sun, but since the apparent diurnal motion of the sun across the sky is irregular, the second is not precise; it must be replaced by a constant interval of the same length. We can create a constant interval by using a mechanical device such as a pendulum or spring constructed to beat out seconds. Instead of using a mechanical oscillator, we can use the vibrations of a crystal or an atom or molecule, and in this way obtain a very accurate clock. Such atomic clocks are so accurate that they do not gain or lose a second in three centuries.

Periodic astronomical phenomena other than the rotation of the earth can also be used to define the second. The period of revolution of the earth around the sun is extremely constant and is used to define the second with great precision. The second can be defined as a definite fraction of the time it takes the earth to revolve around the sun.

Rotating astronomical objects, called pulsars, have recently been discovered; they emit radio signals at extremely precise intervals ranging from a millisecond to about 2 seconds. The period of rotation of a pulsar is so precise that the interval between any two signals from a given pulsar is constant to within one part in a million billion. This precision is equivalent to a clock that does not lose or gain a second in 30 million years.

The time between any two events can now be specified as the number of basic time intervals contained between the two events, and may be expressed as the number of seconds in this interval. A device so constructed that we can count the number of basic periodic intervals contained in a span of time is called a clock. Since many different kinds of clocks can be constructed, we must be able to compare any two clocks to ascertain that they give the same results for a given time span. This does not mean that the basic beat of all clocks must be the same. It means that we must have a way of translating the reading given by one clock to that given by another; when we can do this the two clocks are synchronized. This procedure is simple if the two clocks are together for direct comparison. If they are separated, however, they can be synchronized by a signal sent back and forth from one clock to the other.

Suppose a signal is sent from one clock at the moment that it reads t_1 and reaches the second clock when the latter reads t_2. The signal is then immediately reflected back to the first clock, when it reads t_3. It is then immediately reflected back to the second clock when this second clock reads t_4. The two clocks are synchronized if $t_3 - t_1$ equals $t_4 - t_2$. For this procedure to work, we must use a signal whose speed of propagation from one clock to the other remains constant. As long as the two clocks are fixed with respect to each other, this presents no difficulty, but if they are moving relative to each other, an important question arises. How does the motion of an observer affect the speed of the signal as he measures it? Fortunately, as we shall see later, the motion of light in a vacuum is the same for all moving observers (with certain restrictions) so that light is just the kind of signal we need.

If we have a signal whose speed is the same at all points of its path, we can use this signal to synchronize all fixed clocks with respect to one given clock. Suppose that the time shown on a fixed clock at some moment is t_0, and that at this moment a signal is sent out from the clock to another at a distance l. The second clock is then said to be synchronized with the first clock if it registers the time $t_0 + l/c$ when it receives the signal from the first clock. Here c is the speed of the signal. However, it is helpful if we discuss the concept of speed before we begin talking about the speed of a signal.

In our discussion of distance we spoke of the distance between two events and pictured each event as being associated with a particular point of space. In fact, the very concept of space is associated in our minds with events (or particles) which can be used to set up a frame of reference. However, an important question arises here in connection with the same event (or the same point of space) at different times. Mathematical space can be pictured as having an absolute and unchanging character defined by points whose positions relative to each other do not change. But in actual space as we know it, the concept of the same point at different times is quite ambiguous and has no absolute meaning. The concept acquires a meaning only if we introduce a frame of reference, and then only with respect to this frame of reference. When we say that we return to the same point at different times (e.g., to our beds each night) we certainly do not mean the same point in empty space. We simply mean the same point on the earth's surface, without taking into account the motion of the earth.

It may be argued that if we wanted to, we could return to the same point in space at different times by first carefully ascertaining the earth's motion and then using the knowledge to take the necessary steps to return to our initial point in space at a later time. This would not do the trick, however, because it does not take into account the motion of the sun and, indeed, of the whole solar system, through space. At most, all we can determine is the motion of the solar system relative to the stars, but this does not give us its motion in empty space, which we would need to know to be able to return to the same point of space at different times. This analysis, then, shows us that the concept of the same point at different times has no absolute meaning and can only be defined in some particular frame of reference.

The absence of a uniform axis of time leads back to the problem of measuring the distance between two points. We saw that if the two physical objects that define the points are at rest in our frame of reference, the distance between them can be measured without considering time. If the objects are moving, however, the measurement has meaning only if their positions are noted at the same time. The measurement of distance cannot be entirely disassociated from time and, in particular, from the concept of simultaneity—the simultaneous specification of the positions of two events (or particles) in our frame of reference. We shall see that we cannot define simultaneity independently of our frame of reference (in an absolute way) so that the concept of absolute space itself becomes meaningless. If a change in the motion of our frame of reference, relative to two events, results in a change of our perception of the simultaneity of these events, our measurement of the distance between them also changes.

How are we to determine simultaneity? If the two points at which the events occur are fixed in our frame of reference, we simply see whether a signal detector (a photoelectric cell) stationed midway between the events, receives the signals from the two events at the same moment. If it does, the events occurred simultaneously in our frame of reference. There is no ambiguity when we say that two or three or any number of events occurred simultaneously at different points of space if these points are fixed in our frame of reference.

But if we are observing events relative to which we are moving, the simultaneity concept is ambiguous. If we are located midway between two events but are moving toward one and away from the other, can we draw any conclusions about the simultaneity of the two events from the times at which we receive their signals? It seems that we can if we know how our motion affects the speed of the signals from the events and know the nature of our own motion. The question of the simultaneity of events is a complex one when motion is involved; this point is examined in great detail when we discuss the theory of relativity.

Another difficulty arises in establishing the simultaneity of distant events which stems from the finite time it takes a signal to reach the observer. If a signal existed which traveled infinitely fast, we could become aware of all events occurring everywhere instantaneously. But the fastest signal known is light, which travels at a finite speed, so that the more distant the event is from us, the later we become aware of it. It is meaningless, for example, for us to speak of an event occurring now on a star that is a million light-years away from us, since information about such an event will not reach the earth until a million years from now. We cannot describe a time sequence on such a star. The past and future are not precisely defined for us on distant frames of reference.

We saw previously that with modern types of clocks, very short time intervals can be measured quite accurately. As we go down to intervals of less than a thousandth of a second, however, clocks based on mechanical vibrations such as those produced by springs do not give accurate enough results. Instead of mechanical vibrations, then, we can use electrical vibrations whose frequency can be altered and controlled by using appropriate electrical circuits. If such a circuit is connected to a cathode-ray oscillograph (a television tube), the oscillations can be seen visually and counted, and the time interval corresponding to a single vibration can be determined. Electrical oscillations can be used to measure intervals of time as short as a millionth or a tenth of a millionth of a second. When we go to still smaller intervals, other procedures must be used.

Just how small can physically meaningful time intervals be? In theory, we may picture time as being infinitely divisible with no limit on the shortness of an interval, but we know that we can assign a physical meaning to an interval of time only if we have some way of measuring it. We can, however, deduce the size of the smallest measurable time interval from the basic physical processes that occur in the smallest known physical system. The time intervals associated with atomic processes are as small as one hundredth of a millionth of a second. Such small intervals cannot be measured by means of electrical oscillations, but special atomic clocks can be used.

Still smaller time intervals are associated with the processes inside nuclei and with the creation and disappearance of any one of the newly discovered particles (strange particles) that physicists have found during the last few decades. Such particles are characterized by their lifetimes which are just the intervals of time during which they remain in their created state before changing to some other type of particle. Their lifetimes may be as short as a billionth or a billionth of a trillionth of a second, and the only method we have of measuring them is to observe the length of the tracks the particles leave on a photographic plate during their lifetimes. If we know the speed of a particle, we can then find its lifetime by dividing the length of the track by this speed. The limitations of this method lie in the accuracy of measuring such a small path and in knowing the actual speed of the particle.

We saw that we are not always limited to periodic processes in measuring time intervals but can use the known speed of a particle. Another nonperiodic process for measuring time intervals, particularly those of very long duration, is radioactive decay. This process is based on the fact that certain types of atomic nuclei (e.g., uranium, radium, carbon-14) undergo spontaneous decay and change into other nuclei. The rate at which this happens differs from one type of radioactive nucleus to another, but it is constant (no external condition affects it) for any given type. Half of any given amount of uranium-238, for example, decays in 4 billion years. To find the age of an object containing uranium such as the crust of the earth, we need only compare the amount of uranium it now contains (by observing its radioactivity) with the amount it originally had. In this way, we know that the earth is about 5 billion years old.

Very long time intervals can also be measured or estimated from an analysis of such physical systems as stars, galaxies, and the universe itself. Since the laws that govern the evolution of such systems are known, we can deduce their ages by tracing their evolution. As an example, we can estimate the age of the entire universe (about 20 billion years) if we accept the theory of an expanding universe which originated explosively from a single concentrated globule of matter (the "big bang" theory). The age of the universe can then be estimated by first measuring its present rate of expansion through spectroscopic observations of distant objects, and then extrapolating backwards to the time when the universe was a small, concentrated sphere.

Time can now be combined with distance to obtain a new and important quantity to which we have already referred—the velocity of a particle. In our discussion of time, we spoke of the speed of a signal and of the motion of a frame of reference without defining these concepts precisely. To provide more accurate definitions, we begin by considering just the concept of speed. A particle moves from some point A at time t_0 to point B at some later time t_1 along the straight line connecting these two points. We now measure the distance s between the two points and divide it by the time interval $t_1 - t_0$ to obtain the quantity v which is the average or mean speed of the body along the straight line between the two points. This quantity is a scalar since we are concerned here only with a magnitude—the direction from A to B is of no interest to us at this point. The formula for v shows us what basic measurable quantities go into the makeup of speed—distance and time. Since the time divides the distance, speed is specified as centimeters per second, miles per hour, and so on. We do not have to introduce any new units to measure speed, since speed is a derived quantity and its units are a combination of those that measure distance and those that measure time.

Why do we refer to v as the average or mean speed? The distance between the two points is finite so that the particle spends a finite length of time traversing that distance. During that time, the speed of the particle may fluctuate from one value to another, just as that of an automobile, moving on a highway, does. The mere division of the total distance by the total time smooths out the fluctuations to give us an average value. This average value tells us little about the speed of the particle at any specific moment.

The problem of determining the instantaneous speed of a particle was solved brilliantly by Newton, and its solution led Newton to the discovery of the calculus, which is the single most powerful analytical tool humanity has yet developed. Hardly any of our modern technology or deep insight into the laws of nature could have been obtained without this remarkable and beautiful mathematical instrument.

The difficulties involved in determining the instantaneous speed of a particle can be seen when we note that the particle must be observed at two distinct points along its path (at two different times). The introduction of two distinct times in itself contradicts the idea of an instantaneous speed. In spite of this, however, we may still ask whether we can deduce the instantaneous speed from the measurement of the average speed. We cannot do this if the time interval during which we observe the particle is fairly large, for this leaves room for measurable changes in the speed. The question then becomes what we mean by "fairly large."

Suppose that we carry out the same measurements on the same particle over and over again, but each time we take a shorter time interval. Another way of looking at this situation is to picture a large group of observers, each measuring the average speed but using different time intervals from very short ones to longer ones. In general, each observer gets a slightly different answer. These differences between successive observers grow smaller and smaller, however, with decreasing time intervals until, below a sufficiently small interval, the values for the average speed are the same. This value may then be called the instantaneous speed of the particle at a particular moment. This speed in really not the actual instantaneous speed since a finite time interval is still involved, but our procedure tells us that we can always find a time interval that is so short that our available instruments do not reveal any change in the speed of the particle during that interval. More sensitive and accurate instruments that might detect a change in the speed in such an interval would force us to go to smaller intervals to obtain the instantaneous speed. To obtain the instantaneous speed for a particle, we divide the distance traveled by the amount of time elapsed for the smallest possible interval of time. The actual measured value obtained for the instantaneous speed is determined by the accuracy of our instruments. We can introduce a theoretical instantaneous speed the way Newton did, who pondered the consequence of allowing the time interval to become infinitesimally small. Newton understood that the instantaneous speed can be found only if the time interval is allowed to go to zero. This simple idea, and the mathematical consequences that stem from it, led Newton to the differential calculus.

This procedure of permitting the denominator of a fraction (in this case the distance interval divided by the time interval) to go to zero may appear to be most hazardous (and, indeed, forbidden by the rules of algebra) to the student who has not studied calculus, but Newton saw that no calamity results if the distance interval goes to zero at the same time as the time interval does. It may appear that we would then obtain the indeterminate result 0/0 but in fact the ratio of the distance interval to the time interval has a finite value as the time interval goes to zero, and this value is the instantaneous speed of the particle. Newton's great contribution was to prove that this result is, indeed, so and to show how this value, or limit, as it is called, can be calculated if one knows how the distance a particle moves depends on the time. The method he developed is the basis of the differential calculus and when we apply it, we say that we are finding the derivative of the distance, or differentiating it with respect to time.

The speed concept is extremely general and applies to motion of any kind. We may speak of the speed of a stream of water or air, or a flock of geese as well as light or sound even though we cannot see or feel anything move when light or sound passes us. But we can also extend this idea to economics and speak of the flow of money; the amount of money you receive per week (dollars per week) for work you do (your salary), is an extension of the speed concept.

The average speed of a body is obtained by dividing the distance a body moves in a given time interval by the time it takes the body to move through the distance. But the speed of a body (which is just a scalar quantity) is only one aspect of its motion. If we say that a ship is traveling at 20 miles per hour, we are giving only some of the information needed for plotting its course. When we consider such structures as the solar system, we can obtain a meaningful picture of the structure only if we can plot the course (or orbit) of each of its components (the planets, in the case of the solar system). For this we must know not only the speed of the body at each moment, but also the direction of its motion. When we have both the direction and the speed of the motion of a body, we have its velocity, which is a vector quantity. The speed is the magnitude of the velocity and its direction can be represented by an arrow which points in the direction of the body's motion at each moment.

We can describe the concept of velocity by imagining the path of a particle in some frame of reference. To each point along the path of the particle, such as a point A, for example, we assign a number that gives the speed of the particle at that point. The direction of the motion of the particle can be understood by picturing another point B on the path very close to point A. If we picture the particle as going directly from A to B instead of going along the arc of its true path from A to B, its direction of motion is along a straight line instead of along the actual direction of the particle's path. The closer point B is to point A, the more the direction of the straight line from A to B approximates the actual direction of the motion at A. As a result, we can find the direction of the motion (or of the velocity) of the particle at A by allowing B to come infinitesimally close to A and then taking the straight line from A to B as this direction. When we do this operation, however, this line (and hence the direction of the velocity at A) is just coincident with the tangent line at point A. The tangent line at A is the line (only one such line exists) that just touches the curve of the path at point A but does not cut it.

The velocity of a particle at any given point can be represented as a vector by drawing a line tangent to the curve at that particular point and choosing its length (on some scale) to be equal to the speed of the particle (for example, 1 inch to equal 80 cm/sec). As the particle moves along its path, this vector changes in magnitude (length) and direction. If we know the velocity vector of a particle at each moment of its motion, we can draw the path of the particle. All we need do is pass a curve through the foot of each vector (the end of the vector without the arrow) in such a way that the curve just touches each vector without cutting that vector. Each vector then becomes a tangent of the path of the particle.

As long as we are dealing with the motion of a single particle, we can describe the motion by giving the velocity vectors along a single curve. But in many instances we must deal with the motion of a collection of particles or with that of a liquid. Such cases require that velocity vectors be assigned to many points at the same time. In considering the flow of a river, for example, a velocity vector may be assigned to each point of the river, showing the momentary velocity of the water at that point. The motion of the river is then described by a two-dimensional (or, if depth is taken into account, three-dimensional) collection of vectors. This collection of vectors forms a vector field so that the motion of the river is described by a velocity field.

A coordinate system may also be used to represent the velocity of a particle. If $\mathbf{V}$ is the velocity of the particle at point A we can write down the three components $\mathbf{V}_x$, $\mathbf{V}_y$, $\mathbf{V}_z$ of the velocity $\mathbf{V}$ along the three axes X, Y, Z of the coordinate. These three components define the velocity at the point, the position of which in our coordinate system, at each moment, is given by the vector from the origin of the coordinate system to the point A. As the particle moves from the origin to some other point, its position vector and its velocity both change in magnitude and direction. If we know the magnitudes of each of the components $\mathbf{V}_x$, $\mathbf{V}_y$, $\mathbf{V}_z$, we can obtain the magnitude of the velocity of the particle at a given point from the theorem of Pythagoras which gives the relationship $V^2 = V_x^2 + V_y^2 = V_z^2$. By taking the square root of this sum, we obtain the speed of the particle. This operation shows that the total motion of a particle can be divided into motions along three mutually perpendicular axes and each of them can be treated independently of the others. The total motion of the particle can then be obtained by adding the three mutually perpendicular components vectorially.

This point can be better understood by considering the motion of an object thrown horizontally from a given height. Here we need take into account only the vertical and horizontal directions so that a two-dimensional coordinate system is all that is required. The object is moving both horizontally and vertically at the same time, but the two motions are not the same in the two directions. The object moves horizontally with constant speed (if we neglect air resistance) but its downward speed increases constantly. Owing to this difference in the horizontal and vertical components of the motion, the easiest way of analyzing the motion and determining the path of the object is to treat the constant horizontal motion and the increasing vertical motion separately, and then to combine these two motions to give the total motion.

Since velocity is a vector quantity, two velocities are added vectorially. If we have a kite that is moving through the air (with no wind) at 10 miles per hour due north and the kite is caught in a stream of wind moving due east at 10 miles per hour, the kite will move in a northeasterly direction at a speed of approximately 14 miles per hour. In many cases the two velocities can be added arithmetically. Thus, the velocity of a boat on a river moving in the same direction as the water is obtained by adding the velocity of the boat in still water (c) with the velocity of the river itself (v). The speed of the boat relative to the riverbank, as seen by an observer standing on the riverbank, is the sum of these two velocities. In this case the velocities are added as though they were ordinary numbers. If the boat is moving against the stream, the velocity of the boat relative to the bank is $c - v$. If the boat is moving at right angles to the stream (such as crossing the stream), the velocity of the boat is given by the diagonal of the rectangle formed by the vectors **c** and **v**. Since this is the hypotenuse of a right triangle, the speed of the boat relative to the bank is found by taking the square root of $c^2 + v^2$, according to the theorem of Pythagoras.

The velocity of a particle, as measured by an observer, depends on his frame of reference and, in general, differs from observer to observer. If two automobiles A and B are moving in the same direction, driver B can find his velocity relative to that of driver A by subtracting his velocity from driver A's velocity. If the two automobiles are moving away from each other (in opposite directions along their line of motion), their speed away from each other equals to the sum of their respective speeds. If driver A is driving north at 50 miles per hour and driver B is driving south at 80 miles per hour, the two are separating at a speed of 130 miles per hour.

The apparent motions of the planets among the stars are an example of the importance of relative velocity. To the ancient astronomers, this shifting of the planets indicated a real motion of the planets around the earth, but we now know that this apparent motion can be understood only if we consider the motion of the earth as well as the motions of the planets around the sun. The apparent velocity of the planet is then its velocity relative to the earth.

As an important example of the relative velocity of an object as seen by a moving observer, we can picture such an observer in an enclosed vehicle moving with speed v. A fixed observer outside the vehicle (here we take the earth as our frame of reference) fires a bullet at right angles to the motion of the vehicle (from the side), so that it enters the vehicle at point A. The walls of the vehicle are so thin that they have no effect on the flight of the bullet. What is the direction of the flight of the bullet as seen by the moving observer as it passes through the vehicle at speed c? If the vehicle were stationary, the bullet would leave the vehicle at some point B on its farther side, and the moving observer would find the direction of the bullet's flight (the line from A to B) to be the same as that found by the fixed observer. But the motion of the vehicle alters this result because the bullet then hits the farther wall at some point D closer to the back wall of the vehicle than point B. The distance between the two points B and D on the far wall is just the distance that the vehicle advances during the time the bullet takes to cross the vehicle. To the moving observer it appears that the bullet is moving along the path from point A to point D at an angle with respect to the original path from point A to point B. This is the direction of the relative velocity of the bullet for the moving observer as obtained vectorially. The magnitude of the angle (the apparent change in the direction of the flight of the bullet) depends on the ratio v/c of the speed of the vehicle to the speed of the bullet.

An interesting question arises in connection with the above example. Can the observer in the vehicle deduce anything about his own velocity from his observation of the flight of the bullet? As long as the moving observer is traveling with constant speed in the same direction, he cannot because he cannot know, without additional information, whether he is standing still and the path from point A to point D is the "true" direction of the flight of the bullet (true relative to the earth) or whether he is moving with a fixed speed in some direction and the path from A to D is then the direction of the relative velocity. If the direction of the motion of the vehicle changes periodically (which would be so if the vehicle were traveling in a circle), then the relative velocity (the direction from A to D) also changes periodically and the moving observer may deduce that he is moving relative to the earth.

An important application of this idea is the phenomenon called the "aberration of light," first discovered in 1725 by the British astronomer James Bradley. To illustrate this phenomenon, we can use the above example as a guide and take the earth as our vehicle, and a star situated at right angles to the earth as our gun. The bullet in this case is the light from the star, entering our telescope. Here v is the speed of the earth and c is the speed of light traveling from the star to the earth along a line in space, at right angles to the earth's motion. This line lies along the true direction of the star from the earth. By aligning our telescope to point along this line, we are able to photograph the star on a photographic plate in the telescope. The center of this photographic plate is designated by the point O. If the earth were standing still the image of the star would be at O. But the image of the star is not at the center of the plate because the earth is moving. By the time the light moves from the top of the telescope to the photographic plate at the bottom, the earth advances to the right by a small amount and the light strikes the photographic plate somewhat to the left of the center point. If we want to position the image at the center of the plate, we must tilt the telescope in the direction of the earth's motion through a small angle ϕ with respect to the true direction of the star. To the observer on the earth, the star appears to lie in a direction different from what it would be for an observer in a frame of reference fixed relative to the star. This phenomenon is called the aberration of light and the angle ϕ is called the angle of aberration; its magnitude depends on the ratio v/c where v is the earth's speed around the sun and c is the speed of light.

If the earth were constantly moving in the same direction through space, this effect (the aberration of light) could not be observed. But every six months the direction of the earth's motion (owing to its revolution around the sun) is reversed so that the telescope must also be tilted in opposite directions every six months. The observed star (if it is at right angles to the plane of the earth's orbit) actually appears to move in a small orbit in the sky (the circle of aberration) which is identical in shape to the earth's orbit. The orientation of the telescope following the star changes continuously and this change can be observed and measured. The angle through which the telescope has to be tilted is called the angle of aberration. Its value depends on the direction of the star and the speed of the earth: it is zero if a line from the earth to the star is parallel to the earth's motion and it is a maximum if the line from the earth to the star is at right angles to the direction of the earth's motion. From numerous measurements astronomers have found that the maximum aberration angle (which is called the constant of aberration, and is the same for all stars lying in a direction at right angles to the earth's motion) is 20.49″.

The value of the constant of aberration can be used to calculate the speed of the earth in its orbit around the sun since this constant (actually the tangent of this angle) equals the ratio v/c. All we need to do then to find the speed of the earth is to plug into this ratio the numerical value of c (the speed of light) and then equate it to the tangent of the angle of aberration. Since the speed of light is 299,793 km/sec, the speed of the earth is 29.80 km/sec (or 64,368 miles/hour). Since the earth is moving in an ellipse and not in a circle around the sun, its speed changes continuously, but for all practical purposes the earth's orbit may be regarded as circular and its speed as constant. If the earth's speed is assumed to have the constant value of 29.80 km/sec and its orbit is taken as circular, we can calculate the size of the earth's orbit (the distance from the sun to the earth). By multiplying the length of the sidereal year (365.257 days or 3.156×10^7 sec) and the speed of the earth, we obtain the length of the circumference of the earth's orbit which, when divided by 2π, gives us the radius of the earth's orbit—nearly 150,000,000 km. This number, the mean distance of the earth to the sun, is the numerical value of the astronomical unit; as already stated, it is one of the basic constants in astronomy.

We saw that the speed of the earth in its orbit can be calculated from the aberration of light. But to do this we must know the speed of light, which is one of the most important and basic constants in nature. Various precise techniques are now available for measuring this speed, and it is now known to be $2.99792458 \times 10^{10}$ cm/sec with an accuracy of four parts in a billion (to within 40 cm/sec or 0.004 km/sec). Instead of describing the most precise modern method for measuring this speed at this point, we describe the way it was first measured. In 1675 (half a century before the discovery of aberration) the Danish astronomer Ole Roemer observed that the times of occurrences of the eclipses of the satellites of Jupiter varied from season to season. He correctly interpreted this variation as arising from the finite time taken by the light to travel from the satellites to the earth. We can see how the speed of light is to be calculated from such objects by considering a satellite which suffers an eclipse every t seconds. If the earth were fixed relative to Jupiter, an observer on the earth would then find that an eclipse of this satellite occurs every t seconds. But if the earth is receding from Jupiter at a speed v kilometers per second, how soon after a given eclipse is observed is the next eclipse seen? In the t seconds between eclipses the earth moves vt kilometers farther from Jupiter than it was when the first eclipse was observed, and the light traveling from the eclipse will have to traverse this additional distance to reach the observer, so that the eclipse will appear to be delayed by this additional time. The observed time between the two eclipses is then $t + (v/c)t$ instead of t since $(vt)/c$ is just the additional time that it takes the light, traveling at speed c, to cover the additional distance vt. This is also true for each successive eclipse that occurs during the time the earth is receding from Jupiter. If N such eclipses occur between the times that the earth is closest to and farthest from Jupiter, the total delay for all the N eclipses is $(Nv/c)t$. Roemer observed this value to be 1000 sec. But this delay represents (in time) the total increase in the distance of the earth from Jupiter, as it goes from its closest to its greatest distance from Jupiter. This is just 186,000,000 miles (the diameter of the earth's orbit around the sun) so that $1000 = 186{,}000{,}000/c$ and $c = 186{,}000$ miles/sec. This quantity is very close to the correct value obtained today with the most precise modern methods of measurement.

The addition of velocities played a very important role in the analysis of a famous and crucial experiment on the motion of the earth which was performed in 1887. This experiment can be understood if we imagine a stream, flowing with speed v, and two swimmers, who swim with the same speed c in still water. The swimmers start from the same point at the same time, and one swims a distance L downstream and back and the other swims the same distance cross stream and back. Owing to the stream's flow, the cross-stream swimmer cannot swim directly across the stream to reach his goal; the current pushes him off the straight-line course. Hence, this swimmer has to swim diagonally upstream so that the vectorial sum of his velocity and the stream's velocity will be the correct velocity to carry him directly across. The speed of the swimmer going downstream is $c + v$ (the stream's speed is added to his speed), but his speed back is $c - v$. Using this analysis we can calculate the times taken by the two swimmers for a round trip, and we find that the cross-stream swimmer returns first.

The physicist Albert Michelson, in the late 19th century, devised an experiment based on this analysis, using light beams, to detect and measure the motion of the earth around the sun. He and his collaborator, Morley, directed two light beams at right angles to each other along the same distance L to two mirrors which reflected them back. One beam traveled parallel to the earth's motion and back (like the up-and-downstream swimmer) and the other traveled perpendicularly to the earth's motion and back (the cross-stream swimmer). The motion of the earth was the counterpart of the motion of the stream in our previous example. They discovered that the two beams returned to their starting point at exactly the same moment, contrary to what one expects from the example of the swimmers. Since we know that the earth is moving, this zero result indicates that the motion of the earth does not affect the velocity of light in the same way that the flow of the stream affects the velocities of the swimmers. The zero result was, indeed, very puzzling to the two experimenters, and nobody knew how to account for it at that time; least of all did anybody think that it had anything to do with the special character of the speed of light and that the classical law of addition of velocities based on absolute space and absolute time might not be correct. This idea did occur to Albert Einstein, however, and became the starting point of his theory of relativity. The Michelson–Morely experiment shows that the speed of light in a vacuum is the same for all observers moving with constant velocity with respect to each other. This constancy of the speed of light is one of the basic laws of nature which must be incorporated into all physical theories if they are to be correct.

Using the vector concept of velocity, we can understand why the planets appear to move in the sky as they do, even though they are revolving around the sun in almost circular orbits. The apparent motion of a planet is a vector combination of the motions of the planet and the earth around the sun. This combination is really the motion of the planet relative to the earth and can be obtained vectorially by subtracting the earth's velocity around the sun from the planet's velocity around the sun. If we assume that the planets move in circular orbits around the sun, we can calculate the approximate average speed of any planet in its orbit if we know its mean distance from the sun and the time it takes the planet to revolve once around the sun (which is called its period). Later we shall see, by using planetary theory, that we can calculate the instantaneous speed of a planet in its orbit if we know its mean distance from the sun and its instantaneous distance.

We have seen that we can simplify our analysis of the velocity of a particle by referring the velocity to a Cartesian coordinate system and then treating each of the three mutually perpendicular components of the velocity in this system separately. A particularly important and special case of this occurs when we take as one of the coordinates the line from the origin to the present position of the particle at point P. Here **V** represents the velocity of the particle at some moment and the vector **r** from the origin to the point lies in the plane of the paper. The plane of the paper, then, is the plane in which the particle is moving (the plane of the velocity) at the instant we observe it at P. A moment later the plane of the velocity may alter, but let us, for the moment, just consider the plane determined by the two vectors **r** and **V**. Since the velocity **V** of the particle lies in this plane at the moment that the particle is at P, we need only to introduce a two-dimensional coordinate system (two mutually perpendicular lines) in this plane to describe the velocity of the particle at this moment. We take the line along **r** (from the origin to P) as one of our coordinates and another line from the origin to another point such as T (which is not parallel to the line OP) at right angles to **r**, as the other coordinate. We now break up **V** into a component along **r** which we call the radial velocity $\mathbf{V}_r$ (velocity along the line of sight) and a component along OT which we call the transverse velocity $\mathbf{V}_T$. If we measure $\mathbf{V}_r$ and $\mathbf{V}_T$ separately, we find that the sum of these two velocities equals **V**. As a result, we conclude that $V^2 = V_r^2 + V_T^2$. This equation has an important application to the velocities of the stars because we can find **V** for a star only by first finding $\mathbf{V}_T$ and $\mathbf{V}_r$ separately. For a star, the plane of the motion remains the same since, for all practical purposes, the stars are moving with constant **V**. We can find V_T if we know the distance r of the star and the angle μ (the angle through which r turns in a unit time), which is the angular velocity of the star. The angle μ is also the angular displacement of the star in a unit time. If μ is expressed in radians, we find that $r\mu$ equals $\mathbf{V}_T$. For stars, μ is so small that it is most convenient to express it in seconds of arc; the angle μ, expressed in seconds of arc per year, is then called the ''proper motion'' of the star. If μ is the proper motion expressed in radians per year, and μ'' is the proper motion expressed in seconds per year, we find that $r\mu''/206{,}265$ equals $\mathbf{V}_T$. From this formula we obtain the transverse speed of a star by measuring its proper motion (comparing the photographic positions of the star a year apart) and its distance r. The star's radial velocity $\mathbf{V}_r$ is measured by using an important property of light that was discovered late in the 19th century by Christian Doppler and is now known as the Doppler effect. It is similar to the effect used by Roemer to obtain the speed of light from the delay in the eclipses of the moons of Jupiter. Doppler discovered that the color of light is affected by the radial velocity of the source of the light relative to the observer. The light from a star (or any luminous object) is propagated as a wave, the color of which depends on the wavelength, which is the distance between two successive crests in the wave. The shorter this wavelength is, the bluer the light appears, and the longer the wavelength, the redder the light. If the star is neither receding from nor approaching the earth, the wavelength of the light, as measured by the observer on the earth, is the same as that emitted by the star. But if the star is receding from the earth, the waves are stretched out and the observer on the earth finds that the wavelength of the light he receives is longer than that emitted by the star and the star appears reddened. On the other hand, if the star is approaching the earth, the waves are squeezed together and the observer finds the wavelength to be shorter than if there were no motion, causing the star to appear bluer.

If the wavelength of some particular color in the light from the star is w when the light leaves the star, but is found to be $w + \Delta w$ (where Δw is a small change in wavelength) by the observer on the earth, this observer deduces that the star is receding from the earth at a speed V_r which depends on w, Δw, and the speed of light c: $V_r = c(\Delta w/w)$. By knowing the wavelength w and measuring Δw (the change in the wavelength), V_r can be calculated. As a result, both $\mathbf{V}_T$ and $\mathbf{V}_r$ can be found for a star by direct observations of the star. The actual space velocity $\mathbf{V}$ (relative to the earth) can then be computed by using the theorem of Pythagoras. The actual results of such measurements show that the stars in the neighborhood of the sun are moving almost randomly with respect to the earth at speeds ranging from about 15 to 40 km/sec.

In our discussion of the transverse velocity of a star, we introduced the concept of the angular velocity of a particle relative to an observer by considering the change in the direction of the star from moment to moment. We can imagine a particle moving with constant speed v in a circle of radius r much like a rock tied to a string which is being swung around by a person. The velocity of this particle is changing constantly even though its speed is constant. If we follow the radius vector connecting the particle to the center of the circle (the origin), we see that its direction changes constantly. If the particle moves from one point to another (A to B) along the circumference of the circle in a unit time (so that the distance from A to B equals v), the radius vector to the particle has turned through the angle ω in the same time. The symbol ω is the angular speed of the particle and is expressed as radians per second which gives the equation $\omega r = \mathrm{AB}$ or $\omega r = v$. This equation shows that the linear speed of a particle moving along the circumference of a circle can be obtained from the angular speed, expressed in radians per second, by multiplying this angular speed by the radius of the circle.

Here, we have been talking only about angular speed, which is a scalar quantity, but we are really dealing with a vector quantity, since a definite direction is associated with the circular motion of the particle. This is the direction of the line passing through the origin (the center of the circle) and perpendicular to the plane of the circle. This brings us to the concept of the angular velocity whose magnitude is just the angular speed ω and whose direction is along the axis of revolution of the particle.

CHAPTER 5

Acceleration, Force, and Mass

If a man will begin with certainties, he shall end with doubts; but if he will be content to begin with doubts, he shall end in certainties.

—FRANCIS BACON, *The Advancement of Learning*

In the foregoing chapter we constructed the velocity concept by properly combining space and time, and we thus obtained a new physical entity which plays an important role in the formulation of the laws of nature. But the basic law of nature which we are now leading up to in this chapter cannot be expressed in terms of velocities alone. To see why, we note that we seek a law which gives us an insight into the motions of the bodies in a dynamical structure (such as our solar system) which is held together by a force. Knowledge of the velocities of these bodies at any moment without any other information is not enough to permit us to deduce their orbits and hence the structure of the entire system. If we know the velocity of a body at some particular instant, we cannot say where the body will be at some later moment, because the velocity may be changing. Suppose, for example, that the body is at a certain point A in its orbit. The velocity of this body at time t_A is given by the vector $\mathbf{V}_A$ drawn tangent to the orbit at point A. If we know only the velocity $\mathbf{V}_A$ at time t_A, we cannot conclude that the body will be at another point B in the orbit at some later time. All we can say is that at the moment t_A the body is moving with speed V_A in the direction of the vector $\mathbf{V}_A$, and if there is no change in its velocity the body, at some later time, will be somewhere along the line that coincides with the vector $\mathbf{V}_A$. This conclusion is clearly contrary to what the body is actually doing, since it is moving along the path between points A and B. To predict that the body will be at some point on its orbit after the time t_A instead of somewhere along the line V_A, we must know not only the velocity $\mathbf{V}_A$ at the moment t_A, but also how the velocity is changing—more precisely, the rate at which the velocity is changing at the time t_A.

To analyze this point in detail, let $\mathbf{V}_B$ be the velocity of the particle when it is at the point of its orbit B, at some later time t_B. The velocity of the particle must have changed from $\mathbf{V}_A$ to $\mathbf{V}_B$ during the time interval $t_B - t_A$ it took to travel from point A to B. The rate of change in the velocity is thus the vector difference between $\mathbf{V}_B$ and $\mathbf{V}_A$ divided by the time interval $t_B - t_A$. This quantity is the average acceleration **a** of the particle.

The acceleration is a vector quantity, since it is the vectorial difference between two vectors (which is itself a vector) divided by a time interval (which is a scalar quantity). An important point to keep in mind, which follows directly from the definition of acceleration, is that the acceleration is not necessarily in the same direction as the velocity. It is always in the direction of the change in velocity. If the acceleration is parallel to the velocity (either in the same direction as or in the direction opposite to the velocity), only the speed (the magnitude of the velocity) changes. If the acceleration is at right angles to the velocity, the speed of the particle remains unaltered and only the direction of the velocity changes. If the acceleration is at some angle to the velocity other than 0°, 90°, or 180°, both the magnitude and the direction of the velocity change.

Before we consider the various special cases of acceleration, we see that the definition of acceleration given above is the average acceleration in the time interval $t_B - t_A$. If we wish to obtain the instantaneous acceleration at time t_A, we must again take $t_B - t_A$ as a very short time interval Δt. If we know the particle's instantaneous acceleration at each instant, we can determine the velocity of the particle at each instant and then the orbit of the particle, provided we know the position and velocity of the particle at some initial moment. Since acceleration is a change in velocity divided by time, and velocity is a distance over a time, the dimensions of acceleration in terms of our basic space-time units are distance divided by time squared (abbreviated as cm/sec^2).

The only part of the velocity of a particle that changes when the acceleration is parallel to the velocity is the magnitude of the velocity or the speed. This point can be illustrated by considering the simple case of an automobile moving along a straight road with constantly increasing speed: Its speed is 10 miles per hour when we first see it and 2 seconds later its speed is 14 miles per hour, and 2 seconds after this its speed is 18 miles per hour and so on, until it reaches some final speed. Since the direction of the automobile does not change, the only effect of the acceleration is to increase the speed which, in this case, increases at the rate of 4 miles per hour every 2 seconds. The magnitude of the acceleration is thus 4 miles per hour per 2 seconds or 2 miles per hour per second. Since 2 miles per hour is also 1/30 of a mile per minute, we may also express this acceleration as 1/30 of a mile per minute per second. Finally, since 1/30 of a mile per minute is 1/1800 of a mile per second, we may express the acceleration as 1/1800 of a mile per second squared. The concept of acceleration may be illustrated somewhat differently by considering a person's salary which is a flow of money and is like a velocity. An important aspect of employment is not so much the employee's immediate salary as the raises he can expect; a raise is equivalent to an acceleration. If you are promised a raise of $100 per week every year, the acceleration of money to you is $100 per week per year.

Since acceleration is a vector quantity, the term acceleration is used whether the speed of the body is increasing or decreasing. Decreasing speed simply means that the direction of the acceleration is at an angle of 180° (opposite) to the velocity of the body. Before leaving the simple case of a body being uniformly accelerated in a straight line, we state the relationship between the speed of the body at any moment and its acceleration. Suppose that at some initial moment the body is moving in a straight line with speed v_0. What is the speed v of the body after a time t if it is subject to the constant acceleration a along its line of motion? Since the product of a and t is the total increase in the speed of the body during the time t (the speed is increasing at the rate a) the final speed is the initial speed v_0 plus the increase represented by the product of a and t, or $v_0 + at$.

An interesting and instructive application of acceleration is to the stopping of a car in a given distance. This problem is of particular concern to automobile drivers who often overestimate their ability to stop a car in a given distance and underestimate the importance of the speed of the car in this operation. If the brake is applied to a car that is moving at speed v and this speed then decreases at the rate of a feet per second every second, the distance required to reduce the car's speed to zero is $v^2/2a$. This relationship shows us that the distance in which a car traveling at speed v can be stopped at a constant rate depends on the square of this speed. A car traveling at 60 miles per hour requires four times as great a distance to stop in as the same car traveling at 30 miles per hour. Not knowing this relation of speed to stopping distance for a given acceleration causes more automobile accidents than anything else.

A very important and remarkable example of uniformly accelerated motion is the motion of a body in free fall. By free fall we mean the motion of a body in a vacuum with no interference by other bodies that are not free to move. According to this definition, an unhindered body is in free fall whether it is rising, descending, or moving laterally. Near the surface of the earth all freely falling bodies, regardless of their chemical or physical nature or size, have the same downward acceleration—980 cm/sec^2 or 32.2 ft/sec^2—which means that the speed of the falling body (neglecting air friction) increases by 980 cm/sec or 32.2 ft/sec every second. This phenomenon, which may be considered as a law of nature and which was discovered by Galileo, contradicts Aristotle's idea that freely falling heavy bodies fall faster than light bodies. The full consequences of this law were not understood until Einstein made it the basis of his general theory of relativity.

The acceleration of a freely falling body is called the acceleration of gravity and is designated by the letter g. Although g is a constant to a first (and very good) approximation, it varies slightly from point to point on the earth's surface. It is smallest at the equator and largest at the poles, with a mean value of 980.665 cm/sec^2. This variation is due to two facts: the earth is flatter at the poles than at the equator and the earth is spinning. Another important feature of g is that it gets smaller as we move away from the center of the earth. Thus, it is smaller on top of Mt. Everest than it is as sea level. The way the variation of g is influenced by the distance from the center of the earth was discovered by Newton; it is expressed in his law of gravity, which we shall discuss later.

Several methods that can be used to measure the gravitational acceleration g. From the formula for the distance that a uniformly accelerated body moves in time t, we can write down an expression for the distance s that a freely falling body falls in time t if it starts from rest: $s = \frac{1}{2} gt^2$ or $g = 2s/t^2$. The acceleration of gravity is just twice the distance divided by the square of time. If this body is allowed to fall for 1 second, we place $t = 1$, and obtain $g = 2s$. The numerical value of g equals twice the distance that a body in a vacuum falls in 1 second if it starts from rest; it can be found quite accurately by measuring the value of s. Long before good clocks were available, Galileo timed the descent of bodies rolling down an inclined plane where the inclined plane makes an angle θ with the horizontal surface. If the inclined plane were not present, the body would fall vertically downward with the acceleration g. But the actual motion is along the inclined plane with an acceleration that is smaller than g. Since the vertical acceleration of gravity is a vector, we break it up into mutually perpendicular components. One of these components is along the inclined plane and the other is at right angles to the plane (downward). The component of the acceleration along the incline of the plane gets smaller and smaller as the angle θ is reduced, finally vanishing for $\theta = 0$ because the inclined plane is then perfectly flat. The acceleration along the inclined plane equals g (the vertical acceleration) when $\theta = 90°$. Its value is $g \sin \theta$ for any arbitrary value of the angle θ. If θ is small enough, the body moves slowly enough along the plane to be clocked without difficulty. If s is the distance the body moves (starting at the top of the plane from rest) along the plane in one second, then $g = 2s/\sin \theta$. Since both θ and s can be measured quite accurately, the value of g can be calculated accurately also.

The observed period of oscillation of a pendulum can also be used to calculate g. When the pendulum oscillates, the position of the bob along its arc at any moment is given by the angle θ that the string attached to the bob makes with the vertical at that moment. Since the bob of the pendulum is forced to move along an arc because the string cannot be stretched, only the acceleration along this arc comes into play in producing the motion of the bob. Just as in motion along the inclined plane, this acceleration is $g \sin \theta$, but owing to the constant change in θ, this acceleration is not constant. As a result of this, the motion of the bob of the pendulum is rather complicated. However, if θ is kept small (the amplitude of the swing of the pendulum is kept small), the acceleration $g \sin \theta$ can be approximated by the product of g and θ (where θ is to be expressed in radians) and the motion of the bob can be treated as though it were oscillating at the end of a spring. Motion of this sort is called simple harmonic motion (SHM). The results derived from SHM show that the time required by the pendulum to swing back and forth once (the period P of the pendulum) is $P = 2\pi\sqrt{l/g}$. From the known length l of the pendulum and its observed period P, the acceleration of gravity can be calculated from the equation $g = 4\pi^2 l/P^2$.

Since the acceleration of gravity is always vertically downward and has the constant value 32.2 ft/sec^2, we can analyze the motion of a freely falling body with some simple arithmetic. Consider first a body thrown vertically upward with some initial speed V (we neglect air resistance). After 1 second its speed is $V - 32$ ft/sec; after 2 seconds its speed is $V - 64$ ft/sec and so on, until it has zero upward speed and is at its highest point. Its downward speed is also zero at that moment so that the highest point of the body's path can be defined as the point where its vertical speed is zero. It reaches this point $V/32$ seconds after it begins its ascent. Immediately after this, it begins to descend, gaining 32 ft/sec of speed each second. After $V/32$ seconds it is back at its starting point with the same downward speed V as it had when it started upward. The time spent by a body in its upward journey is exactly the same as that spent in its downward journey, and its downward speed at any height above the ground is the same as its upward speed was at that same height (in a vaccum or neglecting air resistance). This is true whether the body is thrown vertically upward or diagonally upward at some angle to the vertical, so that it has a horizontal as well as a vertical motion. The reason for this is that the vertical motion of the body is entirely independent of its horizontal motion.

If you throw an object at an angle θ to the horizontal with an initial speed V_0 (the speed of the object at the moment it leaves your hand), the initial velocity of the particle is $\mathbf{V}_0$. The moment after the object has left your hand, its velocity begins to change continuously in both magnitude and direction, owing to the acceleration of gravity. We shall see later that the object is really in orbit around the center of the earth, but we can study the characteristics of its motion here without concerning ourselves with the relationship of the motion of the object to the center of the earth. In this example we need only take into account that the particle is subject to a constant downward acceleration g. The assumption that g is constant is quite reasonable so long as the object stays close to the surface of the earth.

This example is very useful because it illustrates the independence of motion in mutually perpendicular directions. We first consider the horizontal motion of the particle and then its vertical motion, and then, by combining the two motions, we obtain the total motion of the particle and its path or orbit. Since the starting speed of the particle is V_0 in the direction θ with the horizontal, we obtain the initial horizontal speed by multiplying V_0 by $\cos\theta$. The object therefore starts its horizontal motion with speed $V_0 \cos\theta$. But this horizontal speed does not change (neglecting air resistance) since there is no horizontal acceleration. This is all there is to the horizontal motion. But what about the vertical motion? The initial upward speed of the object is obtained by multiplying V_0 by $\sin\theta$. But we must now take into account the constant downward acceleration, so that the vertical speed after a time t is $V_{\text{vertical}} = V_0 \sin\theta - gt$. (The minus sign shows that the acceleration is downward.) These two equations for the horizontal and vertical motions of the object tell us that the orbit (path) of the object is a parabola, which is confirmed by observation.

From the expression for the vertical speed (V_{vertical}) we can find the time of flight (the time from the moment the object is thrown to the moment it hits the ground). The range of the projected object can be obtained by multiplying this value for the time by the expression for the horizontal speed $V_0 \cos\theta$. From this result we can conclude that for an object to travel the maximum distance, the angle θ must equal 45°. If a javelin thrower, for example, throws a javelin at an angle of 60°, its initial vertical speed exceeds its initial horizontal speed so that the javelin follows a steeper parabolic path; if the javelin thrower throws the javelin at an angle of 30°, its initial horizontal speed exceeds its initial vertical speed so that the javelin follows a flatter parabolic path. In either case, the javelin does not travel as far as it does when thrown at a 45° angle.

This detailed discussion of the projectile problem shows how much information can be derived from the simple relationship between velocity and acceleration without using any mathematics beyond arithmetic. This simple analysis, however, does not describe the true state of affairs when a body is thrown into the air. To begin with, the air resistance becomes increasingly important as the initial speed of the body is increased. Second, the acceleration of gravity decreases as the body moves away from the center of the earth, so that the above analysis is only approximate and applies only if the total trajectory of the body lies close to the surface of the earth. Finally, the rotation of the earth affects the motion of the body which increases with the initial speed of the body and its range, but the separation of the motion into a horizontal and vertical part still applies.

The acceleration can, in general, make any angle from 0° to 180° with the velocity and both the direction and magnitude of the velocity then change accordingly. With the acceleration parallel or antiparallel (opposite) to the velocity, only the speed of the body changes. If the acceleration changes only the direction of the velocity, the acceleration must always be perpendicular to the velocity so that no component of the acceleration is parallel to the velocity and the speed remains unchanged. Consider an object subject to an acceleration which is constant in magnitude but with its direction always perpendicular to the object's velocity. Since the speed of the object is constant and its direction of motion is constantly changing, the path of the object must be a circle. The acceleration always points to the center of the circle (it lies along some radius of the circle at each instant) so that its direction is constantly changing. You can convince yourself of the relationship between the radial acceleration and the circular velocity of an object by riding a bicycle at constant speed on a large flat area, keeping the handlebars turned at a constant angle. The bicycle will move along a perfectly circular path.

The question now is how large the magnitude of acceleration must be to keep a particle moving in a circle of radius r with constant speed V. This can be deduced in many ways, but without giving an exact derivation, we can obtain a physical insight into the relationship between acceleration and the quantities V and r from a dimensional analysis. It is clear that acceleration can depend only on V and r since these are the only physical quantities that enter into the problem. But we know that acceleration has the physical structure of a length divided by the square of a time. The only combination of V and r that can give such a dimensional structure is V^2/r and this is the exact formula for the acceleration in circular motion.

We see how this formula works when we drive a car. In rounding a small circle or making a hairpin turn (r is small) we are extremely cautious because the car's direction is turning rapidly which means that its acceleration is large; the smaller r is, the larger is the acceleration, as stated by the formula. We also know from our driving experience that it is much more difficult to make a sharp turn when moving rapidly than when moving slowly because the faster we move, the faster our direction changes. All of these considerations are contained in the formula V^2/r.

From our analysis of uniform circular motion, we can now describe another kind of motion which is of great importance in physical phenomena. Before we do that, however, we introduce the concepts of the period and frequency of uniform circular motion. By the period P we mean the time required by the particle (traveling at constant speed) to complete one revolution. Since the circumference of the circle is $2\pi r$ and the speed of the particle is V, $P = 2\pi r/V$; that is, time is distance divided by speed. But if ω is the angular speed of the particle, we have $V = \omega r$ so that $P = 2\pi/\omega$ or $\omega = 2\pi/P$.

The frequency ν of the circular motion may be defined as the number of revolutions of the particle per second. Since each revolution corresponds to 2π radians per second, the angular speed of the particle is $2\pi\nu$. Hence, $w = 2\pi\nu$ and $P = 1/\nu$. Instead of considering the circular motion of the particle P itself, we study the motion of the point p′ on the the Y axis where the perpendicular line from the particle to the Y axis cuts the Y axis. As the particle P moves around the circle, the point p′ oscillates up and down (along the Y axis) between the points A and −A on the Y axis where the circular path traveled by P intersects the Y axis. We call A (which is the radius of the circle) the amplitude of the oscillating motion. Since this oscillating point carries out one complete oscillation when the particle has gone around the circle once, we may assign the same period P and frequency ν to the oscillating motion of p′ as we do to the circular motion. The motion of the oscillating point is simple harmonic motion.

The oscillating point is constantly accelerated and this acceleration causes the speed of the point to change continuously. The point has its maximum speed at the origin (the center of the circle traced by the path of the particle) and its minimum speed at the extreme points A and $-A$. The velocity of the oscillating point at any moment is just the vertical component of the velocity of the particle in its circular motion. The acceleration of the oscillating point is parallel to the Y axis but it always points toward the center of the circle and is proportional to the distance of p' from its equilibrium position and is always directed toward the equilibrium point (the center of the circle). The magnitude of the acceleration changes from zero when the point is at the center of the circle to its maximum value when it is at its maximum distance from the origin. The oscillations of a weight at the end of a spring suspended vertically are an excellent example of simple harmonic motion.

Up to this point we have been dealing with physical quantities that can be derived from our basic elements of space and time, and if we were content with merely describing the motions of the bodies in our universe without attempting to understand why these motions exist, we would have to go no further in our construction or introduction of new physical entities. But the mere description of events does not satisfy scientists—they seek basic laws and relationships that can give them an insight into the reason for the dynamical behavior of bodies. At the same time, such laws permit them to deduce the future behavior of bodies and to predict future events that are not discernible in the description of the present situation and to determine orbits of bodies.

The difference between these two ways of describing nature can be nicely illustrated by considering Johannes Kepler's laws of planetary motion and Isaac Newton's law of gravity combined with his second law of motion. Kepler discovered the laws of planetary motion by carefully analyzing the observational data collected by Tycho Brahe and deriving empirically from these data a set of statements about the motions of the planets. Although Kepler's three laws quite correctly (though not to a high degree of accuracy) describe the orbits of the planets and give a picture of how each planet moves around the sun, these laws tell us nothing about the cause of the planet's behavior nor do they give us an insight into the role played by the sun in this motion. Moreover, from Kepler's laws we can deduce nothing about the precise orbits of the planets which differ from Kepler's orbits, owing to the gravitational interactions among the planets themselves.

These inadequacies are not present in Newton's laws, for these laws are general statements about the way the interactions between bodies determine their motions and their orbits. However, we cannot derive such laws of interaction if we limit ourselves to those basic physical concepts (space and time) which we have introduced up to this point. It would, of course, be a remarkable intellectual achievement if we could do this, for we would then have reduced the laws of nature to a branch of geometry. As we shall see when we discuss the theory of relativity, Einstein did this to some extent with the law of gravity in 1915, but all attempts since then to extend this idea to other branches of physics have failed. In light of this failure, additional basic concepts must be introduced to permit us to go further in our study of structures and forces in the universe.

One possibility is suggested by the very components which define a physical structure, namely the physical bodies that constitute this structure. These are palpable objects which occupy space and move about—they are at the heart of all our efforts to understand nature and its laws and, clearly, we must introduce (or derive) some basic physical quantity that characterizes these objects in a way that goes beyond their purely geometrical features (shape and size). It has been customary in the past to do this by introducing the concept of the mass of a body as the third basic entity (space and time are the other two) which, together with the other two, lead to the construction of the new physical entities which enter into the laws of nature.

In our development here we depart from this procedure because the basic quantities which we have introduced up to now (e.g., space and time) are measurable in terms of procedures that we can readily perceive and understand. It is fairly easy for us to estimate lengths and time intervals. But the concept of mass as a measurable entity is unfamiliar to us. When we look at an object, we are at once aware of its dimensions, but we can tell nothing about its mass or how its mass is to be measured; nor do we know the nature of mass. Even if we are given an object which we are told is of unit mass, we still do not see at once how this unit mass is to be used to measure the mass of any other body. In many texts the mass of a body is introduced and defined as the quantity of matter in the body, but this is as mysterious and incomprehensible an idea as the mass itself, for how does one measure the amount of matter in such a variety of substances as hydrogen, iron, lead, gold, and so on? Another procedure, which is even more confusing, is to state that the mass of a body can be found by weighing it. Although this can be done, certain logical difficulties are associated with it. To begin with, if we are to introduce mass as a basic, undefinable entity, its measurement ought not to involve another concept such as weight. Of course, if mass is to be a derived quantity, then it is permissible to define it in terms of weight. In that case, the weight of a body should be the basic entity and not the mass. Another difficulty stemming from this way of introducing the mass of a body is that the student tends to confuse mass and weight and quite naturally accepts the weight of the body rather than its mass as its basic property. This confusion is quite understandable since we have to contend with the weights of bodies rather than with their masses in most of our daily activities. But weight and mass are different entities.

If we are to follow the procedure we have used up to this point in introducing new physical entities, we should introduce some quantity other than mass as the third of our basic physical elements. This third quantity should be as perceptible and as easy to estimate by means of our senses as are lengths and intervals of time. Moreover, it should lead us to mass as a derived or defined quantity. We do not have far to go to find such a quantity, for it is one that we are constantly involved with, namely force. No matter what we do (and even when we are apparently doing nothing), we are constantly subjected to and must constantly exert forces. Indeed, most of our bodies consist of structures (muscles) whose sole purpose is to exert forces, and the remaining parts of our bodies are designed either to support these muscles or to direct them in their activities.

Owing to this constant muscular activity, we have acquired a fairly precise sense of the magnitude of the forces that are involved in our daily activities. Even though few of us can express the size of a given force numerically, our muscles are so attuned to their functions that we automatically estimate the force required to perform certain acts. Whether we pick up a piece of paper, open a door, hit a tennis ball, or strike a piano key, we exert just the right amount of force to perform each of these acts with the optimum results.

Force may properly be introduced as the third of our undefinable basic quantities. Since we cannot trust our various muscular responses to give us a precise measure of all the different kinds of forces that we deal with in our study of the laws of nature, we introduce some mechanical device with which to measure them. Fortunately, we have a device that has many of the properties of a muscle, namely a helical spring. An important property of such a spring, which was discovered by Robert Hooke, is that if a force is applied to the end of it, the amount by which the spring is stretched is proportional to the force applied. This law, which is known as Hooke's law, permits us to use a helical spring to introduce a unit of force. All we need do is agree to let the unit of force be that force which is required to stretch a specially constructed spring by a definite amount. If the spring is stretched by twice that amount, two units of force are involved, and so on. The magnitude of our unit of force depends on the stiffness of the spring we use and the size of the unit stretch. The unit of force used in most scientific work is the dyne; the muscles in our arm have to exert a force of about 450,000 dynes to lift an object weighing 1 pound off the floor.

Magnitude alone does not give us a complete picture of the force concept, since a force can be exerted in various directions. In other words, a force is a vector quantity, so that we must specify its direction as well as its magnitude (its strength) to define it completely. If a force **F** acts at the origin O of a three-dimensional coordinate system, we represent it graphically by an arrow in the direction of the force; its magnitude is given by the length of the arrow on a scale in which a unit of length is assigned a definite number of dynes (e.g., we may assign to each centimeter of length of the arrow a force of 100 dynes). We can break the force F up into its three mutually perpendicular components along the three axes X, Y, Z so that $\mathbf{F} = \mathbf{F}_x + \mathbf{F}_y + \mathbf{F}_z$, $F^2 = F_x^2 + F_y^2 + F_z^2$. That is, the force **F** in the given direction can be expressed as the sum of the three mutually perpendicular vectors $\mathbf{F}_x$, $\mathbf{F}_y$, and $\mathbf{F}_z$ along the three orthogonal axes X, Y, Z of our coordinate system. The three forces $\mathbf{F}_x$, $\mathbf{F}_y$, and $\mathbf{F}_z$ are the X, Y, Z components of **F**. The square of the magnitude of **F** is then the sum of the squares of the three components in accordance with the theorem of Pythagoras.

Since our bodies respond to the application of a force by muscle flexing, we easily recognize the presence of a force applied to our bodies by the reactions of our muscles. If force is applied to an object that is not part of us, can we still detect the action of the force by studying the behavior of the object or body? If the body is a spring, one end of which is attached to a fixed object, we can watch the spring being stretched. But if the object on which the force is acting is free to move, what feature of its behavior indicates that a force is acting on it? The answer is the motion of the body, which is different when a force is acting on it than when no force is acting on it. Although this answer may seem quite obvious to us now, the full relationship between the state of motion of a body and the forces acting on it was not understood until Galileo and Newton investigated it. Before then, people accepted the Aristotelian notion that a force is required just to keep a body moving at a constant speed in the same straight line. This belief seemed so intuitive a truth that for 2000 years no one questioned Aristotle's statement in his *Mechanics* that "the moving body comes to a standstill when the force which pushes it along can no longer so act as to push it." Physics as a science began when Galileo demonstrated that this intuitive notion is wrong. Galileo did more than merely disprove Aristotle's intuitive ideas about force and motion, for he showed that truths about nature can be obtained only through the observation of nature and scientific reasoning. In the words of Albert Einstein, that "was one of the most important achievements in the history of human thought. This discovery taught us that intuitive conclusions based on immediate observation (without scientific reasoning) are not always to be trusted."

Before introducing the correct statement about the motion of a body that must be substituted for Aristotle's statement, we consider the Aristotelian point of view more carefully to see why his intuition was wrong. According to his idea of motion, a force is required to keep a body moving at a constant speed. This means, according to Aristotle, that if two bodies are moving with different constant speeds, the faster one is being acted on by a larger force. At first this may seem quite reasonable because we know, for example, that if two people pull an object across the floor they can keep it moving about twice as fast as one person can, so that it appears that twice the force is required to maintain twice the speed.

But this apparently correct conclusion does not really give the correct relationship between the application of a force to a body and the state of motion of that body. If we examine carefully the situation of the people pulling or pushing the object across the floor, we find that even when they stop pulling or pushing, the object still continues to move somewhat. But why does it finally come to rest after the pushing stops? Galileo discovered that this happens because of the roughness of the ground. By making the contact between the object and the ground smoother and smoother, he demonstrated that the object maintains its speed for longer and longer times after the push ceases. This convinced him that a push (or a pull) is not required to keep a body moving because he saw that if there were no friction between the ground and the body, the body would continue moving with the same speed it had at the moment the external force on it (the push) was removed. Galileo concluded that the mere velocity of a body does not show whether or not an external force is acting on the body. This idea was later reformulated by Newton and is now known as Newton's first law of motion, or as the law of inertia, which Newton stated as follows: "Every body perseveres in its state of rest or of uniform motion in a straight line, unless it is compelled to change that state by forces impressed thereon." This law is a statement about an idealized situation and is an extrapolation from a real situation in which the condition for the validity of this law can never be fulfilled. The law of inertia cannot be derived directly from a real experiment because we can never eliminate the influence of external forces on a body. We nevertheless accept it because it becomes more and more valid as the external forces are reduced. The continuity of physical phenomena shows us that in the limit of zero external forces the law of inertia would hold exactly.

Another important feature of the law of inertia is that it makes no distinction between rest and uniform velocity so that, in a sense, these two states of motion are equivalent, since they both indicate the absence of external forces. Put differently, the law of inertia states that the behavior of a body when a force acts on it is the same whether the body is at rest or moving with constant velocity.

Before leaving this first law of motion, we discuss the concept of inertia as it is used in this law. The word itself is derived from the Latin *inters* meaning *idle,* and this is the sense in which we apply the word inertia in common usage. When we say that a person has a lot of inertia, we mean that he is an idle, sluggish fellow who gets little done. Some of this sense of the word is contained in its connotation in the first law because it implies that a body possesses some physical attribute which makes it resist a change in the state of its motion. We speak of this property as the inertia of the body and the larger this is, the greater is the force required to set it in motion once it is at rest, or to alter its velocity once it is moving. Although the first law tells us how the state of motion of a body permits us to detect the presence (or absence) of a force, it does not tell us how the force that is applied to a body is related to the way the state of motion of the body changes. This answer is the content of Newton's second law of motion.

As we have seen, the instantaneous velocity of a body gives us no indication of the external forces acting on the body at the moment we observe it. However, by observing the velocity of a body at the beginning and end of a short interval, we can determine whether or not a force is acting because the second law tells us that a force acting on a body changes the body's velocity. This discovery was made by Galileo and described in the following extract from his *Two New Sciences:*

> . . . any velocity once imparted to a body will be rigidly maintained as long as the external causes of acceleration or retardation are removed, a condition which is found only on horizontal planes; for in the case of planes which slope downward there is already present a cause of acceleration; while on planes sloping upwards there is retardation; from this it follows that motion along a horizontal plane is perpetual; for if the velocity be uniform, it cannot be diminished or slackened, much less destroyed.

In this statement Galileo asserts the validity of the first law (the concept of inertia) and at the same time indicates the role of a force (the "external causes of acceleration") in changing the velocity of a body by accelerating it. Although Galileo was aware of the role played by a force in altering the velocity of a body, he never explicitly gave the relation between the impressed force and the change in the state of motion of the body. This task was left for Newton, who expressed it in his second law of motion, which was stated in his *Principia* as follows: "The change of motion is proportional to the motive force impressed; and is made in the direction of the right line in which that force is impressed." Newton goes on to explain his statement more fully in the following:

> If any force generates a motion, a double force will generate double the motion, a triple force triple the motion, whether that force be impressed altogether and at once, or gradually and successively. And this motion (being always directed the same way with the generating force), if the body moved before is added to or subtracted from the former motion. . . .

Newton's second law of motion leads us to the definition of the mass of a body and, at the same time, permits us to measure the mass. Restated somewhat differently, Newton's second law says that if a force is impressed on a body, the body is accelerated in the direction of the force, and the magnitude of the acceleration is proportional to the magnitude of the force. In other words, the ratio of the magnitude of the impressed force to the magnitude of the resulting acceleration is always the same—it is a constant and therefore a measure of some basic characteristic of the body. The second law may therefore be written in the form F/a = constant, which states that force F divided by acceleration a is a constant quantity. To see the physical significance of this constant, we rewrite the above expression in the form $a = F/\text{constant}$. From this equation we see that the larger the constant, the smaller is the acceleration produced by the force F. The constant has the property of resisting the action of the force so that it is essentially what we previously referred to as the inertia of the body. From now on, however, instead of calling it the inertia of the body, we refer to it as the inertial mass of the body and label it with the letter m which gives us the relationship $F/a = m$. We use this equation as the definition of the inertial mass (a scalar) of a body. This equation also tells us exactly how to measure the mass of a body. We subject the body to a known force and then measure the acceleration imparted to the body by the force. We divide the measured magnitude of the force by the measured magnitude of the acceleration; the number we obtain is the mass of the body.

One question still remains: in what units is the mass to be expressed? Since mass is a derived quantity (it is an algebraic combination of force and acceleration), we cannot introduce a unit of mass arbitrarily as we did a unit of length or a unit of time. The unit of mass must itself drop out of the formula for the mass. To see how this occurs, we take a very large variety of particles and subject them one by one to a force of one dyne (a unit force) and measure the acceleration imparted in each case. From our definition of mass, we then have in each case $m = 1/a$ where we have placed $F = 1$, since we are applying one dyne of force to each body. How do we know which of these particles has a unit mass? Our equation tells us that if we want m to equal 1 in the equation, we must have $a = 1$, and this gives us the definition of the unit mass. It tells us that that particle has a unit mass which acquires an acceleration of 1 cm/sec^2 when a force of 1 dyne is applied to it. This unit mass is 1 gram (abbreviated as gm).

Now that we know how mass is defined in terms of force and acceleration, we see that the basic combination of dynes, centimeters, and seconds that gives the definition of mass is dynes sec^2/cm. However, even though we have introduced force as a basic quantity and mass as a derived quantity, in practice we express all physical quantities in terms of mass, length, and time; that is, in terms of grams, centimeters, and seconds. Force, for example, will be expressed as gm cm/sec^2. With this understanding, we can write Newton's second law of motion in the form in which it always appears—$\mathbf{F} = m\mathbf{a}$—probably the most famous equation in the history of science. This expression is a vector equation because it contains both the magnitude aspect and the directional feature of Newton's law.

Since this is a vector equation, we can represent it as three separate equations in a coordinate system. To do this we first note that, in accordance with what we have already said about acceleration, we can study the total motion of a body by breaking this motion up into separate components along the three mutually perpendicular axes of a coordinate system. If $\mathbf{F}$ has the components $\mathbf{F}_x$, $\mathbf{F}_y$, $\mathbf{F}_z$ in our rectangular Cartesian coordinate system and a has the components a_x, a_y, a_z in this same system, then $\mathbf{F}_x = m\mathbf{a}_x$, $\mathbf{F}_y = m\mathbf{a}_y$, and $\mathbf{F}_z = m\mathbf{a}_z$. The total motion of the body can be described as consisting of three separate motions along three mutually perpendicular directions and these motions are independent of each other. These expressions are called the equations of motion of a particle because they permit us to determine the orbit of the particle if we know the magnitude and direction of the force acting on the particle at each instant.

When we attach a body to a vertically suspended spring and the spring stretches, we deduce that the spring is exerting some upward force, say of magnitude F, on the body. But since the body is at rest, we further deduce from Newton's first law that the net force on the body is zero. This null result means that a downward force is also acting on the body and is equal to the upward force F. If we call this downward force W, then $W = F$.

If we call W the weight of the body, we see that it can be measured by the stretch of our standard force-measuring spring. Since weight is a force, it must be expressed in dynes. The weight W of the body is transmitted to the spring as a force which stretches the spring. As long as the body is attached to the spring, it remains suspended above the ground and at rest. But if we cut the spring, the upward force $\mathbf{F}$ of the spring no longer acts on the body which is then subject only to the downward force $\mathbf{W}$ so that, according to Newton's second law, it acquires a downward acceleration g such that $W = mg$ (force = mass $\times$ acceleration). Now we know that g is the acceleration of gravity and equals 980 cm/sec^2. We see that the weight of the body is the force that imparts to it the acceleration of gravity. Again we note that g is the same for all freely falling bodies (in a vacuum) regardless of how much they weigh. Since the weight of a body is the force (of gravity) exerted by the earth on the body, we are thus confronted with a remarkable feature of the force of gravity—it imparts the same acceleration to all bodies regardless of how strongly it pulls on these bodies.

The relationship between weight and mass which we have just established shows us that the weight of a body is certainly not the same thing as its mass, although their numerical values are proportional to each other, since g is the same for all bodies. A body with a mass of 1 gm (unit mass) weighs 980 dynes, for we have $W = 1 \times 980 = 980$ dynes. Since the weight of a body is proportional to its mass, we can now find its mass from its weight. All we need do is place the body whose mass is to be determined in one of the pans of an equal arm balance, and then place on the other pan as many unit masses as are required to balance the scale. The mass of the body then equals the number of unit mass required to balance it.

The best and simplest way to understand the physics of Newton's second law is to apply it to a few examples that stem directly from our daily experiences. In general, the force that acts on a body to set it moving (e.g., the force on a car) changes from moment to moment so that the relationship between the force and the state of motion of the body is a constantly changing one. To escape from such complications, we assume, to begin with, that the force we are dealing with remains constant in magnitude and direction.

As a first example, we consider the net force required to set a body moving with a speed of 20 cm/sec in 5 sec, if the body starts from rest and if its mass is 100 gm (we assume the body acquires its speed at a uniform rate). A few points in the statement of this problem are very important. To begin with, the knowledge of the time during which the body acquires its speed (or the distance over which it acquires its speed) is essential because without it we cannot calculate the acceleration. Second, the only force that can be calculated is the net force—nothing can be determined about all the different forces that may be acting on the body and contributing to the net force.

Since the body acquires its speed in 5 sec, its acceleration is 4 cm/sec^2 (we divide 20 by 5). The net constant force acting on the body is thus its mass (100) times its acceleration (4), or 400 dynes. Now if the body is free to move in empty space, the 400 dynes (acting for 5 sec) is all the force that needs to be applied to give the body a speed of 20 cm/sec in 5 sec. But suppose the body is resting on the ground. There is then a frictional force acting on it which resists your push. It is clear, then, that you will be unable to give the body a speed of 20 cm/sec in 5 sec if you push it with a force of 400 dynes. You must either exert a larger force or push for a longer time to achieve the same results as before. If the opposing frictional force of the ground is 300 dynes, you must exert a force of 700 dynes on the body to give it the speed of 20 cm/sec in 5 sec. If, however, you still push it with a force of 400 dynes, only a force of $400 - 300 = 100$ dynes is available to accelerate the body and the acceleration is 100/100 (force divided by mass) = 1 cm/sec^2. Twenty seconds will be required to give the body a final speed of 20 cm/sec.

As another instructive example, we consider a body with a mass of 100 gm being twirled in a circle at the end of a string whose length is 10 cm. What force does the string have to exert on the body (always toward the center of the circle—the point where the string is held) to keep the body moving in the circle with constant speed of 50 cm/sec? The force (as a vector) is not constant in this case since its direction (along the string) is constantly changing, but its magnitude is constant, so that we can apply the same analysis as in the previous problem. Since the speed of the particle is 50 cm/sec and the radius of the circle in which the particle moves is 10 cm, the acceleration of the particle is $v^2/r = 2500/10 = 250$ cm/sec^2. Since the mass of the particle is 100 g, the force exerted on it must be $ma = (100)(250)$ dynes, which is slightly more than the weight of a 25-gm mass.

We consider now a group of phenomena which may be interpreted in terms of what we will call inertial forces. From one point of view these forces appear fictitious and yet from another, they appear to have the same kind of reality as do gravitational forces (the weights of bodies). To see how these forces arise, we consider an accelerated frame of reference—let us say a train whose speed is increasing at a constant rate of 980 cm/sec^2. Suppose that the floor of the train is quite smooth so that an object is free to move without friction when placed on the floor. How does such an object appear to move as seen by an observer attached to the train? Owing to Newton's first law of motion, this object, as seen by an observer not moving with the train, is at rest (or moving with uniform velocity). Therefore, the back of the train is moving toward the object, as seen by this same fixed observer, with accelerated speed, and there is no apparent force acting on the body. Since the observer on the train is moving in the same manner as the back of the train, it appears to him that the object is moving with accelerated motion—in fact, with an acceleration of 980 cm/sec^2—toward the back of the train. Indeed (as observed by the man on the train), all objects placed on the floor of the train move with the same acceleration of 980 cm/sec^2 toward the back of the train, just as though they were "falling" toward the back. Hence, to the observer in the train, it appears as though a force equal to 980m dynes, where m is the mass of the body, were acting on each body. To stop a body from "falling" toward the back of the train, the observer in the train could attach the body to a spring, the front end of which is attached to the front of the train. The spring would then stretch until it exerted a force of 980m dynes on the body and the body would remain at rest in the train. The man in the train could then say that there is a force acting on every object which is placed on the floor of the train and which is free to move. The stretch of the spring permits him to measure the size of this force. Since there is no physical body in contact with the object as it "falls" toward the back of the train, the observer in the train concludes that either some mysterious force emanating from the back of the train acts on the body (if he were unaware of the train's acceleration) or that the force is not a real force but rather a fictitious force stemming from the forward acceleration of the train and the inertia of the body. The observer would therefore refer to this force as an inertial force. Another example of an inertial force is the "force" that a body experiences when it is in a rotating frame of reference. Objects in such frames of reference (like clothes in a spinning washing machine) appear to be thrown toward the perimeter of the frame. An observer in such a frame then says that these objects experience a centrifugal force, which is really a manifestation of the inertia of objects.

CHAPTER 6

Conservation Principles

Newton's Third Law and the Conservation of Momentum

I do not know what I may appear to the world, but to myself I seem to have been only like a boy playing on the seashore, and diverting myself in now and then finding a smoother pebble or a prettier shell than ordinary, whilst the great ocean of truth lay all undiscovered before me.

—ISAAC NEWTON, Brewster's *Memoirs of Newton*

As far back as we can go in recorded history we find that humanity has always looked for some kind of constancy in the universe and that many of its religious beliefs have sprung from the idea that an unchanging, all-powerful being controls the world of matter, mind, and spirit. These early, naive pictures, which assigned this being to a special place in the heavens and placed the earth in a central position below the heavens, were shattered by astronomical studies which show the universe to consist of bits of matter (the galaxies or clusters of galaxies) distributed very sparsely but homogeneously in vast regions of space. As far out as we can see with our optical and radio telescopes there is no observable difference between our region of space and any other region, so that there is nothing special about the earth, the sun, the solar system, or our galaxy. These observations indicate that the same natural laws that govern our earth and solar system govern all matter in the universe and apply to all regions of space. These laws, then, are the unchanging features of the universe. As we have seen, the statement $F = ma$ is one such law, insofar as we can accept as true the Newtonian concepts of space and time. By this statement we mean that if the same rules for measuring space and time as those we have previously described apply to all regions of the universe, and if mass has the same inertial properties everywhere in the universe and under all conditions, then the law $F = ma$ is valid everywhere. In addition to this law, we have introduced two other laws of nature up to this point: the speed of light in a vacuum is the same for all observers moving with uniform velocity, and all freely falling bodies in a vacuum near the surface of the earth fall with the same acceleration. These laws are the only ones that we have stated thus far but we must know certain other laws before we can understand nature in all its aspects. At this point, however, we consider the general results that can be deduced from the two laws of motion already discussed and leave the discussion of other laws for later chapters.

The second law of motion encompasses the first law because when we state that $F = ma$ we mean that a body for which $F = 0$ experiences no accelerations and therefore moves with uniform speed in the same straight line. The second law, however, does not tell us all we must know about the relationship between the motions of the bodies and the forces acting on these bodies to enable us to understand completely the dynamics of such bodies. Newton knew this and therefore introduced his third law of motion, which, together with the second law, permits us to develop a complete system of Newtonian dynamics. The second law of motion deals with the force exerted on a given body, but it tells us nothing about the source of this force nor about the relationship of the source to the action of the force. To be specific, we again consider the force exerted on a body by a spring. The body, of course, is accelerated by this force, but what happens to the spring at the same time? The spring is stretched, showing that the body on which the spring is exerting a force is, itself, exerting an opposite force on the spring. This is true in all other situations in which a force is exerted on a given body—the body exerts an opposite force on the source of the force. This relationship may be stated in another way by saying that forces of nature occur in pairs, with one member of the pair exactly opposing the other. One part of Newton's third law of motion is a statement of this fact, and the other part of this law states that the two opposite forces are equal in magnitude. This law is sometimes referred to as the law of action and reaction: For every force in nature acting on a body, an equal and opposite force acts on some other body or, alternatively, to every action there is an equal and opposite reaction.

Considered from the point of view of Newton's laws of motion, the sum of all the forces in the universe is zero. One important point must be kept in mind when applying the third law. The two equal forces that constitute the pair of opposites act on two different bodies. The failure to comprehend this point leads to a misunderstanding of this law and to the false idea that the two opposing and equal forces act on the same body. If this were true, a body could never be accelerated. To eliminate this confusion, it is helpful to consider a few simple applications of the third law.

When a person exerts a push against a wall, what forces are involved? We begin with the wall at the moment the person begins to push it, and suppose that it is perfectly free to move, detached from the rest of the room. It then experiences an acceleration under the action of the push and begins to move in the direction of the push; but, owing to its attachment to the floor, the ceiling, and the other two walls, the wall is actually not free to move bodily and the push simply causes the wall to bulge outwardly at the point where pressure is exerted. Though invisible to the eye, the bulge is present, and increases with increase in the magnitude of the push exerted. But under ordinary conditions the bulging ceases very quickly. The reason for this cessation is that the structural forces in the wall are brought into play as soon as the push is applied to it. The harder the push, the larger these opposing structural forces become. If the wall finally stops bulging, it means that the structural forces exactly equal and are opposite to the push so that the acceleration of the wall is reduced to zero. For the sake of simplicity, we may take all of these structural forces as acting at points A and B near the floor and ceiling respectively with their sum just equal to the push F. If the push were strong enough to overcome the structural forces at A and B, the wall would be ripped away from its moorings and would move off with accelerated motion. From this analysis we see that the fact the wall quickly comes to rest means that its structural forces (the reaction) are very strong and quickly balance the push (the action).

But what about the person who is pushing the wall? We know from our own experience that we must brace ourselves against the ground if we wish to push against a wall, so that the person pushing the wall is subjected to two forces. One is the push of the wall against him, which is exactly equal in magnitude to his push against the wall. But the person is not accelerated backward by the push of the wall against him because he is also being pushed forward at his feet by the ground; this push of the ground just equals and opposes the push of the wall, so that the net force on the person is zero. If the person were on roller skates in this case, he would immediately begin to move backward on pushing the wall because the push of the ground would be eliminated and the push of the wall would then be uncompensated and the net force on the person would not be zero.

As another somewhat more complicated example, we consider a pitcher throwing a ball. Essentially all that is happening, as far as the propulsion of the ball is concerned, is that the pitcher's arm is pushing the ball, therefore accelerating it. Suppose that m is the mass of the ball and that a is its acceleration. From Newton's second law, it follows then that the ball is being pushed with a force equal to the product of m and a as long as the ball is in the pitcher's hand. (Here we assume that the acceleration of the ball remains constant during the time the pitcher is throwing the ball.) According to Newton's third law of motion, the ball exerts an equal but opposite force (push) against the arm of the pitcher. The force is communicated to the pitcher's body via his shoulder, but the pitcher is anchored to the ground which pushes against his feet (in the direction of the motion of the ball) with a force which exactly equals ma. The net force on the pitcher is zero. We can look at this situation from a slightly different point of view and consider the pitcher as being an extension of the ground. We may then say that the ground pushes the ball with a force ma and the ball pushes back against the ground with exactly the same force.

Suppose now that the pitcher is standing on skates so that he is completely free to move as he pitches the ball. He then begins to move backward with an acceleration A as soon as he begins to throw the ball, since the ground no longer opposes this backward motion. The quantity A must be such that, when multiplied by the mass M of the pitcher, the product equals the product of m and a so that $ma = MA$. This is another way of stating Newton's third law. From this relationship we can easily find the backward acceleration of the pitcher by dividing the product of m and a by the M of the pitcher, which gives us the relationship $A = ma/M$. This acceleration is as many times smaller than a as the mass of the man is larger than the mass of the ball.

A few more simple examples will clarify the matter still further. Consider a man pushing a box across a floor at constant speed. Since he is certainly exerting a force F on the box, the fact that it is not accelerated may at first puzzle us. But we soon note that the floor is exerting an exactly equal and opposite force F (the force of friction) on the surface of the box that is in contact with the floor. Thus, the net force on the box is zero and it moves with constant speed. The net force on the man and box taken together as a single system is zero. The frictional force F on the floor acting on the box just equals the frictional force F of the floor acting on the feet of the man.

Another application of this same idea is to the motion of a car. Most people believe that a car is pulled in the direction of its motion by its motor, but this is not so. The motor only turns the wheels. The car is pushed forward by the ground. If there were no friction between the wheels and the ground or if the wheels of the car were not touching the ground, the car would not move, regardless of the power of the motor. If the ground is rough enough, the wheels push back against the ground and the ground in turn (by Newton's third law) pushes the car forward. In the same way, a person can walk only because the ground pushes against him at each step. This forward push is the reaction of the ground against the person's backward push on the ground at each step.

This analysis of Newton's third law can now be used to introduce the first of a series of what physicists call conservation principles. These principles deal with certain physical entities, amounts of which can be transferred from body to body or from one system of bodies to another, but which remain unchanged in total quantity during such interchanges. Nature introduces a kind of double entry bookkeeping, as far as such entities go, and keeps them always in perfect balance. We relate the first of these conservation principles to Newton's laws of motion. To obtain this first conservation principle, we introduce a new physical entity called momentum; it is a derived quantity which is defined in terms of mass and velocity.

Consider a particle of mass m moving along an orbit and let $\mathbf{V}$ be its velocity at some point A of its orbit. Suppose, further, that a force is acting on the particle which keeps it moving in its orbit. The magnitude of this force, at any moment, depends on the mass of the particle and its acceleration, according to Newton's second law; and the direction of the force depends on the direction of the acceleration. We now introduce the product $m\mathbf{V}$ (mass times velocity) as a new physical vector associated with the motion of the particle, and call it the momentum $\mathbf{p}$ of the particle. Using the momentum concept we restate Newton's second law as follows: When a force acts on a particle, this force changes the momentum of the particle at a rate which equals the magnitude of the force. The direction of the force is in the direction of the change of the momentum. With this we restate Newton's first law of motion as follows: If no net force is acting on a particle, its momentum remains constant. This statement contains the germ of one of the conservation principles that we discussed above, since it tells us that a certain physical entity—the momentum—remains constant in quantity, and is therefore conserved, when no force acts on a particle. This statement is the essence of the principle of the conservation of momentum. To see its full significance, however, we consider a group of particles interacting with each other, and begin by studying just two such particles moving toward each other. Suppose that one of them has a mass m_1 and is moving in a straight line to the right with velocity $\mathbf{v}_1$, whereas the other has a mass m_2 and is moving to the left along the same line with velocity $-\mathbf{v}_2$ (the minus sign shows opposite directions of motion). What can we say about the velocities of the particles after they have collided? We see that without additional information we cannot specify the particle velocities completely after the collision, but we can derive some important facts about these velocities. To do this we note that before the collision, the momentum of the first particle is $m_1\mathbf{v}_1$ to the right (positive momentum) and the momentum of the second particle is $m_2\mathbf{v}_2$ to the left (negative).

We now introduce the concept of the total momentum of the two particles which taken together constitute a single system. Its total momentum is the vectorial sum of the two momenta (plural of momentum) of the two particles. In this simple case, since the momentum of one particle is opposite in direction to that of the other, the magnitude of the total momentum of the system before collision is $m_1v_1 - m_2v_2$. The direction of this total momentum is either to the left or to the right (along the same straight line as that of the velocities of the particles) depending on whether m_2v_2 is larger or smaller than m_1v_1.

To deduce the situation after the collision, we first examine the collision itself. As soon as the two particles come into contact they begin to push each other, with particle 1 pushing to the right and particle 2 pushing to the left. But by Newton's third law, the magnitudes of these two pushes are equal. It follows from the definition of momentum and Newton's second law that the momentum of each particle is changed by exactly the same amount during the collision. But the direction of the change for particle 1 is to the left and hence negative, whereas the change for particle 2 is to the right and therefore positive. Hence, the two changes cancel each other and the magnitude of the total momentum of the two particles (considered as a single system) after the collision is unaltered.

Thus, if a system of two particles has no external force acting on it (the only force acting on each particle is that exerted by the other particle) the total momentum of the system remains constant, both in magnitude and in direction. This constancy can be stated another way: In a system of two interacting particles, the rate at which one particle is gaining momentum, in any direction at any instant, exactly equals the rate at which the other particle is losing momentum in that direction.

This principle can be extended to a system of any number of particles since we can always concentrate our attention on any one particle in the system and consider its interaction (or collision) with each of the other particles, one by one. In each such interaction this particle exchanges momentum with the other, and there is simply a continual transfer of momentum from particle to particle, with no net change in the total momentum.

The general principle of the conservation of momentum may be stated as follows: In an isolated system of any number of particles (this means no external forces acting on the system) the total momentum of the system remains constant in magnitude and direction. If two systems of particles interact in any way, the momentum gained by one in any direction is exactly equal to the momentum lost by the other in that direction.

Since momentum is a vector quantity, we can express the conservation principle in terms of any component of the momentum. We may say that in any isolated system the component of the total momentum of the system in any given direction remains constant unless a force with a component in that direction acts on the system. The component of the momentum of the system in that direction then increases at a rate which equals the component of the external force in that direction.

The principle of the conservation of momentum permits us to

understand the behavior of elementary particles in nature such as electrons and protons. The revelatory and predictive power of this principle lies in the fact that in all interactions involving the appearance and disappearance of particles, the momentum balance must be maintained. If all the visible particles involved in a reaction do not give a momentum balance, other particles, not yet discovered, must also be involved.

An interesting question is whether there is one particular point between two particles such that if we place an observer at this point and set him moving with this point, the ratio of his distances from the particles always remains the same. The answer is yes. The point divides the distance between the two particles in two parts whose lengths stand in the same ratio to each other as do the inverses of the particle masses; this point, which is called the center of mass of the two particles, is as many times farther away from m_2 than it is from m_1 as the mass m_1 is greater than the mass m_2. This relationship may be illustrated by two people on a seesaw. The heavier person must be closer to the balance point for the seesaw to remain fixed. A child weighing 50 pounds can balance a 150-pound man by sitting three times farther from the balance point than the 150-pound man. In a sense, the center of mass of the two particles is a point in which we may picture the entire mass as being concentrated and which moves in a straight line with such a speed that it carries the total momentum of the system along with it.

Although we have introduced the concept of the center of mass for a system consisting of only two particles, we can extend it to a system containing any number of particles. We can imagine a collection of such interacting particles with each particle possessing a definite momentum at each moment. Owing to the interactions and collisions between particles, the momentum of any one particle changes continuously, but the principle of the conservation of momentum assures us that the total momentum of the system remains constant. To obtain this total momentum, we simply add (vectorially) all the individual momenta and then we find that there is one point in the system (the center of mass of the system) which moves with a velocity $\mathbf{V}$ and has a momentum $M\mathbf{V}$ where M is the total mass of all the particles in the system $(m_1 + m_2 + \cdots + m_n)$ and $M\mathbf{V}$ is the total momentum of all the particles in the system $(m_1\mathbf{v}_1 + m_2\mathbf{v}_2 + \cdots + m_n\mathbf{v}_n)$. This is the most general statement of the principle of the conservation of momentum.

The concept of the center of "something" can be applied to such things as population, wealth, and learning and we find these centers just as we find the center of mass of a collection of masses. Such a "center" is specified by giving its distance from any arbitrarily chosen point. If we seek the center of mass of a number of bodies (five, for example), having different masses, we multiply the mass of each body by the body's distance from any chosen point and add all these products; we finally divide this sum by the total mass of the bodies and thus obtain the distance of the center of mass from the chosen point.

The center of population of a country is found in the same way. We choose a point on the map of the country and measure the distance of each city from the point and multiply this distance by the city's population. We now add all these products and divide this sum by the total population of the country and obtain the distance of the center of population from the chosen point.

Returning to the center of mass, we see that the center of mass of our solar system is close to the center of the sun because the sun contains more than 99% of the total matter in the solar system. The center of mass of the earth–moon system is a point (on the line connecting the center of the moon to that of the earth) which is 81 times as far from the moon's center as it is from the earth's center, which tells us that the earth is 81 times as massive as the moon.

A few specific examples of the application of the principle of the conservation of momentum clarify some points that may still be obscure. We consider first a man standing on a cart that is free to move. The system in this case consists of the man and the cart, which may be treated as two particles, since the mass of each of these bodies may be pictured as being concentrated in a single point (the center of mass of the body). Initially, the cart and the man are at rest, so that the total momentum of the entire system is zero. Since no net external force acts on the man–cart system, its total momentum must remain zero, regardless of what the man does. If his mass is m and he begins to move to the right with speed v, he introduces momentum mv to the right into the system which, if uncompensated, would violate the conservation principle. But another motion compensates this momentum because the cart moves to the left with speed V which is just large enough so that the momentum of the cart to the left exactly equals the man's momentum to the right. If M is the mass of the cart, then $MV = mv$. The speed of the cart to the left is as many times smaller than the speed of the man to the right, as the mass of the man is smaller than the mass of the cart.

This simple example can be applied to the analysis of many familiar phenomena. Just before a bullet is fired from a rifle at rest, the momentum of the rifle and bullet taken as a single system is zero. The rifle must therefore move backward (against the shoulder of the shooter) when it is fired to maintain the zero momentum of gun and bullet since the bullet carries momentum in the forward direction. In the same way, we see that a rocket must be propelled in a forward direction when its fuel is burned rapidly because the gases formed during the burning escape from the back of the rocket and carry off momentum in a backward direction. The whole trick in rocketry of this sort is to use the kind of fuel that releases the maximum amount of gas moving with maximum speed per gram of fuel.

Another example of the momentum principle is that of a man of mass m running with speed v and jumping into a stationary cart of mass M that is free to move. How fast do the cart and the man (now fixed in the cart) continue to move forward? If V is the forward speed of the man and cart together, the momentum of the two is $(m + M)V$. But this amount must equal the initial forward momentum of the man which is just mv. Hence, $(M + m)V = mv$ or $V = mv/(M + m)$. As a final example we consider a stationary cart and a man running toward the cart with constant speed. This time, however, the man runs right across the cart without changing his speed. What happens to the cart? The first response to this question, generally, is to say that the cart moves backwards. But a little thought and the application of the momentum principle show that the cart must stand still while the man is running across it. The reason for this is that the momentum of the man and cart together must remain constant while the man runs across the cart. But this means that the momentum of the cart cannot change, since the man's momentum remains constant; and since the cart was at rest to begin with, it must remain at rest. We can understand this conclusion from a physical point of view by noting that in his first running step onto the cart the man's foot tends to move the cart forward but a moment later this same foot pushes the cart backward as he brings his second foot in contact with the cart. His last step, which is off the cart, cancels the action of his first step onto the cart.

As a final example of the usefulness of the momentum concept, we consider a very rigid, small sphere that is rolled in some direction along a horizontal surface and allowed to bounce off a smooth wall. In what direction does the ball roll after it hits the wall? We cannot answer this question in the general case but if we make certain assumptions about the motion of the sphere before and after the collision, we can then answer the question. We must assume that the speed of the sphere is unaltered while it is rolling along the ground. This assumption is clearly not correct, owing to the friction between the sphere and the ground (but we neglect this detail here for convenience). Moreover, we assume that the speed of the sphere after the collision is the same as it was before. In other words, the collision alters the direction but not the magnitude of the sphere's velocity.

With these assumptions, we can now determine the motion of the sphere after its collision with the wall. To do this, we suppose that before colliding with the wall, the sphere is moving with momentum $m\mathbf{v}$ in a direction that makes an angle ϕ (angle of incidence) with the perpendicular (the straight-line which we designate as OX) to the wall. We now decompose this momentum into a component along the line OX and a component perpendicular to this line. The only component affected during the collision is the one along the line OX—the component of the momentum parallel to the wall remains unaltered. The reason for this is that the force engendered by the collision is perpendicular to the wall—there is no component of this force parallel to the wall. The component of the momentum perpendicular to OX (parallel to the wall) must be the same before and after the collision. As a result, the angle of reflection r of the sphere after the collision is the same as the angle of incidence ϕ, as long as no change occurs in the speed of the sphere because the momentum of the sphere before collision and its momentum after collision must be symmetrical with respect to the line OX.

Newton's second law of motion can be used to calculate the acceleration experienced by a particle if we know the force acting on the particle. As long as we can "see" the force acting on the particle, as in the case of a spring attached to an oscillating particle, there is no mystery about the relationship between the force and the orbit of the particle. But many examples of moving particles occur (such as a particle thrown into the air) in which the force is not obvious. Only the particle's acceleration tells us that a force is acting. If we think about this motion for a moment, we see that there is an element of mystery about it, because all we observe is a particle moving along one particular path (with no apparent force acting on it) rather than along any one of an infinitude of other paths that can be drawn between the starting and end points of the particle's path. The mystery is why the particle chooses its actual path rather than one of the others. If some visible force were directing the motion of the particle, there would, of course, be no mystery, but with no visible agency directing it, we wonder how the particle "knows" which path to choose. This difficulty impels us to examine certain general principles which can be introduced quite independently of Newton's laws of motion and which determine the path of a particle. Newton's laws of motion can, in fact, be derived from these principles, so that they are more general than Newton's laws.

The first of these principles was discovered in the 18th century by the French philosopher, scientist, and mathematician, Pierre-Louis Maupertuis, who was investigating the Newtonian corpuscular theory of light and its relationship to the path taken by a beam of light when it is reflected from a surface or when it passes from one medium into another. Now we know from observation that when a beam of light strikes a smooth surface at a certain angle ϕ, it is reflected from the surface at the same angle. We also know from observation that when a beam of light moves across a surface from one medium such as air into a denser medium such as water or glass, the direction of the beam is bent toward the normal (the perpendicular) to the surface. This phenomenon is called refraction; the angle of refraction is related in a definite way to the angle of incidence and the density of the medium.

Maupertuis sought to explain this behavior of light not by analyzing the manner in which the reflecting or refracting surface affects the light that strikes it, but rather by finding some overall characteristic or property of the entire path, which favors the correct path and, indeed, imposes this path on the light, rather than any other path. To find this property, Maupertuis considered in addition to the correct path any other nearby incorrect paths and then calculated various physical quantities along these other paths. He then found that the correct path is the one along which a certain quantity called the action of the corpuscle of light (he assumed the Newtonian corpuscular theory of light to be correct) is smaller than it is along any neighboring path. Before considering the application of this discovery to a particle of matter, Maupertuis's concept of action must be defined.

Consider a particle of mass m moving with variable velocity along some path (curved or straight) from its starting point A to its final point B. If v is the velocity of the particle at any point of its path, its momentum at that point is $\mathbf{p} = m\mathbf{v}$ as previously defined. If the path of the particle is divided into tiny successive segments, we define the Maupertuis action over a small segment of the path as the product of the length of this small segment of path and the momentum $\mathbf{p}$ along this small segment. We add up all of the small actions segment by segment for the entire path from A to B. This amount gives us the total action over the given path. Maupertuis then proposed the principle that the actual path taken by the corpuscles of light is the one for which the total action is a minimum. This principle is known as the Maupertuis principle of least action, and is closely related to another principle discovered by Fermat, called Fermat's principle of least time, which we discuss later.

The Maupertuis principle of least action fell into disuse as far as light is concerned because the Newtonian corpuscular theory of light was proved to be incorrect. It can be shown that a corpuscular theory of light such as the one offered by Newton, which assigns Newtonian inertial mass and Newtonian momentum to corpuscles of light, requires the speed of light to be greater in a dense medium (e.g., glass) than in a rare medium (e.g., air). Since this is contrary to the experimental evidence, the Newtonian corpuscular theory of light was replaced by the wave theory, which does agree with observation. The Maupertuis principle of least action must then be replaced by Fermat's principle of least time, which states that the actual path taken by a beam of light that experiences multiple reflections and refractions in going from some initial point A to a final point B is the one for which the time of propagation is a minimum.

When Maupertuis first announced his principle of least action, he interpreted it theologically. In his "Essai de Cosmologie," written in 1732, he stated (referring to his principle of least action): "Here then is the principle, so wise, so worthy of the Supreme Being: whenever any change takes place in nature, the amount of action expended in this change is always the smallest possible." Maupertuis was actually too extravagant in these theological claims for his principle, for as we now know, the action along the actual path of a material particle (or the least time for a beam of light) may be either a minimum or a maximum, in other words, an extremum. There is no reason to suppose that the "Supreme Being" is concerned with economy in the events of nature.

The concept of doing things according to a minimal principle or of being guided by such a principle is not strange to us at all for we act on such principles constantly. When we take an automobile trip, we may take the route that requires the minimum time, or one which is of shortest length, or a route which is least expensive. These alternatives are not necessarily minimally equivalent. The shortest route may be very mountainous and entail a greater expenditure of gas than a longer route.

Although Maupertuis discovered the principle of least action by considering the Newtonian corpuscular theory of light, its great importance was first recognized in its application to the motions of particles. This was first delineated in the work of Euler and Lagrange, who extended this action principle to mechanics. The Maupertuis principle of least action implies that a particle carries action along with it, with the action constantly changing as the particle moves. Instead of assigning the action to the particle itself, however, we can picture the action as being a property of the various paths of the particle (we assume that all the paths we are dealing with start from the same point and that the particle is projected along each path with the same speed). We can then say that each point of any of the paths has a certain action associated with it and the particle acquires that action when it reaches the given point on the path.

An interesting speculation about the action of a particle arises when we consider a particle at a point in its path where its momentum is p, and suppose the action of the particle at this point is represented by the variable s. If we now follow the particle for a short stretch q of its path, its action changes (this change may be positive or negative) by an amount equal to $p \times q$. If we follow the particle over a very small interval, its action changes by a very small amount, which is in accord with all the concepts we have developed up to now; classically a particle's action can change by an arbitrarily small amount. But if some kind of restriction governs action, such that the action of a particle cannot change by less than a certain amount (e.g., only by multiples of a given minimum amount h), then the following situation occurs: If we follow the particle over a very small stretch of path, its momentum p cannot remain unaltered but must become large enough to keep the product of p and the small segment of path q equal to the minimum action change h. The smaller the stretch of path is, the larger the momentum of the particle becomes. This inverse relationship leads to the curious situation that the momentum of such a particle in a given stretch of its path depends on how accurately we observe its position in that stretch; that is, the smaller the stretch of path, the larger the momentum. This means that the more accurately we know a particle's position, the less accurately we know its momentum.

CHAPTER 7

Newton's Law of Gravity

The Gravitational Field

Nature and nature's laws lay hid in the night:
God said 'Let Newton be!' and all was light.

—ALEXANDER POPE, *Epitaph for Sir Isaac Newton*

In the foregoing chapters we introduced the basic space-time concepts and derived from them a set of additional quantities that we require to understand the relationship between the structures in nature and the forces that maintain them. In particular, we obtained Newton's second law of motion, which enables us to study the dynamical behavior of the bodies that constitute a particular system, provided we know the nature of the force exerted on any one particle by the others. If we know this force, we can substitute it into Newton's law, $\mathbf{F} = m\mathbf{a}$, and find the acceleration of the particle at each moment that the force is acting on it. Once we find the acceleration, we can determine the orbit of the particle and obtain a dynamical picture of the entire structure. For a structure to exist and remain stable, each particle must move in a closed orbit which means that the motion of each particle must be periodic or very nearly periodic. An example of such a structure is our solar system, consisting, as it does, of bodies that revolve in closed orbits around the sun. If we know the nature of the interaction between the sun and the planets, and between any two planets, we can, in principle, deduce the planetary orbits from Newton's second law of motion.

Newton discovered that the governing force in the solar system is the gravitational force exerted by the sun on the planets; his great contribution to our knowledge of the solar system was his discovery of the mathematical formula for this force of gravity. Since this formula shows how this force depends on the distance between any two attracting masses and on the magnitude of these masses, we can apply it to the sun and any planet to calculate the acceleration of the planet at each point in its orbit.

It may very well be that Newton was inspired by a falling apple to think about gravity and to discover in a flash of genius that the same force that pulled the apple to the ground keeps the planets revolving around the sun in their orbits. But this primitive beginning was a far cry from his final formulation of the law of gravity, which shows how the gravitational force between two bodies depends on the distance between them and their masses, In fact, he came to the mathematical expression for this law by studying Kepler's third law of planetary motion. The important question is whether we can understand why the law that Newton deduced is to be expected from a physical and geometrical point of view.

Newton's genius lay in his ability to look beyond the immediate aspects of a discovery and see its universality in such a way that, for example, the falling apple meant not only that the earth pulls all bodies to it, but that every object in the universe pulls on every other object with a force that depends in a precise way on the distance between these objects and their physical nature. This point may be illustrated by considering two mass points m_1 and m_2 separated by the distance r. The magnitude of the gravitational force exerted on m_2 by m_1 is the same no matter where we place m_2, as long as its distance r from m_1 is not changed. The direction of the force is toward m_1 along the line from m_1 to m_2. This means that if we draw a sphere of radius r about m_1 and place a particle of mass m_2 at each point on the surface of the sphere, the particle m_1 pulls upon each of these with the same force. This model also shows that the presence of other masses does not alter the gravitational force of one particle on another particle. In other words, the presence of many particles of mass m_2 at a distance r from m_1 does not change the gravitational force of m_1 on any one of these particles.

A convenient way to represent the gravitational force (which incorporates this feature) is to introduce the concept of lines of force. We may imagine such lines as emanating from m_1 uniformly in all directions (like the quills of a rolled-up porcupine) and pulling on any particle they meet. Consider now the surface of a sphere at a distance r from the mass m_1 and suppose that N is the total number of lines of force in all directions originating from m_1. These lines pierce the surface of the sphere and their concentration over this surface (the number of lines piercing a unit area of the surface) is just $N/4\pi r^2$, since $4\pi r^2$ is the area of the sphere. The gravitational force of a unit mass placed anywhere on the surface of this sphere depends on the number of these lines of force passing through this unit mass, hence on the concentration of the lines. Thus, the gravitational force of m_1 on the unit mass varies as $N/4\pi r^2$ or inversely as the square of the distance between m_1 and the unit mass. Alternatively stated, as we move away from m_1 the gravitational lines of force spread out more and more, so that the number per unit area diminishes as the areas of the successive spheres through which the lines pass increase. The concentration of lines therefore decreases with the square of the distance. This geometric phase of Newton's law of gravity may be written as $F_{\text{gravitational}} \propto 1/r^2$, where $\propto$ means "proportional to" or "varies as." This equation means that the gravitational force is spherically symmetric; it depends only on distance and not on direction.

To find out how the gravitational force depends on the masses of the two attracting bodies, we again use the imaginary lines of force and suppose that the number of these lines that emanate from a particle of mass m is proportional to m. From this assumption it follows that the gravitational force exerted by a particle of mass m_1 on a particle of mass m_2 is proportional to the number of lines of force emanating from m_1 and, hence, is proportional to m_1 itself. In short, $F_{m_1 \text{ on } m_2} = Q_2 m_1$, where Q_2 is a constant of proportionality that can depend only on some physical property of particle 2 and on the distance between particles 1 and 2. By the same reasoning, the gravitational force of particle 2 on 1 must be proportional to m_2, so that $F_{m_1 \text{ on } m2_2} = Q_1 m_2$, where Q_1 is a constant similar to Q_2 but not equal to it. But by Newton's third law, these two opposite forces must be equal and therefore proportional both to m_1 and m_2. Hence, the force of gravity between the two particles must be proportional to the product of their masses. If we now combine this statement with the geometrical factor, we have for the magnitude of the gravitational force F between a particle of mass m_1 and one of mass m_2, separated by a distance r, the statement that the magnitude of the gravitational force is proportional to the product of the two masses divided by the distance between them squared.

As this expression now stands, it contains all the features of the gravitational force that depend on the distance between the two particles and on their physical attributes. Nothing more about the distance r or the masses m_1 and m_2 influences this force, and yet the above expression is not complete, as Newton himself immediately noted. This incompleteness is evident in two ways. First, if no other factor were introduced into the above expression for F, the magnitudes obtained with this formula for gravitational forces would be about twenty million times larger than they actually are, so that two people could not sit next to each other without being violently pulled together. Second, it is clear that the above formula is not complete because it has the wrong dimensional structure. We know from Newton's second law of motion ($F = ma$) that force has the dimensional structure mass × acceleration. This must be true of all forces but we see that the expression m_1m_2/r^2 has the dimensional structure mass × mass/length^2 so that something is missing: a factor that Newton introduced and which we now call the Newtonian universal constant of gravitation. If we represent this constant factor by G and introduce it into our previous expression for F, we obtain the complete expression for Newton's law of gravity: $F = G(m_1m_2)/r^2$.

We have not said very much thus far about the direction of the force of gravity between two particles, but we have already indicated that it acts along the line connecting them and that the pull of particle 1 on particle 2 is exactly equal and opposite to the pull of particle 2 on particle 1. Taking these features of the gravitational force into account, we can now state Newton's law of gravity as follows: Every particle in the universe exerts a force on every other particle along the shortest line connecting the two particles. The magnitude of this force is proportional to the product of the masses of the two particles and inversely proportional to the square of the distance between them.

This statement contains a remarkably simple and symmetrical law, since, as Newton conceived and stated it, the geometrical aspect of the force depends only on the distance r, and not on the velocities or accelerations of the two particles. One final point worth noting is that the quantities m_1 and m_2 that appear in the formula are the inertial masses of the two particles; that is, they are the same quantities as those that appear in Newton's second law of motion. It is a tribute to his remarkable insight that Newton assigned such an important role to inertial mass because there is nothing that one can observe about a particle that indicates that the gravitational force it exerts depends on its own inertial mass.

Newton had to introduce the constant G as an algebraic factor into his law of force to obtain an expression that has the characteristics of a force. Since the expression m_1m_2/r^2 has the dimensional character mass^2/length^2 and force has the dimensional character mass $\times$ length/time^2, it follows that G must have a dimensional character such that when this character is multiplied by the dimensional character of m_1m_2/r^2 we obtain the dimensional character of a force. We find according to this that G has the dimensions length^3/time^{2}mass. In terms of our basic units, this expression means that G is to be expressed as cm^3 per sec^2 per gm. The numerical value of G has been accurately measured in the laboratory by using very precise torsion balances to measure the gravitational pull on each other of two spheres of known mass, whose centers are separated by a known distance. This value is 6.668×10^{-8} cm^3 per sec^2 per gm. All the measurements of this constant that have been made thus far show that its value is entirely independent of the chemical nature of the spheres used in its measurement, so that chemically speaking, it is, indeed, universal. But all the astronomical evidence strongly favors the conclusion that this constant also has the same value everywhere in space and does not change with time.

The astronomical evidence that indicates that G is everywhere the same in the universe stems from two different kinds of analyses of astronomical phenomena. To begin with, we know that the motions of the planets in our solar system and of the stars in our galaxy are governed by the force of gravity by way of Newton's second law of motion, $\mathbf{F} = m\mathbf{a}$, where for F we must use Newton's expression for the force of gravity. The value of G thus enters into the formula for the orbits of astronomical bodies. If G differed from one region of space to another, these variations would be revealed in differences between the observed and calculated orbits of celestial bodies. Such differences have never been observed.

An even more sensitive astronomical test of the constancy of G in space and time is found in the structure and evolution of stars. Stars are formed from the dust and gas in interstellar space because the molecules that constitute this interstellar medium exert gravitational forces on one another. Even though the gravitational force between any two molecules in space is exceedingly weak, its magnitude on any one molecule, stemming from all the other surrounding molecules, may be large enough to cause all the molecules in a given region of space to contract into a sphere (star) if enough molecules are present. This same gravitational force ultimately causes the star to radiate energy and to pass through a series of evolutionary stages. If the value of G inside stars were different from its measured value here on earth, or if its value now were different from what it was a few billion years ago when the sun was formed, we would obviously find disagreement between the theory of stellar structure and the observed properties of stars.

The very small value of the gravitational constant makes the force of gravity the weakest of all known forces where small and widely scattered bits of matter are involved (such as molecules or atoms). If, however, enough matter is concentrated within a given region of space, the strength of the gravitational force exceeds that of all others. As we shall see later, nothing can stop the gravitational collapse of a given amount of matter once it has been compressed to a sufficiently high density.

Other basic constants of nature play an important role in the behavior of matter. Quantities such as the speed of light in a vacuum, the value of the electric charge on elementary particles (e.g., electrons, protons), the mass of the electron, and so on enter into the laws of nature and are basic factors in the structure of matter. But none of these basic constants bears any relationship to G, as far as is known by scientists today. Since the gravitational force emanates from every particle of matter, we may expect ultimately to find some basic relationship that connects G with the other known constants, but this discovery awaits a better understanding of the interrelationship between the microscopic world of the atom and the structure of the universe.

As noted above, the mass of a particle appears in two different Newtonian laws: the law of motion, where the mass of the particle is called its inertial mass, and the law of gravity, where it is called the gravitational mass. It is quite remarkable that the same quantity that accounts for the inertia of a body also generates the gravitational force which it exerts upon some other body. This equivalence of the inertial and gravitational mass of a body is the reason that all bodies, regardless of mass, acquire the same acceleration at the same point of a gravitational field of force. But another interesting feature of matter stems from the equality of inertial and gravitational mass: the inertial reaction of a body (it is opposite to the acceleration) when the body is being accelerated cannot be distinguished from a gravitational force on the body. Einstein was the first to recognize the importance of this phenomenon, upon which he based his general theory of relativity. He saw in this equality of inertial and gravitational mass more than a coincidence of nature and argued that it has a deep physical significance. He subsequently developed from it his famous principle of equivalence (the equivalence of inertial reactions and gravitational forces) which we discuss in detail in the chapter on general relativity.

To see the physical basis for this equivalence, consider a man standing on a scale in an elevator that is at rest on the ground. The scale registers his true weight W. If the elevator is suddenly accelerated upward with an acceleration a and if the mass of the man is m, his inertial reaction to this acceleration is ma, and he suddenly feels heavier by just this amount. This feeling is not just a subjective reaction on his part because the scale now registers his weight as $W + ma$, instead of W. There is no way for us to tell, just from the reading on the scale, how to distinguish between the amount that is due to the pull of gravity on the earth and the amount due to the acceleration of the elevator.

Before leaving the question of gravitational and inertial mass, we introduce the idea of the gravitational charge on a body by combining G and m. To make this clear, we suppose that the gravitational pull of one body on another is due not to the inertial masses of the bodies but to something we call gravitational charge, which is proportional to the inertial mass. To do this, we introduce the quantity $\sqrt{G}\, m$ as the gravitational charge associated with a body which has an inertial mass m. If we call this quantity g, Newton's expression for the force of gravity between two particles of masses m_1 and m_2 can be written as $F = g_1 g_2 / r^2$. This expression of Newton's law does not present anything new, but it is an interesting illustration of the relationship of the mass of a body to the gravitational force.

Since Newton's expression for the force of gravity between two mass particles was developed for bodies concentrated in a point, it cannot be used as it stands to calculate the force between two extended bodies of any shape. This is an extremely difficult problem for which no simple solution exists. As an illustration, consider two irregular bodies 1 and 2 and suppose that we apply Newton's law, as stated for two particles, to calculate the force between these two bodies. We then have to assume that the total gravitational action of body 1 is equivalent to the gravitational action emanating from a single point somewhere within that body, with its entire mass concentrated at that point. Many people erroneously think that this is so and that there is such a point in each body called the "center of gravity." They believe that the gravitational force between any two bodies can then be found by finding the "center of gravity" of each body and applying Newton's formula as though the mass of the body were concentrated in its "center of gravity." This reasoning is incorrect because there is no physical center of gravity of a body. This erroneous idea stems from the concept of center of mass which we defined previously. As we pointed out, the center of mass of a body or collection of bodies (or particles) is that point in which the total mass of the body (or collection of bodies) may be concentrated for the purpose of dealing with the total momentum of the body or system. We must not, however, assume that this is the point from which the total gravitational action of the body emanates, or the point on which an outside gravitational force acts. For certain special cases, the gravitational action of a body behaves as though it emanates from the center of mass of a body (for example, in the case of a uniform sphere of matter) but this is certainly not true for all bodies, as can be seen by considering a body consisting of two particles of equal mass connected by a thin rod of negligible mass. The center of mass of this structure is at the center of the rod, a point designated as cm, midway between the two particles. Consider now a particle Q placed at some distance from the point cm which does not lie on the structure itself but out in space. If the point cm were the "center of gravity" of the two spheres connected by the rod (in the sense that the gravitational force emanates from this point), the gravitational pull on the particle Q would vary inversely as the square of the distance of Q from the point cm. This means that this force would increase without limit as Q approached cm. But this increase does not happen in reality. In fact, if Q were placed at the point cm, the gravitational force acting upon it would be zero, since the two particles at the ends of the rod would pull equally hard on Q, in opposite directions. As a result, we can conclude that there is no such point as the "center of gravity" in a body. However, we point out later that in the special case of a sphere with the mass properly distributed inside it there is a center of gravity, namely the center of the sphere. With this understood, we can now discuss the gravitational attraction between two extended bodies.

Since we know that we cannot calculate the gravitational force between two irregular bodies by applying Newton's formula directly as though each were concentrated in a single point, we must introduce a procedure that permits us to apply Newton's law. To do this, we divide each body into many small elements of mass (mass particles). Each of these mass particles in each body is to be so small that its size is negligible compared to the distances between it and any one of the particles in the other body. We must then calculate the forces exerted by each of the particles in body 1 on each of the particles in body 2 and then add all the forces in each body vectorially, which in general is a very complex and difficult operation. There is no easy way to calculate the gravitational force of one irregular body on another. This force can certainly not, in general, be represented by a single algebraic expression similar to Newton's formula, as though the gravitational action of each body were concentrated at some point in the body. However, we can use this procedure for special cases. One such case is the sphere, which we now consider in some detail because of its great importance in a variety of physical problems.

A spherical body, inside of which the mass is distributed uniformly or in uniform concentric shells (comparable to the structure of an onion), behaves gravitationally as though all its mass were concentrated at the center. Without proving this, we can see that it is a reasonable assumption if we first consider the gravitational action of a uniform ring of matter on a particle P located in space. For the sake of simplicity, we place P on the axis OP of the ring; the axis OP is the line perpendicular to the plane of the ring which passes through its center. Two equal segments of the ring, A and A′, diametrically opposite one another are of equal mass at equal distances diametrically opposite from the particle P; hence, they exert on particle P equal gravitational forces F_A and $F_{A'}$, one pointing toward A and the other toward A′. Moreover, since F_A and $F_{A'}$ lie in the same plane and make equal angles with OP, the components of F_A and $F_{A'}$ perpendicular to OP are equal and opposite, thus canceling each other. Only the components of F_A and $F_{A'}$ that lie along the axis OP remain. This simply means that the gravitational action of A and A′ on P is directed toward O. But this is true for all pairs of equal mass elements of the ring that are diametrically opposite to each other. Hence, it is true for the entire ring. A uniform ring of matter pulls toward its center any particle lying on the ring's axis.

From the above analysis we deduce that the same gravitational effect must be true for a disk of matter. The reason for this is that a disk can be divided into a series of concentric rings, all having the same axis OP. Since the gravitational action of each of these rings on a particle at P is along the line OP, this is also true for the entire disk. For a uniform sphere and any particle P at some distance from the center of the sphere greater than its radius, we draw a line from the center O of the sphere to the point P and then divide the sphere into parallel disks in planes perpendicular to this line. Each disk pulls the particle at P along the line OP and hence toward O. Clearly, this result occurs no matter where the particle is placed. All the gravitational lines of force of the sphere diverge from its center just as though all its mass were concentrated there. Since this fact can be proved rigorously by applying the integral calculus, we accept as a fact that a uniform sphere behaves gravitationally as though all its mass were concentrated at its center.

The gravitational action of a sphere whose mass is distributed in uniform layers or shells is the same as though all the mass of the sphere were concentrated at its center. The situation does not change for a single uniform shell instead of a complete sphere because we can apply the same reasoning to this example as we did above and deduce that the gravitational pull of such a spherical shell on a particle P is the same as though all the mass were concentrated at its center.

So far we have been considering the gravitational action of uniform spheres and shells on particles lying outside these bodies, but now we consider a particle P′ lying inside a uniform spherical shell. To show that the shell exerts no gravitational force on this interior particle, we consider the gravitational action on it of two segments A and A′ of the shell, lying at opposite ends of two intersecting lines passing through P′. The thickness of each element of the shell cut out by these lines is the same as that of the shell but their areas are different. The area of the element A′ is proportional to r'^2, where r' is the distance of A′ to P′. Similarly, the area of A is proportional to r^2, where r is the distance of A to P′. The volume and hence the mass of the segment A′ are smaller than those of A by just the right amount to cancel the effect of its closer distance on its gravitational pull. As a result, the gravitational pulls of A′ and A on P are equal and opposite so that the net effect is zero. Since this is true for all such pairs of elements of the shell, the total pull of the shell on P′ is zero.

This result is very important in the study of the structure of stars. The reason is that the physical conditions at any point P inside a star at a distance r from the center, are, to a great extent, determined by the gravitational forces at that point. But in analyzing these gravitational forces we can divide the star into two parts: one, a sphere of radius r on the surface of which the point lies, and the other an outer shell of the star above this inner sphere. Only the gravitational action of the sphere of radius r at the point P needs to be considered since the gravitational action of the outer shell at this internal point P is zero. This greatly simplifies the stellar structure problem since we can disregard the outer shell in determining the forces acting on an element of matter at any particular point inside the star.

Since we now know how to determine the gravitational action of a uniform sphere (or a sphere in which the mass is distributed in layers of uniform density), we can determine the mass of the earth. According to Newton's law of gravity for a sphere, the gravitational pull of the earth on a body lying on its surface is described by the equation $F = GmM_E/R_E^2$, where m is the mass of the body, M_E is the mass of the earth, and R_E is its radius. But this pull F is just the weight W of the body of mass m. However, from Newton's second law of motion, the weight W is mg, where g is the acceleration of gravity. By equating F and mg, we obtain the equation $mg = GmM_E/R_E^2$. By canceling the quantity m on both sides of the equation, we obtain the value for g which equals GM_E/R_E^2; this relationship is interesting in itself because it expresses the acceleration of gravity on the surface of the earth in terms of the mass and radius of the earth. One immediately sees that the acceleration of gravity of a body on the surface of the earth is independent of the body's mass. Since g, R_E, and G are all known quantities, we can calculate M_E from this equation by substituting the measured values of g, R_E, and G into the equation. Once these substitutions are made, we find that the mass of the earth equals 5.98×10^{27} gm. From this value of the mass of the earth and from its volume V, we can then calculate the mean density of the earth D_E, which is defined as M_E/V; the result is 5.52 gm/cm^3. On average, a teaspoon of the planet earth has a density of about $5\frac{1}{2}$ gm/cm^3.

When Newton stated his law of gravity as a mathematical expression which gives the force exerted by one mass particle on another, he knew that he had not explained the nature of this force but had merely presented a mathematical description of its effect upon a body. Although this, in itself, was a remarkable achievement, he was unhappy with the physical and philosophical implications of his law because it carried with it an element of mysticism—at least one that could not be entirely understood from a physical point of view. To see this we need only recall that Newton's law refers to the gravitational pull between two bodies separated in space (such as the earth and the sun) which introduces the concept of "action at a distance." Newton found this concept repugnant and was convinced that it would ultimately be replaced by one in which the gravitational force is explained in terms of some kind of physical contact between the two bodies, or in terms of some intermediary that transmits the force from one body to the other. In a letter to Richard Bentley, Newton stated his reservations about action at a distance in the following words: ". . . that one body may act upon another at a distance through a vacuum without the mediation of anything else, by and through which their action and force may be conveyed from one to another, is to me so great an absurdity that I believe no man who has in philosophic matters a competent faculty of thinking could ever fall into it. . . ."

Newton looked upon his law of gravity as a description rather than an explanation; but nonetheless he emphasized the usefulness and power of the law to lead to an understanding of the motions of bodies. His point of view is clearly stated in the following two excerpts from his *Opticks:*

> To tell us that every species of thing is endow'd with an occult specific quality by which it acts and produces manifest effects, is to tell us nothing. But to derive two or three general principles of motion from phenomena, and afterwards to tell us how the properties and actions of all corporeal things follow from these principles would be a very great step in philosophy, though the cause of these principles were not yet discovered.

> I have not yet been able to discover the cause of these properties of gravity from phenomena, and I frame no hypotheses. . . . It is enough that gravity really does exist and acts according to the laws I have explained, and that it abundantly serves to account for all the motions of celestial bodies.

Although the "action at a distance" concept can be dealt with by introducing lines of force, we "know" intuitively that these lines have no real physical existence, serving only as a simplifying device, and so we are really still faced with "action at a distance" and its inherent difficulties. To take another approach to this problem, we picture each particle as surrounded by some kind of field that extends in all directions to all other surrounding particles. We may then define the gravitational interaction between two bodies not as a force between these bodies, but rather as an interaction between each particle and the field of the other particle. In this way we get rid of action at a distance, since the action between a particle and a field is a local phenomenon.

Each particle establishes a field in the space around it which acts as some kind of physical disturbance of the space so that the space itself becomes an active agent which transmits the forces from one body to the other. The interaction of each particle is with the field in its immediate vicinity, rather than with the other particle, which is at some distance away. But a field theory such as this has meaning only if some physical reality can be assigned to the field itself. One way of doing this is to measure the acceleration (the magnitude and direction) that a unit mass experiences when it is placed at any point in the gravitational field. This acceleration, specified at each point, defines the field. Defined this way, the gravitational field is a vector field.

So far we have considered the gravitational field of a single particle only, but we can extend the same reasoning to many particles acting together. The total field may be obtained by adding together the separate fields of the individual particles. This procedure is called the superposition of fields. Since the superposition of the gravitational fields of individual particles is a problem in vector addition, the procedure becomes quite complex with increasing numbers of particles. Therefore, it is useful to introduce a nonvector, or scalar, specification of the gravitational field. This specification is given by the concept of the potential of the field, which is intimately related to the concept of energy which we now derive.

Up to this point, we have introduced a number of basic concepts from which we constructed the derived physical entities such as velocity, acceleration, and mass that enter into the physical laws. As we probe more and more deeply into the world around us, we must either introduce new basic physical concepts or construct additional ones from those we already have, if we are to obtain as understandable a picture of the physical world as possible. Some of these new entities (e.g., momentum) obey conservation principles which provide new insights into the laws of nature and into the relationship between the forces that act on bodies and the motions of these bodies.

There is no limit to the number of new quantities that we can introduce formally since we can always combine old quantities in various mathematical ways to obtain new ones. But few, if any, of these derived quantities are useful or important. The mass of a body multiplied by its volume, for example, is a perfectly good mathematical quantity, as is the speed of a body cubed, divided by the distance it has moved. But these quantities lead to no new laws nor to a better understanding of nature. We must keep the number of new quantities as small as possible and introduce a new one only when it either illuminates what we are doing and thus leads to a better understanding of the physical structures we are dealing with, or when it is necessary for the formulation of a new law. The quantity we now introduce is useful for both of these reasons and, in addition, it leads to a new conservation principle and permits us to present the principle of least action in a more meaningful form.

We can obtain this new physical entity by considering a force acting on a body and studying its effects on the body. Our concern here is with the changes induced in the body by the force. Experience has taught us that the force produces certain changes in the body which can be encompassed in the concept of energy which we must still define. In fact, when a force induces a change in a body, the energy of the body increases. However, until we clearly define the phrase "change in a body'' and ''energy of the body,'' the previous statement remains unclear. The following are the most obvious changes that the action of a force on a body produces in the body: distortion of the shape of the body (bending, stretching, compressing, or shearing); change in the position of the body relative to some other body (such as the earth); change in the state of motion of the body. Although other properties of a body such as its temperature, color and those related to its electrical and chemical nature, can be altered by outside forces acting on it, we consider, for the time being, only the changes in the state of motion of the body and in its position. In any case, when a force produces any of the above changes, the body gains energy.

To express this gain in energy in terms of the basic quantities already introduced, we suppose that the force F displaces a body through a small distance d in the direction of the force. We take the displacement d to be so small that the change in the force F is negligible during the displacement of the body. If a large displacement is produced, the force acting on the body can change appreciably during the displacement and complications arise. The product Fd is defined as the work done on the body by the force F in displacing the body through the small distance d. If W stands for the work done by F, we have the relationship $W = Fd$. This expression indicates that work is a scalar quantity since no direction is specified in the definition of W. All that we said before about a force acting on a body may now be incorporated into the following statement. When an external force does work on a body, the energy of the body increases. If the result of this work is simply to increase the speed of the body, the kinetic energy of the body has increased. If the work results in an increase in the height of the body above the ground (and nothing else), the potential energy of the body has increased. Since, in general, both the kinetic and potential energy of a body are increased by the work done on it, we say that when work is done on a body, its total energy (kinetic plus potential) increases.

Just as we speak of the work done on a body by an external force, we also speak of the work done by the body on its surroundings, or on some other body. In that case (when a body does work) the body loses an amount of energy equal to the work it does. Since the work done on a body is expressed by W, the work done by the body is $-W$, the minus sign meaning a decrease in the energy of the body doing the work.

Having defined the energy a body gains, when an external force acts on it, as the work done on it by the external force, we obtain a formula for this energy by calculating how much work is done on the body. Before doing that, however, we introduce the unit of work, which is therefore also the unit of energy. From the definition of work $W = Fd$ (taking here the force F and the displacement d to be in the same direction) we see that one unit of work is done on a body when a unit of force is applied through the unit of distance so that $W = 1 \times 1 = 1$.

Since the unit of force is called the dyne and the unit of displacement the centimeter, we say that a force of 1 dyne displacing a body through a distance of 1 cm does one unit of work on the body; we call this unit an erg. Since work and energy are equivalent, a force of 1 dyne displacing a body (which is completely free to move) through a distance of 1 cm increases the energy of the body by 1 erg. This statement is correct whether the mass of the body is 1 gm or millions of grams.

Now that we know that the energy gained by a body equals the work done on it, we can write down a formula for the kinetic energy of a body. If a constant force is applied to a body in the same direction as the body is moving, and if the body's speed, owing to the force, increases by the amount v when the body has moved a given distance, the work done by the force can depend only on the mass of the body and the speed v. The more massive the body is (the greater its inertia), the greater is the work that must be done to increase its speed by a given amount; for a given mass, more work must be done to increase the speed by a large amount than by a small amount. This tells us that the kinetic energy of a body depends only on its mass and its speed. The exact formula for the kinetic energy of a mass m moving at speed v is $\frac{1}{2}mv^2$. This formula is very useful since it permits us to calculate the kinetic energy of a particle simply from a knowledge of its mass and speed; we do not have to know anything about the force that gave it its motion. From the formula for kinetic energy we see that the units of energy are gm cm^2/sec^2. What is the kinetic energy of a mass of 100 gm moving with a speed of 50 cm/sec? We square the speed to obtain 2500 and multiply this by one half of 100 gm which gives us 125,000 ergs for the kinetic energy of the body.

The kinetic energy of a body cannot be defined as an absolute quantity but only as a relative one since the speed of a body depends on the frame of reference in which it is viewed. If we say that a body has a speed of 100 cm/sec, we generally mean that this is its speed relative to the surface of the earth. If we were on the sun, however, the speed of the same body relative to us would be about 18 miles/sec, and at the center of our galaxy it would be about 200 km/sec. That kinetic energy has no absolute meaning is not important, however, since only differences in kinetic energy are involved in solving physical problems. This indeterminacy in the definition of energy is present in all forms of energy associated with a body. One generally expresses this by stating that the energy of a body is determined to within a numerical constant.

The concept of potential energy can be introduced by considering a body at rest on the surface of the earth to which we apply a vertical force sufficient to lift it off the ground. To lift the body, we must apply a force equal to the weight W of the body. Suppose we lift the body a distance y above the ground (we neglect air friction) where y is small enough so that the weight of the body shows no measurable change when it is raised through this height. If the body is brought to rest when it is at the height y, the work we have done just equals Wy. But $W = mg$ so the work done $= Wy = mgy$. The work done must equal the increase in the energy of the body when it is raised a distance y above the ground. This amount cannot be the increase in its kinetic energy, however, since the body started from rest on the ground and is at rest at the height y. It must, then, represent an increase in some other form of energy, which we call the potential energy of the body.

Since potential energy has the same indeterminacy as kinetic energy, an arbitrary constant can be added to it without changing its meaning. This is equivalent to saying that we can choose the point of zero potential energy to be at any height above the earth's surface. For the time being, we take all points on the surface of the earth (at sea level) to be at zero potential energy. The potential energy of a body increases as it moves away from the surface of the earth and all bodies of equal mass at the same height above the earth's surface have the same potential energy. As long as we do not go too high above the earth's surface, the force of gravity is about the same as it is at the surface. It follows, therefore, that all bodies of mass m at the same height y above the ground have potential energy mgy.

An important feature of potential energy is contained in its definition. To see this, we consider a body of mass m taken from some starting point A above the ground to some other point B. If the potential energy of the body when at A is $(PE)_A$ and when at B, $(PE)_B$, and if $W_{A \to B}$ is the work done to move the body from A to B, then $W_{A \to B} = (PE)_B - (PE)_A$. If h is the difference in height above the ground between A and B, then $W_{A \to B} = mgy_B - mgy_A = mg(y_B - y_A) = mgh$ where y_B and y_A are the heights of B and A, respectively. We see, then, that the work done in taking a body along a path from one point to another (neglecting air friction) is independent of the path chosen; it depends only on the height difference of the two points. It is not always true, as we shall see later, however, that the work done is independent of the path. There are certain forces called dissipative forces (e.g., friction) for which the work depends on the path. We must then differentiate between a force field such as the gravitational field of force, for which the work is independent of the path, and forces such as friction. The former field of force is called a conservative field and the frictional type is referred to as nonconservative or dissipative.

When work is done on or by a body (or on a system of bodies) the energy of the body (or system of bodies) changes. From this we deduce that the energy remains unaltered if no work is done. Although this statement is essentially the content of the principle of the conservation of energy, it can be formulated somewhat more precisely for the total mechanical energy of a body. A body moving along a frictionless track whose height from the ground varies does no work on its surroundings during its motion (we neglect the friction of air) and no work is done on it. Hence, the total mechanical energy (kinetic plus potential) of the body is the same at all points on the track. Another important application of this conservation principle is to the motion of a body thrown into the air and then allowed to move freely along its natural trajectory. The path of this body is (to a very good approximation) the arc of a parabola. If two successive points in the parabolic path A and A′ are equally high above the ground, the body has the same potential energy at A′ as it had at A. But we know from the principle of the conservation of total energy that the total energy of the body at A′ must also equal that at A. Hence, the kinetic energy at A′ must equal that at A. It follows that the speed of a body at any height in its downward path equals its speed at the corresponding height in its upward path. If we throw a body vertically upward in a vacuum with a speed of 100 ft/sec, it will have exactly that speed when it strikes the ground on its return trip.

Everywhere in nature we find examples of the principle of the conservation of energy. The oscillations of a pendulum are a simple example of this principle. With the bob at rest at some point A in its natural arc, the pendulum's total energy is potential and equals mgh, where m is the mass of the bob and h is its height above the lowest point B in the path followed by the bob when it swings freely. Because B is the point where the bob would remain stationary if it were not subject to any forces, B is the position of zero potential energy. When we allow the bob to move freely, it follows the arc ABC where C is a third point on the opposite side of point B from point A which has the same height as point A. During this swing, the bob loses potential energy and gains kinetic energy. At point B its total energy is kinetic (equal to its initial potential energy at A) and when it is at point C all of its energy is again potential. If there were no friction of the air against the bob and if the string were attached to some point O without friction, the pendulum would go on swinging forever. But the pendulum actually does work against the air and at the point O (where the string holding the bob is attached to its support), so that it slowly loses its mechanical energy and finally comes to rest at point B. The bob does work on the surrounding air by pushing it and the string of the pendulum does work at O by rubbing against the surface to which it is attached. These forces are dissipative forces. The final position of the bob of the pendulum when it comes to rest (at point B) is the one along the arc ABC which has the lowest potential energy.

The example of the pendulum shows that the total mechanical energy of a body is not really conserved in general owing to the frictional forces that the body must constantly work against. There are mechanical systems, however, in which there is hardly any friction present at all. As an example, consider a planet, such as the earth, revolving around the sun. As it moves in its elliptical orbit, its distance from the sun varies, which means that its potential energy relative to the sun also varies. At aphelion (the position farthest from the sun) its potential energy is a maximum, and, conversely, at perihelion (the position closest to the sun) the potential energy of the earth is at a minimum. The kinetic energy of the earth has minimum and maximum values corresponding to the maximum and minimum values of the potential energy, so that the sum of potential and kinetic energies is the same for all positions of the earth. If the total mechanical energy of the earth did not remain constant in time (to a very high degree) the mean diameter of its orbit would either decrease or increase until the earth was drawn into the sun or was out of the solar system entirely. The very existence of the planets in their orbits, illustrating the great stability of our solar system, is excellent empirical evidence supporting the principle of the conservation of mechanical energy. We might say that this principle guarantees the existence of life because the earth would either be vaporized by approaching too close to the sun or frozen by moving too far away from it if this principle were not obeyed. This principle can be used to answer the following question: Can a body like the sun capture an object that approaches it from a very great distance, causing the object to revolve around the sun in a closed orbit, if no other bodies are close enough to interfere and if there is no friction? This hypothetical situation may be explored if we imagine an object approaching the sun from a very distant point A that is so far off that practically all of the body's energy is potential. As the object approaches the sun, its potential energy decreases and its kinetic energy increases until it is closest to the sun at some point B; its total energy at B must be the same as it was at point A (by the conservation principle), but its gain in kinetic energy has now been so great and it is moving so fast that the pull of the sun is not strong enough to keep the object close to the sun. In fact, the object must move along an arc through point B to a third point A′ away from the sun (where A′ is as far away from the sun as A is) such that the total energy of the object at each point of this arc is the same as it was at A. From this it is clear that the path between B and A′ must have the same shape and size as the path between A and B, so that the object cannot be captured by the sun and pulled into a permanent orbit close to the sun.

Simple harmonic motion (SHM) in its simplest form is straight-line oscillatory motion about a point. It differs from all other types of oscillatory motion, however, in that the acceleration of the particle undergoing SHM is proportional to the distance of the particle from the center of oscillation and the acceleration is always directed toward the center of oscillation. This motion can be produced by a perfect spring (a spring in a vacuum in which no friction is present and which does not heat up as it oscillates). We can picture such a spring lying along the X (horizontal) axis with a particle of mass m at its free end. The other end is attached in such a way that the spring can oscillate along the horizontal direction. We now place our origin at the position of the particle when the spring is completely relaxed and not oscillating. In this position, the kinetic energy of the particle is zero. The gravitational field can be disregarded since the motion of the particle is horizontal and we may take the gravitational potential energy as always equal to zero. But the spring contributes potential energy to the particle. To take into account the potential energy arising from the forces in the spring, we let k be the spring constant, which, by definition, is the force required to stretch the spring one centimeter. For a perfect spring, k remains the same regardless of how much the spring is stretched. If we stretch the spring slightly, displacing m by a small amount, we do a small amount of work which is stored in the spring as potential energy. If the spring is oscillating, m has both kinetic and potential energy at each moment. The sum of these energies is its total energy which remains constant though the kinetic and potential energies are changing separately.

The principle of the conservation of energy, like the principle of conservation of momentum, places a restriction on the possible motions that a particle can experience under a given set of conditions. But neither one of these principles by itself can be used to determine completely the motion of a particle. As an example of this, we note that a freely thrown particle, if it were governed only by the principle of the conservation of energy, could oscillate up and down, or move in a circle or series of loops—any one of a number of different complex trajectories—and still satisfy the energy principle. But such orbits conflict with the momentum principle. Applying both the energy and momentum conservation principle, we see that the orbit of the particle is limited to only one of the many different orbits permitted by just one of these principles and that the orbit is therefore completely determined.

Another point in connection with the determination of orbits is the concept of action. We saw that Maupertuis had introduced the idea that a particle takes that particular path along which the Maupertuis action is a minimum. Maupertuis defined the action for the path as the sum of the quantities $p\Delta q$ for the entire path, where Δq is a small piece of the path and p is the momentum of the particle within that small piece. This Maupertuis definition of the action deals only with the momentum, but we have seen that the energy must also be taken into account in dealing with the dynamics of a particle. We therefore surmise that the energy should also be related to the action in some way, which is, indeed, the idea expressed by Hamilton in what we now call the Hamiltonian action of a particle (or system of particles). This concept extends the Maupertuis definition of action by including energy.

To see how Hamilton related the energy to the action, we recall that the Maupertuis action of a particle with momentum p moving a distance Δq is altered by an amount $p\Delta q$. If we introduce a coordinate system and let p_x, p_y,p_z be the x,y, z components of p, and Δx, Δy, Δz the components of Δr, then $p_x\Delta x$, $p_y\Delta y$, and $p_z\Delta z$ are the contributions to $p\Delta q$ from each coordinate component of the motion. The change in the Maupertuis action s may be written as $\Delta s = p_x\Delta x + p_y\Delta y + p_z\Delta z$. This expression, however, does not take into account changes that can occur in the action with time, and therefore does not give a complete picture of the most general kind of motion that a particle can experience. Hamilton corrected this deficiency by extending the concept of action to include the effect of the energy of the particle on the action by showing that the energy of the particle is related to the change of action with time. To see how Hamilton did this, we consider a particle with energy E moving for a very short time Δt. Since we defined the change in the Maupertuis action of a particle over a small stretch of path Δq as $p\Delta q$, it is quite natural to define the change in its action with time alone (assuming there is no change in position or that it is disregarded) as $E\Delta t$, where E is the energy of the particle and Δt is the small time interval in the life of the particle, during which it has the energy E. The only question that now remains is whether this value should be taken as negative or positive.

Suppose we consider the change in the action of a particle resulting from both the change in its position and the change in time. According to Hamilton, we take $p\Delta q - E\Delta t$ as the total change in action (or, more precisely, the change in the Hamiltonian action) during the time Δt and over a stretch of path Δq. The remarkable thing in this expression is the appearance of the minus sign in front of $E\Delta t$, since one might naturally expect the total change in action to be $p\Delta q +$

$E\Delta t$ (the sum of space and time contributions) rather than $p\Delta q - E\Delta t$. But Hamilton's choice of the expression with the minus sign is a tribute to his genius because this expression contains the basic space-time concept of Einstein's special theory of relativity. Since Hamilton knew nothing about relativity theory or space-time when he introduced his action principle, his use of this expression shows great foresight.

To justify the minus sign, we now introduce the idea that action is associated with a point in space occupied by a particle at a certain time. As the particle moves, the action of each point along the path of the particle changes from moment to moment. If a particle moves with constant momentum along the x direction and at time $t = 0$ the particle is at the origin $x = 0$, we may say that its Hamiltonian action at that moment is s. As the particle moves along its path, this same value of this action should occur at the new points occupied by the particle, but at later times. If the change in action were given by $p\Delta x + E\Delta t$, the action would always grow and we could not have the same value at succeeding points at later times. If we want the same value of action to appear at points farther along the path of the particle, we must define the action in such a way that its change remains zero if we follow the particle with a certain speed. But if the change in action in a distance Δx and time Δt were $p\Delta x + E\Delta t$, there would be no speed which would make this quantity zero. If, however, we define the change as $p\Delta x - E\Delta t$ we find that there is a definite speed for which the change in action is zero.

When we introduced the Maupertuis action, we saw that it obeys a certain minimal principle which Maupertuis discovered, and which may be stated as follows: Consider all possible paths connecting two points, A and B, in a conservative field of force (total energy of a freely moving particle along any one of these paths conserved) and consider a particle which can move from A to B along any one of these paths, but always with the same total energy (that is, starting out with the same initial speed). Then the path along which the particle actually moves is that along which the total Maupertuis action is a minimum. This principle of least action contains an important restriction because the total energy of the particle must be conserved. Hence, this principle cannot be applied to situations in which the energy of the particle does not remain constant, but the Hamilton principle of least action (or stationary action) does apply to such situations.

Hamilton's principle may be stated as follows: Consider all possible paths connecting two points, A and B, in some force field. Let these paths be such that the time taken by a freely moving particle to go from A to B is the same along each path. Then the path along which the particle actually moves is the one for which the sum of all the quantities $p\Delta q - E\Delta st$ taken over the entire actual path and for the entire duration of the motion is a minimum (more accurately, is stationary) as compared with this sum for any nearby path. In this principle, no restriction is placed on the energy, which, in general, may vary from moment to moment.

The Hamilton principle of least action now says that the value of this sum along the actual path is smaller (or larger) than its value along any neighboring path. The power and beauty of this principle is found in its elimination of the need to introduce forces, so that the repugnant concept of action at a distance is discarded. From this point of view, the particle is not directed along its path by a force but rather by the value of the action at each point of space. The action is intimately related to another important concept in modern physics—the wavefunction of a particle.

We can clarify the Hamilton principle of least action by picturing a salesman who has to meet a certain number of customers. He will minimize his expenses by traveling along the shortest routes between customers. Suppose he is also guaranteed a certain hourly pay in addition to his commission. He would then subtract this amount from his hourly expenses and try to minimize that difference so as to conform with what we might call the principle of least expense.

We introduced the concept of angle by noting that a body can have two general kinds of motion: translation and rotation. Translation changes the position of every point in a body by the same amount relative to the observer's frame of reference. In this type of motion all the points of the body move along lines that are parallel to each other; associated with this motion is a vector (parallel to the lines of translation) called the displacement of the body. Its magnitude is the distance through which the body has been translated.

Rotation of a body (as distinct from bodily revolution around a point) occurs when all the points on one particular line through the body remain fixed, whereas all other points in the body change their positions relative to the origin of the observer. In this kind of motion, each of the moving points of the rotating body moves in the arc of a circle whose plane is perpendicular to the line of fixed points and whose radius equals the distance of this moving point from the line of fixed points. This "line of fixed points" is called the axis of rotation. Associated with rotational motion is a vector whose direction is parallel to the axis of rotation and whose magnitude equals the angle through which the body is rotated. If the rotation about a vertical axis, as seen from below, is closewise, the vector representing this rotation is (by agreement) arranged to point in the direction along which a right-handed screw parallel to the axis advances.

It is important to distinguish carefully between the rotation of a body about an axis and the revolution of a body about a point. We first consider the revolution of a body and, to simplify our discussion as much as possible, we start with a mass particle. Since a particle has no extension, it cannot rotate in the sense defined above. The particle itself would have to lie on the axis of rotation and there would be nothing left over to rotate about this axis. We then see that rotation, in the usual sense of the term, has no meaning when we are dealing with a point mass, or a particle. We can, however, introduce a motion of the particle that is, in a sense, equivalent to rotation. This type of motion, called revolution, was previously described when we considered a particle moving in a circle around a point. Although the particle itself is not rotating (it is really undergoing a translation from point to point in the circle), its motion can be defined as rotation if we treat the radius drawn from the center of the circle (described by the particle) to the particle as a massless rod of zero thickness. The rod and the particle can then be treated as a rotating body, with the axis of rotation passing through the end of the rod that coincides with the center of the circle. Despite this method of enlarging the rotation concept, however, we always refer to the motion of the particle in a circle as revolution and reserve the term rotation for an extended body.

With this understood, we can discuss the important physical characteristics of revolutionary motion by picturing a particle of mass m revolving in a circle of radius r around the point O with constant speed v. Is this motion associated with a physical entity that remains unchanged under the proper conditions? We saw that in linear (i.e., translational) motion of a particle, such a quantity (the momentum) does exist if no force acts on the particle. This is expressed by the principle of the conservation of momentum. But the momentum of a particle revolving in a circle with constant speed is not constant. The magnitude of the momentum of the particle is always mv and this quantity does remain unchanged, but the direction of the momentum changes constantly. From this it follows, according to Newton's laws of motion, that a force acts on this particle constantly. Because the particle experiences an inwardly directed acceleration, we know from Newton's second law of motion that the revolving particle constantly experiences a force. This force, which always acts along the radius, from the particle to the center of the circle, is called the centripetal force; it is always at right angles to the momentum of the particle. This is why the direction of the momentum of the particle changes continuously, whereas the magnitude of the momentum remains constant. The magnitude of the momentum would change only if the force acting on the particle had some component in the direction of the momentum. In any case, we see that momentum is not conserved in uniform circular motion.

An inwardly directed force must constantly act on a particle to keep it moving in a circle. This force can be applied to the particle by means of a string of length r, one end of which is attached to the particle and the other end of which is held fixed at the center of the circle. The inward pull of the string on the particle is the centripetal force we have just described. But by Newton's third law of action and reaction, the particle exerts an outward force on the string which tends to stretch the string. This outward force, which is equal and opposite to the centripetal force, is called the centrifugal force, and is usually thought of as acting on the particle itself. Actually, as we see from the physical arrangement, no outward (or centrifugal) force acts on the particle; the only force on the particle is the inward pull of the string. Nevertheless, an observer moving along with the particle feels a real outward (centrifugal) force.

The momentum of a particle moving with constant speed in a circle is not conserved. However, another quantity is conserved in circular motion. Like momentum, this quantity is also a vector, and we can see, from our discussions of angular velocity and momentum, how to construct this new vector quantity. We saw previously that there is a constant angular velocity $\boldsymbol{\omega}$ associated with the constant speed v of a particle moving in a circle. As long as the speed v remains constant, the angular velocity remains constant. This is not quite the vector quantity that we are seeking, since the angular velocity gives a picture only of the kinematics of the revolutionary motion and tells us nothing about the dynamics of this motion. The same situation exists in uniform translational motion in a straight line, which is characterized by the linear velocity $\mathbf{v}$. This vector does not tell us the full story of translational motion, but the vector $m\mathbf{v}$ (momentum) does, because it involves the mass of the particle as well as its velocity.

In the same way as we constructed a dynamical vector $m\mathbf{v}$ from mass and velocity for linear motion, we must now construct a dynamical vector for revolutionary motion from the angular velocity $\boldsymbol{\omega}$ and the other dynamical elements associated with the circular motion of a particle. These elements are the mass m of the particle and the radius r of the circle in which it is moving. But if we simply multiply the three quantities m, r, and $\boldsymbol{\omega}$, we obtain $mr\omega$ or mv, since $r\omega = v$. But this is just the magnitude of the momentum and we get nothing new. If we take the quantity $mr^2\omega$, however, which is the magnitude of the momentum mv multiplied by r, we obtain a quantity that remains unaltered as the particle revolves in the circle; this quantity is the magnitude of the vector that we seek. This vector is called the angular momentum or spin $\mathbf{J}$ of the particle and $J = mr^2\omega$. The vector $\mathbf{J}$ plays the same role in rotational motion as the momentum vector $\mathbf{p}$ plays in translational motion.

At this point one may ask why we cannot use $mr^3\omega$ or $mr^4\omega$ or any other such combination for the quantity that remains constant. The reason is that $mr^2\omega$ is the only combination (besides $mr\omega$) that is the magnitude of a vector. To emphasize the conceptual parallelism of $mr^2\boldsymbol{\omega}$ with $m\mathbf{v}$ in linear motion, we note that just as the momentum of a particle is the product of its mass and its velocity, the angular momentum as defined above is the product of the mass of the particle and another velocity. To find this other velocity, consider the radius r drawn from the center of the circle to the particle. As the particle moves from point A to point B in its circular orbit, this radius sweeps out an area. When the particle moves from point A to point B in its circular orbit, the radius sweeps out the area AOB (where O is the center of the circle), which is the sector of the circle bounded by the arc AB and the two radii OA and OB. The faster the particle moves and the larger the radius of the circle, the more rapidly a given area is swept out.

We now consider the rate at which the area is swept out if the speed of the particle is v and the radius of the circle is r. To find this we note that the rate at which something is happening, if the rate is constant, equals the total amount that happens in a given time, divided by this time. The rate at which the area is swept out is found by dividing the total area of the circle by the time it takes the particle to make one revolution, since this is the total area swept out in one revolution of the particle. The total area of a circle is πr^2 and the time it takes the particle to go around once is computed by dividing the circumference of the circle by the speed of the particle, or $2\pi r/v$ ($2\pi r$ = the circumference of a circle). Hence, the rate at which the area is swept out is $\pi r^2/(2\pi r/v)$ or $rv/2$. This quantity is called the areal velocity of the particle and is represented by $\dot{A}$. Thus, $\dot{A} = rv/2$ or $rv = 2\dot{A}$.

The angular momentum of the particle can now be obtained by multiplying twice the areal velocity by the mass of the particle so that $\mathbf{J} = 2m\dot{\mathbf{A}}$. Since the areal velocity is a vector in the same direction as $\boldsymbol{\omega}$ and $\mathbf{J}$, we can write $\mathbf{J} = m(2\mathbf{A})$. Here we see the close analogy between linear momentum $m\mathbf{v}$ and angular momentum $m(2\mathbf{A})$. It also shows the justification for treating the product $mr^2\omega$ or mrv as the physical quantity in rotational motion that is equivalent to the product mv in linear motion.

The angular momentum $\mathbf{J}$ is a constant vector for a particle moving with uniform velocity in a circle. The question which naturally arises at this point is whether an angular momentum vector is associated only with uniform circular motion and, if not, under what conditions it is constant if and when it is associated with other kinds of motion. We shall see that angular momentum is associated with every kind of motion and that it remains constant only if a certain important condition is fulfilled, which we discuss now.

Before considering motion in general, we describe what must be done to change the angular momentum of a body moving in a circle, e.g., the rim of a rotating wheel; as long as the wheel is spinning with constant angular velocity (this means that the angular speed of the wheel and the direction of its axle or axis of rotation does not change), the angular momentum of the wheel is constant. To change it, we must either change the direction of the wheel's axle or the spinning rate of the wheel. We know that the application of just any force to the wheel does not necessarily do this and therefore does change its angular momentum. As we noted above, although a centripetal force is constantly acting on every particle in the revolving rim, the rim's angular momentum remains constant. The forces acting on a particle or system of particles, then, need not be zero for the angular momentum of the system to remain constant. A force may be applied to the wheel which has no effect on the angular momentum if the line of action of this force passes right through the center of mass of the wheel; it only accelerates the wheel as a whole along this line of action translating it along that line. This force has no effect either on the direction of the wheel's axle or the rate of revolution of the wheel's rim. In fact, any force which passes through the center of the wheel, regardless of the direction, can have no effect on the angular momentum of the wheel—it can only accelerate the wheel as a whole.

If a force $\mathbf{F}'$ is applied which does not pass through the center of the wheel, it may alter the speed of rotation and also tilt the axle or axis of rotation. In other words, such a force $\mathbf{F}'$ changes the wheel's angular momentum. To measure the effect of this force on the angular momentum, we draw a vector from the center of the wheel to the point p on the wheel at which the force is applied to the wheel. The effect of $\mathbf{F}'$ on the angular momentum is greater the larger r is, and the closer the angle between r and $\mathbf{F}'$ is to 90°. In other words, the effectiveness of $\mathbf{F}'$ in changing the angular momentum depends on the product of $\mathbf{F}'$ and the perpendicular distance from the center of the wheel to the line of action of $\mathbf{F}'$. This product is called the moment of the force $\mathbf{F}'$ relative to the center of the circle, and we refer to the product as a torque. From this definition we see that the moment of a force or the torque is a vector quantity since it expresses the turning action of a force. An example of this quantity is the turning action of a wrench applied to a pipe, which can occur in either a clockwise or counterclockwise direction.

The moment of a force or torque can be expressed as the vector $\mathbf{T}$. If a force $\mathbf{F}$ is applied to the end of a wrench at a distance r from the screw, the screw experiences a torque of magnitude $r \times \mathbf{F}$ and its direction is along the screw. From this example we see that the larger r

is, the larger the torque. This is why long-handled wrenches are used to turn heavily embedded screws.

The principle of the conservation of angular momentum can now be expressed as follows: Given a collection of particles (or, in a special case, a single particle) moving in any way whatsoever, the total angular momentum of the system remains constant unless a torque acts on the system. When this occurs, the rate at which the angular momentum changes equals the torque, and the change in the direction of the angular momentum is in the direction of the applied torque.

Torque is far more important in our daily activities than is force by itself. Every motion of our arms, legs, head, and torso involves one or more torques within our bodies and when we carry out our daily activities, whether it is opening a door, unscrewing the cover of a jar, steering a car, or swatting a tennis ball or a golf ball we are applying torques.

Although we have introduced the principle of the conservation of angular momentum for a system of particles, we have not yet defined the angular momentum for such a system. We have only defined the angular momentum for a single particle moving in a circle, but this leads to a definition of the angular momentum for a system of particles. We first consider particles revolving in circles that lie in parallel planes. The different axes about which these particles are revolving do not coincide but they are parallel to each other. Each particle has its own angular momentum **J** which lies along its axis of rotation. Since all of these axes are parallel, all the angular momentum vectors are parallel and we can simply add them numerically. This addition gives us the vector **J** for the total angular momentum of the system of particles. This vector is parallel to the individual angular momentum vectors of the particles and its magnitude equals the sum of the magnitudes of these separate angular momenta.

To understand the concept of the rotation of a rigid body, we can picture a plate with several concentric circles. The plate is held fixed at its center point O so that it cannot be translated as a whole. A point i placed on one of the concentric circles can move even though the plate as a whole cannot be translated. In fact, the point i can revolve in a circle around the point O, which is the only kind of motion that can occur there. This ability to revolve around point O is of course true for every other point on the plate. Suppose now that point i is moving with constant speed v_i about O so that it traces out the circle with radius r_i (the distance of i from O). Since the plate is rigid, all the other points on the plate must be moving in their own circles around the point O. But the speeds of these different points are not the same since they are, in general, at different distances from point O. Therefore, the speed of any one particular element of the body cannot be used to specify the motion of the rigid body as a whole; it requires a certain dynamical parameter.

To see this parameter, we consider not only the point of the plate at i, but also another point j whose distance from O is r_j. Since the body is rigid, the angle ϕ between r_i and r_j must remain constant as the body rotates. This is so only if the lines r_i and r_j turn around O at the same rate. The angle ϕ_i must increase at the same rate, then, as does the angle ϕ_j, which are the angles made by r_i and r_j with respect to a fixed direction on the plate. In other words, a constant angular speed ω is associated with the motion of the plate such that the angle ϕ_i for any point i increases at the rate ω. From this relation we deduce that if v is the speed of the particle at point i, we must have $v_i = r_i\omega$, because the plate is rotating with constant angular speed ω about an axis through the point O at the center of the plate.

Angular momentum can be associated with this rotational motion by returning to point i on the plate and assuming that this minute portion of the plate has mass m_i. The magnitude of the angular momentum of this point is $m_i r_i v_i$ or $m_i r^2 \omega$; the direction of this angular momentum is along the axis which passes through point O and is perpendicular to the plane of the motion of the element m_i. This is true for every point on the plate (excluding point O at the center) since they are all moving in circles (of different radii) about the axis. If we now picture any rigid body, rotating with angular velocity $\boldsymbol{\omega}$ as divided into n elements 1, 2, . . . , n with masses $m_1, m_2, \ldots, m_n$ and at distances $r_1, r_2, \ldots, r_n$ from the axis of rotation, these elements have angular momenta with magnitudes $m_1 r_1^2\omega$, $m_2 r_2^2\omega$, $m_3 r_3^2\omega, \ldots,$ $m_n r_n^2\omega$, respectively. Since all of the angular momenta point in the same direction, we can add them together as we would ordinary numbers to obtain the total angular momenta because ω is the same for each element of the rigid body. If we introduce the symbol **J** to represent the total angular momentum, we have for a rigid body the relation $\mathbf{J} = I\boldsymbol{\omega}$. The quantity I is called the moment of inertia of the rigid body relative to the axis of rotation. Notice that if the position of the axis of rotation is changed, the magnitude of I changes so that the magnitude of angular momentum also changes. The direction of the angular momentum of the rotating body changes if the axis of rotation changes its orientation. But neither of these changes can occur spontaneously. A torque must be applied to the rotating body for this to happen.

Since the magnitude of the angular momentum is $I\omega$, it can be altered by changing ω which requires the application of a torque to the spinning body. There is an interesting feature of the conservation of angular momentum as applied to a spinning body which contains particles (or points) that may move about inside the body. If the body is spinning, with no external torque applied to it, and some of the matter in the body rearranges itself relative to the axis of rotation, the moment of inertia I changes. But this change means that the magnitude of the angular momentum also changes, unless something compensates for this change in I. The only other factor that can change, however, is ω. In fact, ω has to change by just the right amount to offset the change in I and thus keep $I\omega$ constant.

Some interesting examples of this application of the principle of the conservation of angular momentum can be found in different physical phenomena around us. The earth is a spinning body whose angular momentum vector lies along its axis of rotation. Careful observation of the motion of the earth's axis as well as the period of rotation of the earth reveals two things: the earth's axis changes its orientation in space continuously and the earth's period of rotation fluctuates from month to month. We show later that part of the change in the direction of the earth's angular momentum (its axis of spin) arises from the action of an external torque exerted on the earth by the moon and sun, but some of this change and the fluctuation in the earth's period of rotation occurs because of the continual rearrangement of matter within the earth itself.

Another example is a figure skater pirouetting with her arms outstretched. As long as her arms remain extended, she spins at about the same rate, but as she brings her arms close to her body, her rate of spin, or her angular speed ω increases. The angular momentum of the skater equals $I\omega$ and, by the principle of conservation of angular momentum, this quantity must remain the same as long as no torque acts on the skater. But by bringing her arms close to her body, she reduces her moment of inertia I, because I depends on the distance between the skater's axis of rotation and the various parts of her body. When the skater's arms are extended, these distances are larger than when her arms are brought in toward her body. If I is reduced, ω must increase by a corresponding amount to keep $I\omega$ constant.

Other examples of this kind of phenomenon can be found in the twirling of acrobats and divers. By bringing their legs up to their chests, they can spin rapidly as they move through the air, and then they can slow themselves down as they straighten out their bodies. In fact, all acrobatic stunts and gymnastic feats depend on the principle of the conservation of angular momentum.

As an example of the application of the principle of the conservation of angular momentum, when the speed of rotation ω does not change but the direction of the angular momentum changes, we consider a man standing on a turntable, holding a spinning wheel with its axle horizontal. The platform is free to rotate about the vertical axis only but it cannot be tilted in any direction. The man is holding the wheel so that its axis points to the right and the wheel (as seen by the man) is rotating in a clockwise direction. Treating the man, the wheel, and the platform as a single dynamical system with no external torque acting on it, we know that its total angular momentum must remain constant. But the total angular momentum of this system is just the angular momentum of the rotating wheel, which is represented by the vector $\mathbf{J}$ pointing along the X axis of a three-dimensional coordinate system. The dynamical system of man, wheel, and platform has no angular momentum parallel to the vertical direction—along the Z axis. Since there is no external torque tending to rotate the system about the vertical direction, we conclude, by the conservation principle, that no matter what the man does with the wheel, the total angular momentum along the Z axis must remain zero.

Suppose that the man begins to tilt the axle of the wheel upward until its axis is vertical. The angular momentum of the wheel is then entirely along the vertical direction. What happens to the platform? It begins to rotate to oppose the motion of the wheel, so that the total angular momentum of the system remains zero along the Z axis as it was initially, because there is no external torque about the Z axis. This can be so, however, only if the upward vertical angular momentum of the wheel is compensated by the downward vertical angular momentum of the entire system. Since the moment of inertia of the entire system (man, platform, and wheel) with respect to the vertical axis is much larger than that of the wheel, the angular velocity of the platform is considerably smaller than that of the wheel.

One other interesting and amusing example of the conservation of angular momentum is seen in the behavior of a cat that is dropped from a given height. As soon as the cat starts to fall, it begins to twirl its paws in such a way that its whole body rotates about an axis parallel to its spine, thus permitting the cat to land on its paws when it hits the ground. Tightrope walkers and children who walk on fences know intuitively about this principle when they twirl their outstretched arms (or, in the case of the tightrope walker, an umbrella) to keep their balance.

We must now consider an important example of a rotating body to which an external torque is applied in such a way that the magnitude of the angular momentum remains unaltered but the direction of the angular momentum changes. A top spinning on a horizontal surface with its axis tilted at some angle ϕ with respect to the vertical axis illustrates this phenomenon very well. The first remarkable thing about this spinning system is that it does not fall to the ground but remains with its axis always tilted at the same angle as though it were defying gravity. However, gravity does not stop acting on the top when it begins to spin. Gravity is pulling down on the top with a force Mg, which acts at the center of mass of the top. If M is the mass of the top, Mg is simply its weight. If we suddenly brought the top to rest by stopping its rotational motion, it would fall over on its side in response to this vertical pull of gravity, but as long as the top is spinning, it does not fall over. The other remarkable aspect of the rotating top is that its axis describes the surface of an imaginary cone whose apex coincides with the point where the top is in contact with the ground. The vertical line through this point is the altitude of the cone traced out by the axis of the top, and the angle between the axis of the top and the vertical line remains constant as the top spins. This phenomenon is called precession; it can be understood if we apply the principle of the conservation of angular momentum. This principle tells us that unless a torque is applied to a spinning body, the angular momentum ($\mathbf{J} = I\boldsymbol{\omega}$) remains constant, which means that both the quantity ($I\omega$) and the direction of the axis of spin remain unaltered. But if a torque is applied to the body, its angular momentum changes at a rate equal to the applied torque. If a torque is applied, the magnitude $I\omega$, the direction of the axis of spin, or both, may change. For the spinning top, we have just noted that the axis of rotation precesses about the vertical but the magnitude $I\omega$ of the angular momentum does not change. It is clear that a torque is acting on the top. Only the direction of the axis of the top changes. Hence, the torque must always be acting at right angles to the top's axis. To see how such a torque arises, we note that the force of gravity acts vertically downward at the center of mass of the top. But an equal and opposite upward force acts on the top at its point of contact with the table. The upward push of the table on the top just equals the downward pull of gravity, since the top has no acceleration in the upward or downward direction. We thus have acting on the top two equal and opposite forces, separated by a distance.

Two such forces constitute a torque and we see that this torque tends to rotate the top in a counterclockwise direction. The torque can be represented by a small vector **T** at right angles to the plane of the paper and pointing away from it. Since **T** is at right angles to **J**, it cannot change the magnitude of **J** but only its direction. The axis of the top moves in the direction of **T**, toward the reader, which is exactly what happens.

So far we have limited ourselves to a special case of angular momentum—a particle (or a collection of particles) moving in a circle with constant speed. By considering a collection of such particles whose distances from each other remain constant, we were led to the concept of a rotating rigid body. But we must now consider the general case of a particle moving with variable speed in an orbit of any shape. If we have a particle at point P of its orbit and its momentum is $m\mathbf{v}$, how can we define the angular momentum of this particle? The answer to this question requires that we return to the particle moving with constant speed in a circle, where we obtained the angular momentum by multiplying the radius of the circle by the momentum of the particle. The center of the circle thus became the reference point for our definition of angular momentum. To proceed in the same way and define the angular momentum of the particle at P relative to the observer at O by multiplying the momentum of the particle by its distance r from O is incorrect. For the particle moving in a circle, the line from the center of the circle to the particle is the radius of the circle and hence perpendicular to the momentum of the particle; this is not true of the line from O to the particle in our example. We must take the perpendicular x from O to the particle's momentum. To be consistent and in complete analogy with circular motion, we define the angular momentum of the particle at P as mvx which is the general definition of angular momentum $J = mvx$. The angular momentum of a particle moving in any orbit relative to any observer is then twice the mass of the particle times the rate at which a line from the observer to the particle sweeps out an area. In short, the areal velocity is just the angular momentum divided by twice the mass of the particle. The principle of the conservation of angular momentum may now be expressed somewhat differently. If a particle moves in such a way that its angular momentum relative to an observer is constant, the line from the observer to the particle sweeps out equal areas in equal times; we call this the law of areas.

CHAPTER 8

The Earth as a Gravitational Structure

'If everybody minded their own business,' the Dutchess said in a hoarse growl, 'the world would go round a deal faster than it does.'

—LEWIS CARROLL, *Alice in Wonderland*

We began this book by outlining the relationship between the various structures in nature and the forces that exist between the various bodies that constitute the structures. Up to this point we have considered in detail only one force—the force of gravity—but we have developed the basic principles and laws that reveal how any kind of force gives rise to a structure. To get a preliminary look at how a structure can evolve when forces come into play, we first consider a collection of particles moving about freely without interacting with each other (no forces between them). From Newton's first law of motion we know (if we neglect collisions) that each particle continues moving without change of speed in the same straight line so that these particles in time fill all regions of space uniformly. Of course, collisions among these particles constantly alter the speeds and directions of the colliding particles but these random collisions do not change the overall picture. The absence of forces between particles leads to a structureless, completely random and undifferentiated universe.

To take into account the actual situation, we first assess the effect of the gravitational force between pairs of these particles. Although the gravitational force between any two particles is, in general, quite small, if enough of them are concentrated in a given region of space, they pull on any outside particle with a large enough force to draw it into their region of space. Whether this particle enters this concentrated region of space and remains there to increase the number of particles already there or passes by and goes off again into the outer regions of space, depends on certain energy relationships which we discuss below. But if the particle does remain (we then say that it has been captured), it becomes a member of the concentrated group of particles. Since this can happen to many of the particles outside the concentrated group, a well-defined gravitational structure results. This account gives us a rough picture of how the gravitational force leads to a structure in an otherwise undifferentiated distribution of particles.

We now consider more carefully the energy conditions that must be fulfilled before a structure of the sort described in the previous paragraph evolves. In our previous discussion of some of the consequences of the principle of the conservation of energy, we saw that a single particle, however massive it may be, cannot capture another particle approaching it from a very great distance if the total energy of the two particles is conserved. If the two particles were initially at rest with respect to each other and separated by a very large distance, the total initial energy was zero. Since the total energy must always remain zero to satisfy conservation of energy, the two particles cannot remain attached to each other no matter how close they approach—they must separate again. If we now apply this same idea to our collection of randomly moving widely dispersed particles, with the force of gravity operating among them, a permanent structure consisting of a group of these particles can be formed only if this group of particles loses energy in coming together. If their total energy after they have come together to form a concentration is the same as it was before coming together, the particles disperse again and the concentration is dissipated. Assume now that somehow or other this subgroup of particles loses energy so that a stable configuration is formed. This event must then be the first step in the formation of all the other structures such as the sun, the earth, complex molecules, living organisms, and so on that may evolve from this initial, structureless subgroup of particles.

The first step in the formation of the various complex structures all around us results from the gravitational force; this force has just the right properties to bring together particles spread over vast distances. This phenomenon is called the gravitational contraction phase of the origin of such bodies as the sun, the moon, the planets, and the solar system itself.

To go beyond this first gravitational phase in the formation of structures, forces other than gravity must come strongly into play since the gravitational force under ordinary conditions is much too weak to account for the structural strength of the solid bodies that surround us. At this point in our story, however, we do not introduce these strong forces; we limit ourselves to the structural properties arising from the force of gravity. With the aid of Newton's law of motion and his law of gravity, we can now understand many of the phenomena associated with such structures as the earth, the moon, the sun, and the planets.

We know that the earth is (to a very good approximation) a sphere (actually an oblate spheroid because of the flattening action of the rotation of the earth), which is just what we expect from the spherical symmetry of the force of gravity. According to Newton's law of gravity, the gravitational force exerted by a particle is the same in all directions. Such a particle tends to draw all things toward it along lines radiating from it; since no radial direction is preferred over any other by the gravitational force, all bodies at the same distance from the attracting particle fall toward it with the same acceleration, thus forming an imploding spherical surface of matter. Such spherical surfaces of matter falling in toward an attracting center ultimately form a sphere of matter. This is a highly simplified picture of the way a sphere of matter like the earth was formed, but the important point here is that a spherically symmetric force like gravity leads (under appropriate conditions) to spherical structures when particles are free to move under the action of gravity and arrange themselves in geometrical patterns. We consider the exact conditions under which gravitational spheres arise in a later chapter and merely note here quite generally that the force of gravity accounts for the shape of the earth.

Although the earth is described in a general way as a sphere, it is not quite spherical, but is an oblate spheroid flattened at its poles. At the earth's equator we must walk 68.7 miles for our vertical to turn through 1° whereas we must walk 69.4 miles for it to turn through 1° at the poles. Before the advent of artificial satellites, the exact shape of the earth was estimated from the variation in the acceleration of gravity from point to point on the earth's surface and with surveying measurements. The observations of the motions of artificial satellites now give us much more accurate pictures of the earth's shape, and we now know from such observations that although the earth is flattened at both poles, the amount of flattening is not quite the same at both poles so that the earth is slightly pear-shaped. We shall see the reason for the polar flattening of the earth later when we discuss the earth's rotation.

The simplest way to study the structures of the earth is to consider it as consisting of concentric shells of matter in different states and with different densities. The outermost shell starts at the solid (and liquid) part of the earth (referred to as sea level) and extends out into space for a few thousand miles. It is a mixture of various gases: nitrogen, 78.09%; oxygen, 20.95%; argon, 0.9%; carbon dioxide, 0.03%; with traces of such rare gases as neon, krypton, and xenon also present. Water vapor is present in variable amounts which depend on climatic conditions; an ozone layer starting at a height of about 10 miles and extending out to about 100 miles is also present.

The atmosphere is not of uniform composition but shows some chemical and physical stratification. For that reason it has been divided into concentric shells which are referred to as "spheres." The troposphere which extends from sea level up to a height of about 10 miles is the usual familiar atmosphere where the daily weather phenomena occur. The ozone layer starts at the top of this shell and extends into the next shell which is called the stratosphere. The ozone layer is extremely important for life here on the earth because the ozone molecule (the triatomic oxygen molecule O_3) readily absorbs ultraviolet radiation so that the ozone layer protects the living organisms from exposure to this harmful solar radiation. Of course, not all the ultraviolet radiation is absorbed, but the amount that gets through the ozone layer is quite small and harmless. The stratosphere continues up to a height of about 50 km where the mesosphere begins. The ozone layer ends about halfway through the mesosphere and in the mesospheric region above this layer ionization of such atoms as oxygen and nitrogen occurs quite readily. The highest known (noctilucent) clouds are found at the top of the mesosphere. Here, too, we observe the trails of meteors that are vaporized as they speed through the atmosphere from interplanetary space.

The ionization of atmospheric atoms, first met in the mesosphere, is due to the unscreened ultraviolet radiation and high-energy particles from the sun that penetrate down to the mesosphere. This ionization of oxygen leads to the formation of the ozone. Under the ordinary conditions which exist in the troposphere of the earth's atmosphere, only the diatomic molecule O_2 of oxygen exists. Such diatomic molecules are quite stable and do not interact to form ozone. But when oxygen is ionized, O_3 (ozone) can be formed. This is why the typical odor of ozone can be detected immediately after a lightning discharge which ionizes oxygen atoms along its path.

The ionization that occurs in the regions above the mesosphere also accounts for the ionosphere—a layer of ions (positively and negatively charged particles) that extends to a height of about 200 km. The ionosphere plays an extremely important role in radio communications on the earth since the electrically charged particles that constitute the ionosphere reflect radio waves. The region above the ionosphere is called the exosphere. It extends out to a distance of about 5000 km which is the greatest height at which auroras have been observed.

Before artificial satellites were introduced, the maximum height of the earth's atmosphere could not be accurately determined. At best one could only say that the atmosphere is certainly as high as the highest observed auroras, but the study of the motions of artificial satellites shows that these objects experience atmospheric drag out to distances of some 10,000 miles. Moreover, the ion detection instruments placed in such vehicles show belts of ionized (electrically charged) particles (the Van Allen belts) which circle the earth and extend out to about 15,000 miles. These Van Allen belts are due to the earth's magnetic field which captures and traps energetic charged particles emitted by the sun during solar storms and flares.

Not only do the atmospheric layers differ in their composition, but they also differ in their physical properties (temperature and density). The density of the air at sea level (mass per unit volume) is 1.2×10^{-3} g/cm^3, which means that 1 cm^3 of sea level air contains about 4×10^{19} molecules. The air density falls off quite rapidly with height, and is only 10^{-13} (one ten billionth) g/cm^3 at a height of 125 miles. The number of molecules per cubic centimeter at this height is about 4 billion.

We now consider a column of atmosphere having a cross-sectional area of 1 square inch and extending from sea level up to the very top of the atmosphere. Although the density in this column diminishes as we go up, we can easily measure the total weight and, hence, the total mass of this column of air since the weight of this column is just what we call atmospheric pressure. First, we define pressure, which we now introduce as a new derived quantity, constructed from force and area. We can apply a force at right angles to a surface (e.g., we push against a wall) by bringing various objects in contact with the wall and pushing against them. The effect this push has on the wall at the point of contact between the object and the wall depends on the cross-sectional area of the object (let us say a rod). The thinner this object is (e.g., a spike), the greater the impression it makes on the wall even though the push exerted on the object is the same in all cases. From this we see that the impression that an object makes on a wall when it is pushed against the wall is determined not alone by the total force exerted on the object, but by this force and the area of contact between the wall and object; the smaller this area is for a given force, the larger the impression. We introduce a new quantity called pressure to describe this and define it as F/A, where F is the force exerted and A is the area of contact. The pressure is expressed as dynes per square centimeter or pounds per square inch.

To get back to the earth's atmosphere, we define the atmospheric pressure (in pounds per square inch) as the weight of a column of air having a cross-sectional area of 1 square inch and extending from the surface of the earth to the top of the atmosphere. Such a column weighs 14.7 pounds. From this number we can now find the total weight and hence the total mass of the atmosphere. All we need to do is to multiply 14.7 by the total number of square inches on the earth's surface. We then find that the weight of the total atmosphere of the earth is 5.10×10^{21} g, which is only about one millionth of the earth's mass, which we calculated previously. Most of this weight arises from the air lying in a shell whose lower surface touches the earth and whose upper surface is less than 20 miles above sea level.

The variation of the temperature in the earth's atmosphere is quite complex. Instead of dropping continuously with increasing height, the atmospheric temperature alternately increases and decreases as one passes through the various layers. At first the temperature decreases steadily with increasing height through the troposphere. But then it begins to rise again, reaching about -8°C at the mesosphere. Here again the temperature begins to drop, going down to -80°C at a height of about 90 km, where another reversal occurs, with the temperature increasing steadily through this zone called the thermosphere. The temperature increase is rapid here and it reaches a maximum of 1220°C at a height of 500 miles.

Although the earth's atmosphere is primarily of biological interest to us because its oxygen content supports the variegated forms of life on the earth, it is of geological and astronomical interest as well. We discuss the geological effects of the atmosphere later and consider only the astronomical aspects here. Since most of the information the astronomer can obtain about celestial bodies is contained in the light that reaches us from these bodies, we must know how the atmosphere affects this light before we can correctly interpret its message. There are three such effects: (1) The atmosphere absorbs light and thereby changes the apparent brightness of celestial objects such as stars; (2) the atmosphere scatters blue light more readily than it does red light so that celestial objects appear redder than they are in reality (this effect is especially pronounced when the celestial object is near the horizon such as when the sun is rising or setting); (3) the speed of light is diminished in the atmosphere as compared to its speed in a vacuum. As a result of this third effect the path of a beam of light from a star (if the star is not directly overhead) is bent so that the star appears to be displaced from its true position. Since these three effects depend on the density of the atmosphere at each point, we must know the atmospheric conditions quite accurately before we can interpret the message of starlight properly.

The hydrosphere is the thin layer of water, soil, and rock outcroppings that separates the atmosphere from the rocky crust of the earth. It is essentially a discontinuous shell of fresh and salt water consisting not only of the oceans but also of continental ice, lakes, rivers, ground water, and water in biological combinations. The total mass of the hydrosphere, most of which is ocean water, is 1.430×10^{24} g. This mass is about 0.03% of the earth's total mass. The waters of the ocean, which have an average depth of about 12,500 feet and cover 71% of the earth's surface (about 140 million square miles), have a mass of about 1.4130×10^{24} g, which is 98% of the total hydrosphere mass. The total volume of the oceans is 330 million cubic miles. The amount of fresh water and ice in the hydrosphere is quite small compared to the ocean waters. There are about 0.005×10^{24} g of fresh water and 0.0228×10^{24} g of ice. Most of this ice lies on Greenland and on Antarctica. Since there is no land mass around the north pole, the ice in the north polar regions of the earth is part of the ocean; it is just a fairly thick frozen layer of ocean water. If all the ice in the north polar regions were to melt, the ocean levels would be the same as they are now because this ice is already part of the ocean. But if all the ice on Greenland and on Antarctica were to melt, the ocean levels would rise by about 200 feet.

Chemically speaking, ocean water is a weak electrolytic solution of various salts with sodium chloride the most abundant. In addition to ordinary salt the oceans contain the salts of magnesium, calcium, potassium, and aluminum in measurable quantities. Smaller amounts of the compounds of such atoms as carbon, silicon, fluorine, iodine, strontium, bromine, iron, and boron are also present. Indeed, all the chemical elements that are found on the earth can be found in the oceans. The salt content of the oceans (taking into account all elements) is about 3.5% of the total ocean mass. The oceans have an enormous effect on the earth's climate and play a major role in the hydrologic cycle. Most of the water vapor in the atmosphere and the rain and snow stem from the oceans.

The rocky crust upon which the hydrosphere and the earth's soil rest is called the lithosphere. It is a thin layer of rock which in turn rests on the mantle or barysphere (so called because the material below the crust is the heavy material of the earth). The only part of the earth's crust that can be studied directly is the superficial part which is exposed as outcroppings by complex folding, mountain building, or, where the soil has been worn or swept away, by the erosive action of air and water. In particular, such deep clefts as the Grand Canyon and such deep valleys as those in the Alps permit geologists to construct a fairly complete history of the progressive layers of the crust that have been built up for billions of years.

The crust of the earth is divided into two parts by geologists: the continental crust and the oceanic crust. The continental crust has a mean thickness of about 45 km and its chemical composition is mainly silica–aluminum-rich rocks (called "sial"). The mean height of the crust above sea level is 2707 feet with a maximum elevation (Mt. Everest) of 29,028 feet. The mean density of the continental crust is 2.7 g/cm^3. The oceanic crust has a mean depth of about 12,500 feet with a maximum depth (below sea level) of about 36,000 feet. Its composition is mostly silica–magnesia-rich rock (primarily basalt) and is referred to as "sima." The mean density of "sima" is about 3.2 g/cm^3, which is about 25% higher than the density of "sial."

Although we have described the two parts of the earth's crust as though they were sharply separated, this is not really so because there are transition regions where the continental shelves meet the ocean waters. These areas are intermediate regions which seem to be quite mobile (geologically speaking) in the sense that either the oceanic crust is slowly being transformed into continental crust or segments of the continental crust are breaking off and becoming part of the oceanic crust.

Interesting features of the oceanic crust are the deep-sea trenches. These trenches are wedgelike clefts in the ocean floor (often twice as deep as the average ocean depth) that run parallel to the continents and island chains and are marginal to these land masses. They also run parallel to the continental mountain chains, often extending for more than 500 km. They are about 50 km wide at the top but narrow down to about 100 m at the bottom.

The generally regular features of the surface of the earth's crust above and below the hydrosphere do not remain the same at all times but change slowly over a long time. Occasionally, violent disturbances such as earthquakes and volcanos bring sudden and drastic changes. The slow variations are caused by weathering processes which are due to rains, rivers, and the atmosphere over the continental crust. Since erosion and weathering processes of this sort have been going on continuously since the Paleozoic era (about 500 million years ago) a layer about 17 km thick must have been worn off the continental shelf of North America (taken as a typical continent). There must be some compensating process that keeps the crust of the earth as wrinkled as it is today. If this were not so, the earth would have been worn quite smooth by this time and there would be no dry land at all. However, the constant shifting of the earth's crust has prevented the planet's surface from approximating a billiard ball. The variation in the density of the earth's crust as one moves from the continents to the ocean floor indicates that a constant vertical adjustment occurs to maintain pressure and density equilibrium in the earth's crust. This equilibrium is called isostasy.

Immediately below the earth's crust is a very thick layer called the mantle or barysphere. This shell extends to a depth of about 3000 km and ends where the core or centrosphere begins. It is separated from the core by a discontinuity (a jump) in the density. Specimens of mantle material are present on the earth's surface in the form of slabs of very dense rock that has been dredged up in the lava of deep-seated volcanos. This rock is a greenish-black material consisting of the silicates of iron and magnesium. Its density is 3.2 g/cm^3. The upper mantle, the region right under the crust, has the properties of a rigid, brittle solid; earthquakes are probably generated in this part of the mantle down to a depth of about 700 km. In the very top region of the mantle and in the crust itself earthquakes are probably due to displacements of the earth layers along rock faults. But at depths of a few hundred kilometers the earthquakes may originate in sharp density discontinuities.

The rigid brittle part of the mantle extends down to a depth of about 75 km below the crust, where a rather sudden change in the rigidity occurs. A kind of plastic zone (a so-called "zone of no strength") is present, which flows under high pressure. This plastic zone is indicated in the behavior of certain types of earthquake waves (called P waves) as these waves pass through the mantle. The speed of earthquake waves (they are really acoustical waves) depends on the density and rigidity of the rock through which they are propagated. The higher the rigidity (or strength of the material), the greater is the speed of the waves (increasing density decreases the speed). The speed of the P waves increases steadily (from 8 km/sec to about 14 km/sec) as they pass through the mantle, but a sudden decrease in speed occurs below the 75-km zone. This slowdown suggests a plastic region. The presence of such a plastic region is important for the maintenance of equilibrium between the continental and oceanic crust. As the ocean basins are depressed by the deposition of layers of silt that are carried into the oceans from the land masses by erosion, equilibrium ("isostasy") is maintained by the displacement of an appropriate amount of material back under the continents via the plastic regions of the mantle.

The density of the material in the mantle increases steadily from 3.5 g/cm^3 right under the crust to about 6 g/cm^3 at a depth of 3000 km. At this depth the mantle gives way to the core of the earth which is divided into two zones: the liquid outer core, also called the centrosphere, and the solid inner core. The liquid character of the outer core is detected by the cutoff of the S-type (shear type) earthquake waves at the bottom of the mantle and the drop in speed of the P waves as they enter the outer part of the core. We shall soon see why S waves cannot pass through the outer core and how this tells us that this part of the earth's core is liquid. This liquid region is thought to consist of an iron–nickel alloy in a liquid state. The rotation of the earth sets up currents in this material which generate the earth's dynamo-type magnetic field.

The density of the inner core is about 17 g/cm^3 and the pressure there is about 3.64 million atmospheres. The temperature near the center is estimated to be about 5000°C. These figures are quite tentative since they are not based on direct measurements but on the analysis of the propagation of elastic waves.

Although we have just given a fairly detailed description of the principal layers of the interior of the earth, the reader is justified in asking how we obtained this picture. The only way the interior of a physical structure can be studied is to subject it to some kind of force and to study the results. This force may be mechanical, gravitational, electromagnetic, or nuclear. In the earth's interior the forces are mechanical and gravitational and the results are acoustical waves (or earthquake waves). The analysis of the interior structure of the earth is based on a study of the way these waves are propagated through the earth.

When a solid is struck, the disturbance (or sudden distortion) caused by the blow is propagated in all directions in the solid in the form of two waves. One of these waves is a compressional or P wave. In this kind of wave the displacements (alternate compressions and expansions of the solid) are in the direction of the propagation. P waves can be transmitted through liquids and gases (sound in air) as well as solids. In addition to P waves there are also distortional or S waves (shear waves) in which the direction of the displacement of the solid material is perpendicular to the direction of propagation of the wave; these are also called transverse waves. Transverse waves can only be transmitted through solids; they cannot be transmitted through liquids or gases. Both P and S waves are propagated through the earth from the point where an earthquake occurs. The velocities of both P and S waves depend on the density and the rigidity (the reciprocal of the compressibility) of the solid but not exactly in the same way; the speed of the P waves is always larger than the speed of the S waves. The P waves move about 1.7 times faster through a solid than do the S waves. Owing to this difference in speed, we can distinguish between them and determine in this way the internal structure of the earth by studying the propagation of earthquake shocks. Another important and useful property of earthquake waves in this analysis is the change in speed and direction of both P and S waves when they pass from a region of one density to a region of another density. In seismic stations set up at different points on the earth's surface, observers monitor the P and S waves that emanate from an earthquake at a point p in the earth's interior. By measuring the difference between the times of reception of P and S waves at these different stations, the variations in the density of the earth's interior can be determined.

The data obtained from the S waves at these stations also permit one to calculate the point below the earth's surface at which the liquid core of the earth sets in. Since S waves are not transmitted through a liquid region, these waves cast an acoustical shadow of the interior liquid sphere, and seismic stations within the shadow (on the side of the earth opposite the point of origin of the earthquake) do not detect S waves. From the size of this shadow the size of the liquid core can be found.

Earlier in this chapter we dealt with the physical structure of the earth and pictured it as consisting of a series of concentric material spheres which differ in density and chemical composition. There is, however, another sphere attached to the earth which is not material: a force field. We have seen in our discussion of the gravitational force that we can describe this force as a field surrounding the mass which is exerting the gravitational force. In addition to the gravitational field, other force fields exist among which is the magnetic field. Since a magnet exerts a magnetic force on other magnets in the space around the magnet, we say that a magnet is surrounded by a magnetic field of force. Here we consider only the magnetic field that surrounds the earth. The ancient Greeks, the early Chinese, and most early navigators knew that a rod of a special type of iron ore (the lodestone) orients itself along the north–south direction if suspended above the ground by a string. This property of the lodestone enabled these earth navigators to explore the oceans since they could use the lodestone to find the north–south direction. The modern magnetic compass is an outgrowth of the lodestone.

The compass needle does not point exactly to the geographic north but rather to a point slightly off the north pole that is called the north magnetic pole; this point shifts about from year to year. Just as there is a north magnetic pole, there is also a south magnetic pole so that the earth as a whole behaves magnetically like a bar magnet whose axis is somewhat tilted (about 15°) with respect to the earth's axis of rotation.

The earth's magnetic field may be pictured as consisting of magnetic lines of force that emerge from the earth's surface at the north magnetic pole and reenter the surface at the south magnetic pole. The earth's magnetic field is similar to what one would find if a large bar magnet at the earth's center were lined up at an angle of about 15° to the earth's axis of rotation. The origin of this magnetic field is not completely understood. If the temperature at the center of the earth were not of the order of 5000°C, one might argue that the earth's magnetism is due to the permanent magnetism of the solid iron–nickel core that is thought to exist at the earth's center. But this cannot be because iron and other magnetic substances such as nickel cannot remain permanently magnetized above a temperature of 500°C. Aside from this, the core cannot be treated as strictly solid since the high temperature and pressure in the core make it behave to some extent like a liquid.

To account for the earth's magnetism we must therefore call upon the dynamic processes in the core. Owing to the rotation of the earth there is differential motion between the central liquid inner sphere and the solid central core. This motion induces electric currents in the highly conducting inner mass and these currents generate the earth's magnetic field. This dynamic theory can account for the strength of the earth's magnetic field and also for its observed slow westward drift (about 22 km/year). Since the earth is rotating eastward, there is a westward lag of the core (owing to the inertia of the core).

An interesting and geologically important application of the study of geomagnetism is to the analysis of magnetic remnants of ancient iron-bearing rocks. This analysis has produced a whole new field of geology called paleomagnetism in which investigators have shown that the direction of the earth's magnetic field must have changed drastically since ancient times. The residual magnetism found in very old rocks containing iron ores shows that the direction of the earth's magnetic lines of force was quite different when these rocks were formed (either from volcanic lavas or from iron-ore sedimentation) from the present direction. A worldwide survey of the old magnetic imprints on such rocks permits geologists to construct paleomagnetic maps of the earth. Such maps show that the earth's magnetic pole lay considerably to the west and south of its present position so that it must have been somewhere in the neighborhood of Hawaii about 1 billion years ago. We believe today that the change in the direction of the earth's magnetic field cannot be described as a drift of the magnetic pole along a well-defined orbit but rather as a reversal phenomenon. Paleomagnetic evidence and modern measurements of terrestrial magnetism show that the intensity of the earth's magnetic field is decreasing at a steady rate and will diminish to zero (pass through the zero point) in about 2000 years. The magnetic field will then reverse its polarity and its intensity will increase but in the opposite direction so that the north and south magnetic poles will be reversed. Magnetic polarity reversals of this sort have occurred many times in the past and will continue in the future.

Before leaving the structure of the earth we briefly describe one other important geological feature of the earth's crust. An examination of the outline of the earth's continents shows that they seem to be the fragments of a large jigsaw puzzle and that if the continents were brought together along appropriate shore lines they would form a continuous land mass extending southward from the north pole to the tip of Africa. The evidence from rock formations shows that Africa and the American continents began drifting apart about 200 million years ago.

Although the gravitational force is the dominant force that accounts for the earth's overall shape and its internal structure, the detailed geological features of the earth can be understood only if we also take into account the structural forces that produced the rock formation of the crust and mantle. These structural forces come into play when the atoms and molecules that constitute matter are moving slowly enough so that they can form the large aggregates of the solid structures that we call rocks. If these molecules are moving too rapidly, solids are not formed and the materials remain either in gaseous or liquid form (depending on the chemical nature of the molecules) as in the case of the earth's atmosphere, the earth's hydrosphere, and the earth's outer liquid core. The question, then, as to whether certain parts of the earth are solid, liquid, or gaseous depends on how rapidly the molecules of these parts of the earth are moving, which, in turn, depends on the temperature of these terrestrial regions.

The surface of the earth receives its thermal energy from two principal sources—the sun (solar radiation), and the decay of radioactive elements in the earth's crust. Most of the heat that the earth's surface receives comes from the sun, and this energy amounts to 2 calories/min per cm^2 at the top of the earth's atmosphere. This number, which is called the solar constant, does not represent the total thermal energy that each solid or liquid square centimeter of the earth's hemisphere that faces the sun receives every minute, for two reasons. First, a good deal of this radiation (about 35%) is reflected by the earth's atmosphere and some more of it is absorbed in the atmosphere. The amount absorbed by the atmosphere contributes to the heating of the earth but the reflected radiation contributes nothing to this thermal effect. Second, because of the curvature of the earth's surface, only those portions of the earth's atmosphere whose areas are normal (perpendicular) to the solar rays receive 2 cal/min per cm^2. The unit areas of the surface of the other regions are slanted with respect to the solar rays and thus receive fewer than 2 cal/min. This difference accounts for the seasons and the variation in climate as we go from one zone to another on the earth.

Owing to atmospheric reflection and absorption, approximately 51% of the solar radiation striking the upper atmosphere of the earth reaches the hydrosphere, and this radiation keeps the average global temperature of the earth at 15°C. That the solar radiation received by the earth per second must have remained fairly steady for hundreds of millions of years is indicated by the continuity of life. If the solar constant had undergone a gross change in the period during which the present forms of life were evolving, many temperature-sensitive organisms would have been wiped out and the biological continuity that we now observe would have been destroyed.

Just as there is a steady flow of heat from the sun to the earth, there is also an outward flow of heat from the interior of the earth to the earth's surface. The evidence for this heat flow from the earth's interior is that the temperature in mines and deep earth borings exceeds the mean annual temperature at the earth's surface. Moreover, the interior temperature increases with depth so that a definite temperature gradient exists. The rate of increase of temperature with depth varies from place to place but its average is about 20°C/km. This downward increase in temperature means that an upward flow of heat occurs by conduction. From the measured increase in temperature with depth and the knowledge of the heat conductivity of rocks, one can calculate the rate of heat flow from each square centimeter of the earth's surface. The mean value is about one millionth (10^{-6}) of a calorie per second per square centimeter. If this rate holds for the entire earth, the total amount of energy conducted from the interior to the earth's surface every second is 6 trillion (6×10^{12}) cal. This amount is about 25,000 times less heat than is radiated to the earth by the sun every second, so that the temperature (and hence the climate) of the earth's surface is controlled by the sun.

Although the heat conducted from the interior to the surface of the earth per second is a negligible fraction of that received from the sun, it is quite large when taken over geological time. In a million years it amounts to 40,000 cal/cm^2 which is equivalent to the heat supplied by burning a seam of coal 40 m thick. This outward-flowing terrestrial heat cannot come from chemical reactions within the rocks of the earth's crust or mantle.

We also know now that this heat cannot still be the slow escape of the original heat that was supposed to have been trapped in the earth's interior when it was in a molten state. If that argument were correct, the earth would have had to be entirely molten no more than 30 million years ago, for otherwise the solidified outer rock layer which increases in thickness with time, would be too thick to allow the amount of heat that we now observe to pass from the interior out to the surface even if the temperature inside of the earth is still thousands of degrees. In other words, the mantle of the earth has become so thick during the 3×10^9 years of the earth's existence that hardly any heat can pass through it from the molten core unless one assumes that the heat conductivity of the mantle increases with depth and is about the same as that of metal right below the earth's crust. From these considerations we conclude that no more than 20% of the observed outward heat flow can be accounted for as stemming from the original heat of the earth.

Since we have discarded the original heat of the earth as the source of the observed heat flow, we must look for another source and we find it in the radioactivity of the rocks. All rocks contain measurable quantities of such radioactive elements as uranium, thorium, and potassium. A chemical analysis of the rocks in the outcroppings of the earth's crust shows that enough uranium, thorium, and potassium are present in granite and basalt to account for the observed heat flow. The radioactivity in the crust (down to a depth of 35 km) contributes about 70% of the observed heat flow, the radioactivity in the mantle contributes about 7%, and the original heat contributes about 23%.

From the observed radioactivity of the surface rocks on the earth we deduce that most of the radioactive material in the earth has migrated to the surface of the earth. If this were not true, we would have to assume that each gram in the mantle and core of the earth is as radioactive as a gram in the crust and this would lead to extremely high internal temperatures for the earth.

Previously we discussed the earth as a gravitational structure without regard to its relationship to the other bodies that surround it, which together, with the earth, form the still larger gravitational structure we call the solar system. Knowledge of the structure of the earth can help us evaluate the information that comes to us in the radiation from these bodies, but we must also know how the earth is moving within the solar system before we can construct a correct picture of this system. Before we can describe the earth's motion meaningfully, however, we must define motion in general. As we shall see when we discuss the theory of relativity, the concept of absolute motion (motion that is not relative to some frame of reference in which the observer is fixed) is meaningless. Hence, we must choose some frame of reference and discuss the earth's motion in this frame. We may choose any frame of reference we please, but the most natural frame that suggests itself is that determined by the "fixed stars" since the stellar configurations remain unaltered (or change extremely little) over long periods of time. If we use the fixed stars as our frame of reference, we find that in it the earth has the following principal motions: (1) rotation (spin) about an axis that precesses very slowly (the period of precession is 26,000 years); (2) revolution around the sun once every $365\frac{1}{4}$ days; (3) precession of the axis of rotation already mentioned in point 1; (4) a translational motion among the stars.

Although we have listed the four principal types of motion together, we distinguish between the rotational motion of the earth and its revolutionary motion. All motion is relative but rotational motion (and, for that matter, accelerated motion in general) appears to have a special quality which distinguishes it from uniform translational motion. This special quality stems from the manner in which Newton's laws of motion were introduced and the manner in which these laws are to be interpreted. If we go back to Newton's first law and consider it very carefully, we see that it has an unambiguous meaning only if we stipulate that it holds only in unaccelerated frames of reference. In other words, Newton's first law should be restated as follows: A frame of reference exists in which the velocity of a body (as observed in this frame) remains constant if no force is acting on the body. This conclusion was implicit in Newton's thinking because he was well aware that not all frames of reference are on the same footing vis-à-vis his laws. His second law is also to be considered as valid only in those frames in which the first law is valid.

If we now consider the nature of the frames of reference in which Newton's laws are correct, we see that they must be unaccelerated frames (frames that are either at rest or moving with constant velocity). Such frames are called inertial frames of reference and we see that Newton's laws of motion enable us to differentiate between such frames as a group and accelerated frames. All we need to do to determine whether we are in an accelerated frame of reference or in an inertial frame is to observe the motions of freely moving bodies. If all such bodies obey Newton's first law (move with constant velocity), we are in an inertial frame. But if such bodies move with accelerated motions and we can find no force acting on them, we are in an accelerated frame. We can also use the second law of motion $F = ma$ and say that we are still in an inertial frame but there is a special kind of force ma (an inertial force) acting on the body we are observing to make it behave the way it appears to behave. This approach is equivalent to saying that all frames of reference are inertial frames and that bodies do not move according to Newton's first law because inertial forces act on them.

Although the above argument is a logically correct and self-consistent point of view, we adhere to the classical Newtonian point of view and say that if the peculiar motions of a freely moving body, as seen from the earth, cannot be accounted for by invoking gravitational forces, these motions are evidence of some kind of nonuniform (accelerated) motion (rotation or revolution) of the earth. It is clear from the behavior of ordinary bodies on the earth, whose motions can be understood and explained quite accurately by Newton's first law of motion and the force of gravity, that the earth is almost an inertial frame and that its nonuniform motion (rotation or revolution) is very small. If this were not true, there would be considerable departure from straight-line motion of bodies moving freely on the earth's surface. Since such gross departures are not observed, the rotation of the earth must be very small; this is why it took humanity a long time to accept and understand the earth's rotation even though direct evidence for this rotation is found in the daily rising and setting of celestial bodies.

Since we now accept the observed daily movements of the stars as direct evidence of the earth's rotation, we may wonder why the Europeans of the medieval period rejected (or perhaps did not even conceive) this idea. The reason is that they had no concept of the laws of motion or of the relationship of force to motion. It was not necessary, according to their point of view, to invoke a force to keep any star revolving in a fixed orbit around the earth (as it appeared to do so in its daily rising and setting). This motion was divinely ordained as was the sun's "motion around the earth." Goethe presents this medieval picture of the heavens very succinctly in the single stanza of the Angel Raphael's prologue from Faust:

> The sun makes music as of old
> Amid the rival spheres of heaven,
> On its predestined circle rolled
> With thunder speed: The angels even
> Draw strength from gazing on its glance,
> Though none, its meaning fathom may;
> the world's unwithered countenance
> Is bright as at creation's day.

But if we accept Newton's law of motion, the rising and setting of the moon, the sun, the planets, and the stars is evidence that the earth is spinning since to believe otherwise means that a force is acting on these bodies to keep them revolving the way they do at present. But since there is no such force, we conclude that the apparent diurnal rotation of the sky, which carries all celestial objects from east to west across the sky, is really due to the earth's rotation from west to east.

Though the diurnal rotation of the heavenly bodies may seem to be proof enough for us that the earth is spinning, it really is not. A very skeptical, hard-to-convince person can argue that some cosmic force keeps all the celestial bodies turning around the earth in unison. To convince such a person that the earth is spinning, we must search, instead, for apparent departures from Newton's laws of motion in the motions of bodies on the earth stemming from the earth's rotation. Since the earth is rotating very slowly (only 15°/hour) the effects of this rotation on earthly bodies are quite small and therefore not easy to observe in the everyday behavior of bodies. But a few quite pronounced phenomena of this sort can be easily observed under the proper conditions. The most famous of these is the Foucault pendulum experiment carried out in 1851 by Foucault. In this experiment a very long pendulum is suspended from a high ceiling (Foucault used the ceiling of the Pantheon in Paris) and set oscillating in a particular direction (e.g., north–south). One then observes that the direction of the swing of the pendulum changes slowly; the plane in which the pendulum is swinging appears to rotate in the same direction as the stars do in their daily rising and setting (opposite to the direction of the earth's rotation). If this kind of pendulum is mounted at the north (or south) pole, its plane of oscillation rotates westward and goes around once in 23 hours and 56 minutes, the time for one complete rotation of the earth. But if the pendulum is placed south of the north pole, the time required for one complete rotation of the plane of the pendulum is longer than the time it takes the earth to rotate once. As the pendulum is brought closer to the equator, the time required for its plane of oscillation to rotate once increases, until at the equator this time becomes infinite.

We can understand why the pendulum's plane of oscillation rotates more and more slowly as the pendulum is brought closer to the equator by studying a Foucault pendulum mounted at the north pole and another one mounted at a point of latitude Ω. At the north pole the string of the pendulum coincides with the earth's axis of rotation. Therefore, the floor of the room turns under the pendulum at the same rate as the earth is spinning. Hence, the plane of the pendulum appears to spin in the opposite direction once every 23 hours 56 minutes. Consider now the pendulum mounted in a building at latitude Ω which begins oscillating in the north–south direction. After the earth has spun half way around, the new north–south direction makes an angle 2Ω with the old north–south direction. But the old north–south direction is the original direction in space of the swing of the pendulum. Hence, the original direction of the swing of the pendulum makes an angle 2Ω with respect to the north–south direction after half a day. The pendulum thus appears to alter its direction of swing relative to the room, but, clearly, not as rapidly as at the north pole. The angle between the new north–south direction and the original north–south direction is not 180° (as it would be after half a day at the pole) but 2Ω which is smaller than 180° for any point not at the pole. Since the original north–south direction was the direction of oscillation of the pendulum, we see that this direction of swing changes with respect to the north–south direction of a room in which the pendulum is suspended, but the rate of this change decreases as we move to the equator. This behavior of the Foucault pendulum demonstrates that the earth is rotating.

Other simple observations of the behavior of moving bodies on the earth show that the earth is spinning. The general behavior of winds can be accounted for by the rotation of the earth. For the north temperate zone the warm winds come from the southwest (the prevailing westerlies) whereas the cold winds come from the northeast (the trade winds). The reason for this pattern is that the warm winds are air masses that drift to the north from the equator and hence start from the equator with a west-to-east speed of about 1000 miles/hour (the speed of rotation at the equator). This speed is greater than the west-to-east speed of air masses north of the equator. Hence, the warm equatorial air masses drift eastward with respect to an observer in the north temperate zone. Just the opposite is true of cold air masses drifting southward from the polar regions. The rotations of hurricanes, cyclones, water spouts, and whirlpools can all be explained by the rotation of the earth.

The earth is slightly flattened at the poles so that one has to walk somewhat farther at the poles to change the plane of one's horizon by 1° than at the equator. The flattening of the poles is also indicated by the decrease in the period of the oscillations of a pendulum. The reason this occurs is that the period of a pendulum equals $2\pi\sqrt{l/g}$, where l is the length of the pendulum and g is the acceleration of gravity. Since one is closer to the center of the earth at the poles than at the equator, g is larger at the poles than at the equator and the pendulum's period is thus smaller, as can be deduced from the above formula.

The flattening of the poles and, hence, the bulging of the earth all along the equator can also be deduced from the behavior of artificial satellites moving in orbits around the earth. If the earth were a perfect sphere, the orientation of the plane of the orbit of any such satellites would remain the same for all times (if we neglect the gravitational actions on the satellite of all the other bodies in our solar system). But this is not so for any satellite whose orbit does not lie exactly in the earth's equatorial plane. In that case the plane of the orbit precesses (it wobbles around).

If there were no equatorial bulge, the gravitational action of the earth on the satellite would be exactly along the line from the satellite to the earth's center and there would be no torque acting on the orbit and hence no precession since the angular momentum of the satellite would then be conserved. But the equatorial bulge introduces a component **F** of the gravitational force which does not point to the center of the earth but rather to the center of the bulge. This component of the gravitational force itself has a component that is at right angles to the plane of the satellite's orbit and hence one that produces a torque (it tries to reorient the plane of the orbit and set it parallel to the plane of the equator). This torque produces the observed precession (wobble) of the satellite's orbit.

The bulge at the earth's equator leads to phenomena that may at first appear paradoxical and contradictory to the laws of gravity, but we shall see that the very property of the earth that causes the bulge also explains these phenomena. Note that a unit mass at the north pole is about 13 miles closer to the center of the earth than a unit mass at the equator. We may therefore compare this situation to a unit mass at the foot and a unit mass at the top of a 13-mile-high mountain. Now we know that we must do a considerable amount of work to take the unit mass from the foot of such a mountain to its summit. Moreover, if the slope of the mountain were smooth (covered with ice, for example), a man placed anywhere on the slope would slide right down to the bottom. If water were placed on the slope of the mountain, it would flow down. We now consider the oceans of the world near the equator. They bulge away from the equator and form a "mountain" of water 13 miles high with the foot of this "mountain" at the poles. How does such a mountain maintain itself? Why do not all the waters flow down from the equator which, of course, would have a catastrophic effect on the present continental shelves.

We may consider this question differently by supposing the earth's surface to be perfectly smooth (no friction), and an object, placed at the north pole, were given a slight push (just enough to start it moving). Would it continue moving (apparently uphill) to the equator and then beyond it and back to the north pole or stop? The answer is yes; it would continue moving and simply circle the earth (along a circle of longitude) endlessly. Moreover, if a particle on such a smooth earth were placed at the equator, it would not slide down to the pole but would stay there.

The reason for all these curious phenomena which seem to defy gravity is the rotation of the earth, which is itself the cause of the earth's bulge. The spin of the earth introduces a centrifugal force which is just sufficient to keep a particle from sliding down the 13-mile-high "mountain" at the equator and to keep the "mountain of water" piled up at the equator. If the earth stopped spinning, the waters would recede from the equator and the land bulge at the equator would slowly flatten out until the earth became perfectly spherical.

This discussion can be clarified by considering a perfectly spherical earth that is not spinning. The force of gravity $\mathbf{F}$ then acts perpendicular to the surface at each point along a line toward the center of the earth. The spherical surface of the earth is then an equipotential surface; no work is required to displace a given mass from any point on the surface to any other point. We now set this perfectly spherical earth spinning with angular velocity $\boldsymbol{\omega}$ around an axis and consider a particle that is free to move north or south on the earth's surface (which we assume to be perfectly smooth) but which is carried around from west to east as the earth spins. Since the particle is moving around in a circle of latitude, it experiences a very small centrifugal force whose direction is perpendicular to the axis of rotation. We now decompose this force into a vertical component, whose effect is merely to decrease the weight of the body slightly, and a horizontal component directed toward the equator. This horizontal component of the centrifugal force accelerates the particle toward the equator, so that all objects on the surface of a perfectly smooth rotating spherical earth would drift toward the equator and there would then occur a piling up of such objects at the equator.

Consider now the surface rock structure of the earth. Any given part of this structure is not free to move so that all the centrifugal force can do is to stretch the rock toward the equator. Thus, the entire structure of the earth is distorted. If the earth had begun as a perfect sphere initially and had then begun to rotate, it would have changed its shape slowly until an object placed on it at any point would no longer tend to move toward the equator. This statement means that the earth would thus ultimately reach its present shape, which is such that the gravitational force and the centrifugal force combine to give a resultant force that is perpendicular to this nonspherical surface at each point. As a result, the nonspherical surface is the equipotential surface for the rotating earth. That is why the waters of the oceans near the equator appear to defy gravity by heaping up and why the Mississippi River appears to defy gravity by flowing toward the equator and apparently "uphill."

The motion of the earth in its orbit around the sun causes the light from a star whose true position is perpendicular to the direction of the earth's motion to appear to come from a direction that is tilted (in the direction of the earth's motion) with respect to the true direction of the star. The observer on the earth thus "sees" the star "displaced" in the direction of his own motion. Since the amount of this apparent displacement (called the aberration of light) depends on the speed of the observer and the speed of light, the observer can calculate the speed of the earth from this aberration phenomenon. This orbital motion accounts for the apparent annual motion of the sun among the stars and for the difference in length between what the astronomer calls the sidereal day and the solar day.

The sidereal day which is 23 hours 56 minutes long is the actual time it takes the earth to rotate once on its axis, or, put differently, the time it takes a star to get back to the observer's meridian measured from the moment it first crossed it—the interval of time between two successive meridian transits of a star. The solar day, which is 24 hours long, is the time between two successive meridian transits of the sun and is given by the time from noon to noon. This period of time (the solar day) is about 4 minutes longer than the sidereal day because the sun (as seen from the earth) appears to drift to the east among the stars by about 1° per day and thus appears to trace out a great circle in the sky called the ecliptic. This great circle whose plane (the plane of the earth's orbit around the sun) is tilted $23\frac{1}{2}°$ to the plane of the earth's equator passes through twelve different constellations which the ancients called the signs of the zodiac. Because of the tilt of the earth's orbit with respect to the equatorial plane (also referred to as the tilt of the earth's axis of rotation) the annual motion of the earth around the sun produces the change of seasons on the earth.

As the earth moves eastward in its orbit, the sun appears to drift eastward along the ecliptic. Owing to this apparent motion of the sun along the ecliptic, the stars rise about 4 minutes earlier each night and different constellations are visible at the same hour at night time at different times of the year.

The seasons change from quarter year to quarter year because of the earth's motion in its orbit and the tilt of the earth's axis with respect to the plane of its orbit. On June 21 the north pole of the earth is tilted toward the sun and the direct rays (the vertical rays) from the sun strike the circle of latitude which is $23\frac{1}{2}°$ above the equator, and summer begins in the northern hemisphere. On September 21 the direct rays strike the equator and fall begins. On December 21 the south pole is tilted toward the sun and the direct rays strike the circle of latitude $23\frac{1}{2}°$ below the equator so that winter begins. On March 21 the direct rays again strike the equator and spring begins. The earth's axis of rotation is assumed to remain parallel to itself always pointing to the north celestial pole (which is pictured as a fixed point in the sky) as the earth revolves around the sun. But this is not exactly so because the earth's axis precesses slowly so that the north celestial pole shifts its position in the sky slowly, tracing out a circle as it does so.

Before leaving the earth's annual revolutionary motion, we mention one more direct observation of the stars that demonstrates this motion—the Doppler effect for stars that lie along a line (as seen from the earth), that coincides with the line along which the earth is moving at any one moment. Consider the stars toward which the earth is moving in June such as the stars in the constellation Aries. The spectral lines in the spectra of these stars are shifted, on the average, toward the blue end of the spectrum relative to their position when the earth is in its March 21 orbital position. On the other hand, the spectral lines of the stars in Scorpio and Sagittarius are shifted toward the red end of the spectrum. Six months later (on December 21) just the reverse is true. This occurrence shows us that the direction of the earth's motion is reversed every 6 months, resulting in a periodic approach toward and recession from the same star.

We have seen that three distinct stellar observations indicate that the earth is revolving around the sun: the annual parallactic shift of the stars; the aberration of light from stars whose direction is perpendicular to the direction of the earth's motion around the sun; and the annual variable Doppler shift in the spectral lines of stars lying along the line of the earth's motion. A very careful and precise analysis of these effects shows that the earth's orbit around the sun is not a perfect circle as Nicolaus Copernicus first suggested, but an ellipse, as was first proposed by Johannes Kepler. Note that the ecliptic, which is a great circle and is the apparent orbit of the sun among the stars, is not the orbit of the earth around the sun. It is the projection on the sky of the earth's orbit. We obtain a fairly good replica of the earth's orbit around the sun by observing the gradual change in the apparent size of the sun as the earth revolves around it. By the apparent size of the sun we mean the angle formed at the eye of the observer on the earth by two lines drawn from two diametrically opposite points on the sun's disk. This is about $\frac{1}{2}^\circ$ (about the same as the apparent size of the moon), but it changes from day to day because it depends on the distance from the earth to the sun, which changes over time. If the apparent size gets larger from day to day, the earth is approaching the sun; if it gets smaller, the earth is receding from the sun. One thus finds that the earth is closest to the sun in January and farthest from the sun in June, and by comparing the apparent size of the sun over the entire year one finds that the changes in the earth–sun distance are just those that correspond to an elliptical orbit.

We come now to the third component of the earth's motion to which we alluded when we noted previously that the north celestial pole does not remain fixed on the sky but drifts westward (opposite to the direction of the earth's revolution around the sun) very slowly. This phenomenon is called the precession of the equinoxes because, owing to it, the two diametrically opposite points on the sky where the celestial equator (the great circle that is the projection of the earth's equator on the sky) and the ecliptic cut, called the vernal and autumnal equinoxes, move westwardly. The vernal equinox is the point on the ecliptic where the sun appears to be on March 21 and the autumnal equinox is where it appears to be on September 21. These two points drift westward very slowly because of the slow precession of the earth's axis of rotation. This westward drift of the equinoxes was first discovered by Hipparchus in 125 B.C.; he noted that the tropical year (the year from the beginning of spring to the beginning of spring) is shorter than the sidereal year (the actual time it takes the earth to revolve around the sun once and return to its starting point, or the time it takes the sun to come back to the same star on the ecliptic) by about

20 minutes. Hipparchus explained this difference by saying that the vernal equinox moves westward by about 50″ of arc every year. This phenomenon was also known to the ancient Egyptians.

The earth is like a huge spinning top in space revolving very slowly (15°/hour) about an axis. If the earth were a perfect sphere, there would be no torque on it and the principle of the conservation of angular momentum would keep the earth's axis always pointing in one direction in space. But the earth is not perfectly spherical; instead it bulges at the equator. Owing to this nonsphericity of the earth, the sun and moon exert a torque on the earth by pulling on the bulging parts of the earth in a different direction and with a different magnitude (per unit mass) from that with which they pull on the center of the earth. This torque causes the earth's axis to precess once every 26,000 years which means that the north celestial pole shifts its position on the sky by about 47° every 13,000 years.

CHAPTER 9

The Solar System and the Motions of the Planets

In questions of science the authority of a thousand is not worth the humble reasoning of a single individual.

—GALILEO GALILEI

Having established the basic laws of Newtonian physics we can see how a structure such as our solar system can exist for billions of years. It is always something of a puzzle to people who do not understand the laws of mechanics how it is possible for the planets and the thousands of other bodies that belong to our solar system to continue to move in their well-defined and unaltered orbits year after year without catastrophic collisions. As we shall see, the solar system has a remarkable stability which can be deduced from the basic principles we have introduced in the previous chapters.

Before we discuss the planetary motions and the structure of the solar system, we will review briefly the basic laws and principles that guide us in solving the problem of the solar system. At the very top of these principles we must place Newton's second law of motion $\mathbf{F} = m\mathbf{a}$, from which all deductions must flow once we specify the nature of the force $\mathbf{F}$. We note first that this law is a general statement that relates any force, regardless of its nature, to the acceleration imparted to a body by this force. Second, we see that this equation is a vector equation, which means that it can be replaced by three independent scalar equations—one equation for each of the three mutually perpendicular directions of the coordinate system in which the motion of the body is to be described. This equation holds only in an inertial frame of reference, that is, a frame which is itself not accelerated.

We must now specify the nature of the force F that must be placed on the left-hand side of Newton's second law and which accounts for the orbits of the planets and the stability of the solar system. As we have seen, this expression is just Newton's law of gravity which we write as $F = GMm/r^2$, where M is the mass of the body (let us say the sun) that is pulling the body of mass m (let us say some planet) and keeping it moving in its orbit. The direction of $\mathbf{F}$ is along the line r from M to m and Newton's gravitational law of force is completely symmetrical between M and m. This symmetry means that the gravitational pull of M on m is exactly equal to but opposite in direction to the pull of m on M so that there is no net gravitational force on the system consisting of the two bodies M and m. Put differently, this statement means that the motion of the center of mass of M and m is in no way affected by the gravitational interactions of M and m.

Since Newton's law of motion and his law of gravity are the source of our understanding (within the limits of Newtonian concepts of space and time) of the gravitational structures in our universe, we place the two laws together and label them as Newton's classical gravitational theory: $F = ma$ and $F = GMm/r^2$. We need to introduce no relationships other than these to explain the general features of the motions of the planets and, indeed, Newton derived Kepler's three laws of planetary motion by a straightforward application of these equations to the pull of the sun on a planet. However, the general analysis of planetary motions (and the behavior of any group of objects in a gravitational field) can be greatly simplified by using the other general principles that we have deduced from Newton's laws. These are the three conservation principles that deal with the momentum, the energy, and the angular momentum of the interacting bodies, respectively.

The conservation of momentum is directly related to Newton's laws of motion when we are dealing only with the translational motions of bodies. This principle tells us that given a system of bodies moving about randomly, the total momentum of this system remains unaltered as long as no external force acts on this system. Associated with this system is a point (the center of mass) which moves through space with a velocity **v** such that the total mass of the system multiplied by the velocity is a constant and equal to the total momentum of the system. This conservation principle thus assigns a constant vector to the system of moving bodies that is conserved.

The conservation of angular momentum deals primarily with the rotational motions of rigid bodies (or a collection of bodies) and with the orbital motion of bodies. It assigns a vector to such a motion (the angular momentum vector) which is conserved as long as the body or system of bodies is not subject to an external torque.

Finally, the conservation of energy assigns a scalar quantity (the total energy: kinetic plus potential) to a system of moving bodies in a conservative force field (a field in which the work done on a body in taking it from one point to another is independent of the path along which the body is transported) which is conserved as long as no work is done on or by the system.

Although one can derive the orbits of the planets by a straightforward application of Newton's laws of motion and his law of force, the conservation principles greatly simplify the analysis. To see just how this happens we return to the motion of two masses M and m (sun and planet) interacting gravitationally, and set up the procedure for solving this problem if we were to do it without the aid of the conservation principles, which is probably the way Newton did it, although he certainly was aware of but had no exact concept of the conservation of energy and momentum. We introduce a nonaccelerated (inertial) coordinate system with the origin at some appropriate point and set down and then solve the equations of motion for each of these masses. These equations are just the X, Y, Z components of Newton's second law (which is a vector equation) and the quantities $\mathbf{F}_x$, $\mathbf{F}_y$, $\mathbf{F}_z$ on the left-hand side of these equations are the X, Y, Z components of Newton's gravitational force which we need not write down explicitly for our present discussion. But we must keep in mind that these three components depend on the separation between M and m and thus vary from moment to moment. The simplest way to do this calculation is to split the motion of each mass into three distinct motions, one along each of the three directions of our coordinates. We thus obtain three equations for each mass, or six equations for the two interacting masses m and M.

The problem one faces, then, in deriving the orbits of the two masses M and m by a straightforward application of Newton's law of motion, is to solve the six equations. However, these are not algebraic equations such as one meets in a course in algebra; they are what are known as differential equations, which leads to great complexities. To see why these complications arise, we briefly discuss differential equations in general to indicate how they differ from algebraic equations. Finally, we also discuss briefly why the equations of motion are differential equations.

To see the difference between the information given by an algebraic equation and a differential equation, we consider the very simple case of a particle that is moving in a straight line which has a slope of 45° (with respect to the x axis of our coordinate system) and cuts the x axis at the point $x = 2$. We can express this straight line algebraically as $y = x - 2$. Now this algebraic equation refers only to one line and therefore gives us very specific information about the motion of the particle. But if the motion of this particle were described by a differential equation, the only information that we could get from this differential equation would be that the particle is moving in some straight line; the differential equation would not specify any particular straight line. The solution of the differential equation would be expressed algebraically as $y = ax + b$, where a and b may be any two numbers which are unknown and cannot be known from the differential equation itself. From this result we see that when we solve a differential equation for the motion of a particle the actual orbit is not determined; all that we discover from this solution is the kind of orbit (e.g., straight line, parabola) that is involved but not which one this particular kind of particle is actually moving along. The reason for this indeterminacy is that a differential equation gives us information only about the way some particular quantity (the velocity or the position) associated with the motion of a particle changes from moment to moment. This limitation means that the solution of such a differential equation contains undetermined constants (such as a and b in the above equation for a straight line) and these constants can be specified only if some initial information about the motion of the particle is available. Actually, one must know the initial position and the initial momentum (or velocity) of the particle for each component of the motion of the particle if the orbit of the particle is to be found from the differential equations of motion. This means that six unknown constants (one initial position and one initial momentum for each coordinate axis) appear in the solution of the three equations of motion for a particle.

Here, where we are dealing with two interacting particles, the solution of the six equations of motion contain twelve constants. These constants are known as constants of integration because they are not present in the equations of motion themselves (the differential equations) but appear only after the differential equations have been "integrated" (solved).

Since these twelve constants may be treated as unknowns, we must seek some kind of relationships that permit us to specify them. These relationships are found in the conservation principles discussed previously. Each of these conservation principles permits us to eliminate some of the constants of integration and thus to reduce the complexity of the equations. Without going into all the details of the technical and mathematical aspects of the problem, we show below in a very general way how the conservation principles permit us first to replace the six equations of motion for the two particles by just two equations and then how a further application of the same principles enables us to solve these two remaining equations.

When we write down the six equations of motion of m and M we may use an arbitrary inertial coordinate system with its origin at some point O, which we assume to be at rest or moving with uniform speed in a straight line. However, if we use the principle of the conservation of momentum, we gain a great advantage by placing the origin of our coordinate system at a point determined by this conservation principle. We recall that the principle of conservation of momentum tells us that the total momentum of the two particles must remain unchanged in magnitude and direction because no net force acts on the two particles taken as a single system; the force of M on m is exactly equal and opposite to the force of m on M. Hence, the change in the momentum of M during any short interval of time must be opposite and equal in magnitude to the change in the momentum of m during that same time interval. Put differently, we may state that $m\mathbf{v} + M\mathbf{V}$ is constant at all times where $\mathbf{v}$ is the velocity of m and $\mathbf{V}$ is the velocity of M. This constancy means that some point c lying on the line connecting M and m (the center of mass) moves with a constant speed v' in a fixed direction. This point c divides the line connecting M and m into two parts which are always in the same ratio to each other regardless of where M and m may be in their orbits. If d_{m} is the distance of m from c and d_{M} is the distance of M from c, then $d_{\mathrm{m}}/d_{\mathrm{M}} = M/m$. The principle of the conservation of momentum thus tells us that the actual orbits of m and M about the point c are entirely independent of the uniform motion of c (i.e., of the total momentum of the system). We can now use this to simplify our problems as follows: Instead of placing the origin of our coordinate system at any arbitrary point, we place it (and keep it) at the point c. Since we are then always at the center of mass of our system, the total momentum of the system in our new coordinate system (with the origin at c) is zero, and we need no longer concern ourselves with the motion of c. The only thing that matters now is the way m and M move relative to c. In this way we eliminate three equations (corresponding to the three coordinates of the center of mass).

We can now show by a simple argument flowing from the conservation of momentum principle that the orbit of m about c is identical in shape to the orbit of M about c; but the orbit of m is smaller (or larger) than that of M by exactly the same factor that m is larger (or smaller) than M. Moreover, the speed of m, in its orbit relative to c at any moment, must be smaller (or larger) than that of M by this same factor.

To see that the orbit sizes relative to c are in the inverse ratio of the masses of the two particles we note that no matter where m and M may be in their orbits, the line connecting them must pass through c and the distances d_{m} and d_{M} of m and M from c must obey the equation $d_{\mathrm{m}}/d_{\mathrm{M}} = M/m$. If we are situated at c we always see m as many times farther away from us than M is, as m is smaller than M. To be specific, if m is ten times smaller than M, we always find m ten times farther away from us than we find M. This means that the orbit of m around c is ten times larger than the orbit of M around c.

We now consider the velocities of m and M as seen from c. Since the total system can have no momentum relative to an observer always stationed at c (he is moving along with the entire system) the momentum of m relative to c must be equal and opposite to the momentum of M relative to c. Thus, $mv_{\mathrm{c}} = -MV_{\mathrm{c}}$ where v_{c} and V_{c} are the velocities of m and M in the center of our mass coordinate system. From this it follows that v_{c} is always opposite to V_{c} and the magnitude of v_c is to the magnitude of V_c as M is to m. Thus, $v_c/V_c = M/m$. From the above analysis we see that by placing the origin of our coordinate system at the center of mass of the two particles we reduce the problem of the motion of the two bodies to the motion of a single body since the orbits of both bodies relative to the center of mass are completely similar so that if we discover one of the orbits we have them both. But since the motion of a single body can be described by just three equations (one of each of the three coordinates) we have reduced our original six equations of motion for the two bodies to three equations. This reduction shows us the usefulness of conservation principles in analyzing dynamical problems and simplifying such problems.

So far we have used only the principle of the conservation of momentum. We still have two other conservation principles to help us along in our analysis and we now use one of these (the conservation of angular momentum) to reduce the three remaining equations of motion to two equations. We can split the principle of the conservation of angular momentum into two parts. One part tells us that the direction of the angular momentum vector of an isolated system (a system, such as our two bodies, that experiences no external torque) must remain unchanged. The other part tells us that the magnitude of this angular momentum vector (under the same condition of no external torque) must remain unchanged.

We now apply the first part of this conservation principle to the motion of the two bodies, leaving the second part for later use. If we place ourselves at the center of mass of the two interacting bodies, we know from the principle of the conservation of momentum that at any moment the two velocity vectors of the two bodies must lie in the same plane because they must be opposite in direction. This plane then must be the angular momentum plane of the two bodies at that moment, and the angular momentum vector of the two bodies must be perpendicular to this plane at that moment. But we know that the direction of the angular momentum vector cannot change with time. Hence, the plane in which the two bodies are moving cannot change. This tells us at once that we are dealing with a two-dimensional problem (motion in a single plane) and that our three equations of motion can be reduced to two equations (one for each of the two coordinates in a plane).

Knowing from our conservation principles that we have only two equations, we can simplify our problem considerably by following the motion of only one of the particles as it moves around the center of mass under the gravitational attraction of the other. We may use any two-dimensional coordinate system we please for this purpose and we may place the origin wherever we please; e.g., either at the center of mass of the two particles or to coincide with one of the two particles. In the first case, we obtain two similar orbits (identical in shape but different in size) around the origin (the center of mass). In the second case, we obtain a single orbit of one particle around the other. This single orbit is called the relative orbit of the two particles. The relative orbit is identical in shape to that of the two individual orbits around the center of mass but its size is the sum of the sizes of the two orbits relative to the center of mass. We shall try to demonstrate these statements below by elementary means using simple algebra combined with the angular momentum and energy conservation principles. We shall see that the energy conservation principle determines the size of the orbit and the angular momentum conservation principle determines the shape of the orbit. Later we shall see how the results obtained here are related to and, indeed, are a restatement of Kepler's laws of planetary motion. Since the results in this section are derived from Newton's law of gravity, this proves the equivalence of Kepler's and Newton's laws.

We begin by considering first the simple case of circular orbits with our coordinate origin at the center of mass of the two particles. The only dimension that enters into this special circular case is the size of the orbit which then poses the following problem: What must the speed of particle m be (or of particle M) if it is moving in an orbit or radius r (or of radius $R = mr/M$ for particle M) about the center of mass of the two particles? We start with the two particles m and M separated by the distance $r + R$ and with the center of mass at the distance r from m and R from M. If we simply started with the particles at rest and then let them move, they would be accelerated toward each other along the line connecting them, and we would not have circular orbits. To obtain circular orbits and, in fact, any kind of orbits, we must give each particle a velocity that is transverse to the line connecting them. If we want the orbits to be circular, the direction of the velocity of m (call it v) and that of M (call it V) must be exactly perpendicular to the connecting line $r + R$. This condition exists because, as we know from our discussion of circular motion, the velocity of a particle at any point moving in a circle must always be perpendicular to the radius of the circle drawn to that point.

Consider the velocity of mass m moving in a circle of radius r about the center of mass c. We know the direction of the velocity v (perpendicular to r) but we still have to determine its magnitude. We call the magnitude of this constant circular velocity v_{cir} where the subscript stands for circular. To calculate v_{cir} we now apply Newton's two laws: $F = ma$ (law of motion) and $F = GMm/(r + R)^2$ (law of gravity) to the particle m. Note that the denominator in this formula for the force is not r^2 but $(r + \mathrm{R})^2$ since $r + R$ is the distance of M from m, where F is the force acting on M and a is m's acceleration. The acceleration of a particle moving with constant speed v_{cir} in a circle of radius r_c always points inwardly along the radius and equals v_{cir}^2/r_c. But from Newton's law of motion $a_{cir} = F/m$ and from his law of gravity $F/m = GM/(r_c + R_c)^2$, where we have used R_c in place of R to show that M is moving in a circle of radius R_c. Hence, a_{cir} must equal $GM/(r_c + R_c)^2$. But a_{cir} is also v_{cir}^2/r_c. This equation gives us $v_{cir} = [1/(1+R_c/r_c)]\sqrt{GM/r_c}$. This equation permits us to calculate the speed with which we must launch m exactly perpendicular to r_c if its orbit is to be exactly circular. We call this speed the circular speed of m relative to an observer at c. Of course, from what we have already said, we know that M also moves at exactly right angles to R_c with its own circular speed which is smaller than v_{cir} by the factor m/M. This procedure gives the exact speed relative to an observer at c with which a body of mass m at a distance $r_c + R_c$ [where $R_c = r_c(m/M)$] from a body of mass M must be thrown if it is to move in a circle of radius r_c about the center of mass of M and m. Moreover, the body must be launched exactly perpendicular to the line $r_c + R_c$ connecting m and M. Even the minutest departure from either of these two conditions results in a noncircular orbit. In any event, the circular speed for M about c is $V_{cir} = \sqrt{Gm/R_c(1+M/m)^2}$.

We notice that the smaller m is compared to M, the less important is the motion of M in this problem. If one of the bodies is the earth (mass M) and the other a satellite (mass m) being launched into circular orbit, M is so large compared to m that m/M is practically zero so that we may drop m/M in the previous expression for v_{cir} for the satellite and the expression for v_{cir} becomes $\sqrt{GM/r_c}$. The total energy of the two bodies m and M moving in circular orbits of radius r_c and R_c in the center of mass coordinate system can now be calculated. We recall that the energy of a body is not an absolute quantity but depends on the frame of reference in which the energy is calculated; one such frame in which the energy of a system of bodies is of special importance is that at the center of mass system.

To see why this statement is true we consider the relative orbit of two gravitationally interacting bodies, where the relative orbit is defined as the orbit of either one of the bodies as viewed from the other (the orbit of the earth or any one of the other planets around the sun when the observer is on the sun). This relative orbit has a definite size (its largest diameter through the sun) and a definite shape. We shall see later that the size of the orbit is determined by the total energy of the two bodies as measured in the center of mass system. The energy associted with the motion of the center of mass has no influence on the size of the orbit; whether the system as a whole is moving slowly or rapidly with respect to some outside point, the size of the relative orbit is the same. If we wish to understand how energy is related to the geometry of the relative orbit, we must deal with the energy in the center of mass frame of reference.

In the center of mass system the square of the speed of body m (for a circular orbit) is $v_{\text{cir}}^2 = GMr_c/(R_c+r_c)^2$. Hence, the kinetic energy of m in the center of mass system is $(\frac{1}{2})mv_{\text{cir}}^2 = (\frac{1}{2})GMmr_c/(R_c+r_c)^2$. The square of the speed of M in this center of mass system is $V_{\text{cir}}^2 = GmR_c/(R_c+r_c)^2$ and its kinetic energy is $(\frac{1}{2})MV_{\text{cir}}^2 = GmMR_c/2(R_c+r_c)^2$. The total kinetic energy is thus the sum $(\frac{1}{2})(mv_{\text{cir}}^2+MV_{\text{cir}}^2)$, or $(\frac{1}{2})GMm/(R_c+r_c)$. If we place $r_c + R_c = a$ and call this the size of the relative orbit, we see that the total kinetic energy, measured in the center of mass system, is $(\frac{1}{2})GMm/a$. On the other hand, the mutual potential energy of m and M is $-GMm/(R_c+r_c)$ or $-GMm/a$, that is, negative, so that the total energy (kinetic plus potential) which is just $(\frac{1}{2})GMm/a - GMm/a = -GMm/2a$ where we have placed $a = r_c + R_c$. This result is very important because it tells us that the total energy for a circular orbit is negative and the total energy (for a given M and m) of the system of two bodies is determined only by the size of the orbit a—the larger the value of a, the larger the energy. We shall see the full significance of these two statements later when we discuss the most general kinds of orbits.

One other point is of interest for circular orbits: the kinetic energy is half the numerical value of the potential energy. This point is important for the following reason: We consider a very massive body like the sun, with numerous other bodies of very small mass revolving around it in circular orbits. Since the bodies of small mass hardly affect each other gravitationally, these bodies continue to move in circular orbits so that the entire system does not alter its overall dimensions. It neither contracts nor expands. In other words, the system is certainly in dynamical equilibrium if all the orbits are circular. But this statement is equivalent to saying that the system is in dynamical equilibrium if the total kinetic energy (the sum of the kinetic energies of the individual particles) equals half the numerical value of the total potential energy of the system. This statement is a special case of a very general theorem known as the theorem of the virial which says that if any system of bodies is in gravitational equilibrium, regardless of the kinds of orbits they are moving in, the total kinetic energy of the system equals half the numerical value of the potential energy; circular orbits ensure that this condition is fulfilled, but the theorem is true no matter how the individual particles are moving. As long as the total kinetic energy numerically equals half the numerical value of the potential energy of the system, the size of the system remains constant.

Previously we discussed the circular orbits of two gravitationally interacting bodies about their common center of mass, but in studying the motions of the planets around the sun the orbits relative to the sun are important. For that reason we now consider the same problem as above but with the origin of our coordinate system at the particle M. The orbit is now a circle of radius $a = r_c + R_c$; and the velocity of m relative to M is the relative velocity V which is just the sum of the velocities v_{cir} and V_{cir} of m and M relative to the center of mass. Thus, $V = v_{cir} + V_{cir} = \sqrt{G(M+m)/(R_c+r_c)}$. The relative circular speed of a particle of mass m about a particle of mass M is the same as it would be about a fixed particle (a particle that is kept anchored to one spot) of mass $M + m$ and the radius of the orbit is just the distance between the two particles which we call a. We see then that if we want a particle of mass m to move in a circle of radius a (where a is the distance between the two particles at the moment under consideration) about a particle of mass M, when the two particles are interacting gravitationally, the particle m must be thrown exactly at right angles to the line connecting M and m with a speed (we call it the circular speed from now on) exactly equal to $\sqrt{G(M+m)/a}$. Because everything must be exact for a circular orbit, we see that the probability of achieving such an orbit by throwing a body at random is zero.

If a body of mass m is thrown exactly at right angles to the line connecting it with a body of mass M, it moves in a circular orbit about M only if its initial speed is the circular speed appropriate to its distance at that moment. If its distance happens to be r_1, the circular speed is $\sqrt{G(M+m)/r_1}$ as already discussed and any speed other than this one (even though the velocity is at right angles to r_1) results in a noncircular orbit. We can deduce the shape of the orbit from the initial speed which we assume to be v_1, when m is farthest from M. We still take the initial velocity to be exactly at right angles to r_1 at this point, but place no restriction on the speed which may have any value from zero to infinity. To deduce the shape of the orbit we must deduce the distance r_2—the distance of m from M when m is on the other side of M and close to it and thus again moving at right angles to the radius from M to m. If we take $v_1 = 0$, the particle m will simply move toward M and then collide with M and bounce back again. We must therefore take $v > 0$ if we wish to have any orbit at all. As long as $v > 0$, m moves in an orbit around M and $r_2 > 0$.

To deduce the shape of the orbit we now use the two basic conservation principles that apply to this system—the conservation of energy and the conservation of angular momentum (rotational motion). Actually, we apply these two principles to the motion of the particle m only since we take M as fixed at the origin of our coordinate system. Notice that the principle of the conservation of momentum does not apply to m because the force of gravity is constantly acting on it and is thus changing its momentum, but conservation energy and angular momentum do apply. The reason that the principle of conservation of energy applies even though the force of gravity does work on m is that the increase in kinetic energy of m that results from the work done on m by the gravitational force exactly equals the loss of potential energy of m so that the total energy is constant. Note that even though the gravitational force does work on m from moment to moment, the total work done on m during one complete revolution is zero. Conservation of angular momentum applies to the motion of m because the gravitational force acting on m passes through M and thus exerts no torque on m relative to M.

The principle of the conservation of energy applied to m tells us that its total energy E (kinetic plus potential) must remain the same at all points in its orbit. If it is at some point in its orbit at the distance r from M and its speed there is v, the total energy of m at that point is $(\frac{1}{2})mv^2 - GMm/r$. This quantity (the total energy of m) does not change as m moves in its orbit even though m's speed v and distance r from M do change continuously. By placing $v = v_1$ and $r = r_1$ in this expression we obtain the energy of m at r_1, and by placing $v = v_2$ and $r = r_2$ we obtain its energy at r_2. Since these two energies are equal, we equate these two expressions (conservation of energy) and obtain a relationship among the four quantities r_1, v_1, r_2, v_2. We now apply the principle of conservation of angular momentum, which tells us that the angular momentum mv_1r_1 at r_1 must equal the angular momentum mv_2r_2 at r_2 so that $mv_1r_1 = mv_2r_2$ or $v_1/v_2 = r_2/r_1$. We thus obtain a second relationship among the same four quantities; by combining these two relationships (actually by using one of them to eliminate v_2 from the other) we find how r_2 is determined by r_1 and v_1. The second relationship $v_1/v_2 = r_2/r_1$ already tells us something important about the motion of m (e.g., of a planet). Its speed at perihelion (when it is closest to the sun) which, as we shall see, may be either at r_2 or r_1, depending upon whether v_1 is smaller than or larger than the circular speed at r_1 given by $\sqrt{Gm/r_1}$, is as many times greater than its speed at aphelion (the most distant point of orbit from the sun) as its aphelion distance is greater than its perihelion distance. Thus, the speed of the earth in January when it is at perihelion (about 90,000,000 miles from the sun) is about 9.6/9 times (about 7%) greater than its speed in June when it is about 96,000,000 miles from the sun. Applying some elementary algebra to eliminate v_1 and v_2 from the expression for the total energy of the particle m we obtain $-GMm/(r_1+r_2) = -GMm/2a$ for the total energy where we have placed $2a = r_1 + r_2$. We see from this expression that the energy E is a constant for all points of the orbit since it depends only on the quantities G, M, m, and a that are the same for all points of the orbit. We also note that E is negative, which means that m stays attached to M so that the two bodies form a gravitationally bound structure. Finally, we note that if we go from a smaller to a larger orbit, a increases and the total energy of m increases so that it is less negative. Thus, the size of the orbit determines the total energy of m. The only way we can get m to move in a larger orbit is by giving it more energy (doing work on it). This relationship between energy and orbit size tells us why it is more costly to put a space vehicle into a large orbit around the earth than in a small orbit—more energy or fuel (which is a great part of the cost) is required for larger orbits than for smaller orbits.

So far we have spoken only about the size of the orbit of body m about M (which we assume to be fixed) but not about its shape and we saw that the size of the orbit determines the total energy of m in its orbit. We now consider an orbit of a given size (a given mean distance a of m from M) so that the total energy of the particle m is fixed (equal to $-GMm/2a$); what can we say about the motion? We know from the principle of conservation of angular momentum that the angular momentum of m in its orbit relative to M must be fixed. If we call $\mathbf{J}$ the orbital angular momentum of m, then $\mathbf{J}$ must be expressible in terms of unchanging orbital characteristics as is E. But if we keep a fixed, the only thing we can change about the orbit is its shape which may range from a circle to a narrow ellipse, which we define below.

Since the energy E is the same for all such orbits, a being the same, the only thing that can differ from one orbit to the next is the angular momentum $\mathbf{J}$. Thus, $\mathbf{J}$ must depend on the shape as well as on the size a of the orbit. We shall see that the angular momentum m is largest in a circular orbit of a given size as compared to all possible elliptical orbits of the same size.

Before we treat the angular momentum $\mathbf{J}$ of m we consider first just how the shape of the orbit of m depends on the speed v, with which m is projected into its orbit at the distance r_1 from M (note that the direction of v_1 is perpendicular to r_1). We saw that only one value of v gives a circular orbit, and the value of this v_{cir} is just $\sqrt{GM/r_1}$. If v_1 just equals this value, r_2 equals r_1 and the orbit is a circle. But if v_1 is less than or larger than this value, the orbit is noncircular since r_2 does not equal r_1. If v_1 is less than v_{cir}, r_2 is less than r_1 and r_1 is thus the aphelion point (if M is the sun) and r_2 is the perihelion point. Just the reverse is true if v_1 is larger than v_{cir} but less than $\sqrt{2GM/r_1}$. To see exactly how all this works out we deduce a relationship that expresses r_2 in terms of v_1 and v_{cir}; we do this algebraically by combining the expression for the energy and the angular momentum and obtain the equation $r_2 = r_1/[2(v_{cir}/v_1)^2 - 1]$. This is a very simple but quite instructive relationship because it permits us to determine r_2 as a fraction (or multiple) of r_1 if we know v_1. This fraction (or multiple) is entirely determined by the value of the denominator on the right-hand side of the above expression. We note that if $v_1 = v_{cir}$ the value of the denominator is just $2 - 1 = 1$ and $r_2 = r_1$, as it should be. If v_1 is less than v_{cir}, the denominator is larger than 1 and r_2 is less than r_1. If v_1 is larger than v_{cir}, the denominator is less than 1 and r_2 is larger than r_1. This expression simply verifies algebraically what we have already stated.

An interesting situation arises when v_1 is such that the denominator vanishes. This occurs when $2(v_{cir}/v_1)^2 = 1$ or $v_1{}^2 = 2v_{cir}^2$ or when $v_1 = \sqrt{2GM/r_1}$. If v_1 has this value, then the denominator in the expressions for r_2 is zero and r_2 becomes infinite (the body m never returns to its starting point). For that reason we call $\sqrt{2Gm/r_1}$ or $\sqrt{2}v_{cir}$ the speed of escape from the body M at the distance r_1. The orbit is then a parabola so that $\sqrt{2}v_{cir}$ is also called the parabolic speed. We return to this point later.

To see how the value of v_1 determines the shape of the orbit, we first define the shape more precisely than we have up to now. This is best done by comparing r_1 and r_2 with the size of the orbit (i.e., with $a = (r_1+r_2)/2$). The actual quantity we use is called the eccentricity e of the orbit and is defined as $e = (r_2-r_1)/(r_2+r_1)$. It is useful to have a different expression for e as well so from the above equation we obtain $r_1 = a(1-e)$ which gives e in terms of a and r_1. The reason e is called the eccentricity is that it is a measure of the extent to which the orbit departs from circularity, and hence, a measure of the shape of the orbit. If r_2 is close to r_1, e is small and the orbit is almost a circle. If $r_2 = r_1$, the eccentricity is exactly 0 and the orbit is perfectly circular. If r_2 is infinite (for a given finite r_1), e is 1 since the numerator and denominator in the expression that defines e are equal (they are both infinite). In this case, the orbit is a parabola, as we have noted above. When e is larger than 0 but less than 1, the orbit is finite (closed) but noncircular.

The angular momentum J of a particle in an orbit of given size $(r_1+r_2) = 2a$ depends on the shape of the orbit e. From the definition of e we obtain $e = (v_1/v_{cir})^2 - 1$. This equation immediately shows us how the speed v_1 with which m is projected at right angles to r_1 determines the eccentricity e and therefore the shape of the orbit. Note that this expression for e is correct as long as v_1 is larger than v_{cir}, so that r_2 is larger than r_1, but if v_2 is less than v_{cir} (so that $r_2 < r_1$) we must write $e = 1 - (v_1/v_{cir})^2$. We can now use this last expression or the first one to express J in terms of e and a. From the first expression we obtain $J = m\sqrt{GMr_1(1+e)}$. If we use the result, obtained previously, that $r_1 = a(1-e)$, we finally have $J = m\sqrt{GMa(1-e^2)}$. The smaller the value for e (the more circular the orbit), the larger is the value of J for a given value a (given size) of the orbit. Another way to look at this is to note that the angular momentum of a particle in any orbit is measured by the rate at which a line drawn from M to m sweeps out area as m describes its orbit. Since the time it takes m to complete its orbit once depends only on the size of the orbit (on a), the rate at which the area is described depends on the area enclosed by the orbit and hence on the shape of the orbit. Hence, the angular momentum is largest for a circle because the circle surrounds the biggest area so that more area is swept out per unit time in a circle than in a noncircular orbit.

We saw above that the motion of a planet in the gravitational field of the sun (if we neglect the gravitational actions of the other planets and the motion of the sun) is determined entirely by two dynamical quantities: the total energy E of the planet and its angular momentum J relative to the sun, both of which quantities are conserved in the sense that they are the same at all points of the orbit of the planet. We saw further that each of these two quantities is related to a geometrical property of the orbit: the total energy is determined by the size a of the orbit and the angular momentum (for a given size orbit) is determined by the eccentricity e (the shape) of the orbit. As we have seen, these two results were derived entirely from Newton's law of motion and his law of gravity. We shall now see that they are equivalent to (but more general than) Kepler's three laws of planetary motion which, as we have already stated, Kepler obtained empirically after many years of ingenious guesswork, wild speculations, and very long and tedious arithmetic calculations. That we can derive the same results in a matter of minutes by applying simple algebra to Newton's laws shows the tremendous power of theory and how superior it is to mere fact gathering and to empiricism in general. Moreover, the results obtained from theory are general and applicable to any two or, for that matter, to any number of gravitationally bound bodies and not only to the motion of a planet around the sun. The superiority of Newton's laws over Kepler's laws manifests itself in one more way. Kepler's third law of planetary motion (as Kepler stated it) is not quite correct, but the results obtained with it (when applied to the motions of the planets) differ from those obtained by applying the correct law by so small an amount that Kepler's data were not accurate enough to reveal this discrepancy. But the correct third law of planetary motion is obtained directly from Newton's laws so that the discrepancy in Kepler's law is revealed immediately even though it is very small for the motions of the planets.

Newton's laws can predict very tiny effects whereas Kepler's empirical procedure cannot because general laws such as those formulated by Newton lead to all possible consequences (however small) of a given set of conditions whereas an empirical procedure is limited by the accuracy of one's observational techniques and instruments. This limitation makes it difficult to test the full implications of the theory because it may predict results which are far too small to be measured. Another result which stems from Newton's laws but cannot be deduced from Kepler's laws directly is that not only ellipses but a whole class of orbits called conic sections are available to m moving in the gravitational field of M. The particular orbit from this class that m actually follows is determined by the speed v_1 given to m when it is launched at r_1.

To see how the results that we have deduced up to this point are related to Kepler's three laws of planetary motion we first state these laws as Kepler did when he discovered them: (1) Each planet moves around the sun in an ellipse one focus of which (there are two foci in an ellipse) is occupied by the sun; (2) the radius vector drawn from the sun to a planet sweeps out equal areas in equal times; and (3) the squares of the periods of any two planets stand in the same ratio to each other as the cubes of their mean distances from the sun. Stated algebraically this "harmonic" law says that if P_1 is the period of one planet (the time for one complete revolution of the planet around the sun) and if P_2 is the period of any other planet, then $P_1^2/P_2^2 = a_1^3/a_2^3$, where a_1 and a_2 are the mean distances of the two planets, respectively, from the sun.

Kepler called this third law the harmonic law because to him it represented the great "harmony of the world" and fulfilled his expectation that the "ultimate secret of the universe" would be found in an "all embracing synthesis of geometry, music, astrology, astronomy and epistemology." What an all-embracing concept and glorious vision it was in itself, and what a sense of exaltation it must have inspired to lead him to his final vision of the truth! To him this ''handiwork of his creator'' is, indeed, the ''music of the spheres'' for, as he stated, ''the heavenly motions are nothing but a continuous song for several voices.'' The final discovery of the third law drew from Kepler this ecstatic paean:

> The thing which dawned on me twenty-five years ago before I had yet discovered the five regular bodies between the heavenly orbits . . . which sixteen years ago I proclaimed as the ultimate aim of all research; which caused me to devote the best years of my life to astronomical studies, to join Tycho Brahe and to choose Prague as my residence—that I have, with the aid of God, who set my enthusiasm on fire and stirred in me an irrepressible desire, who kept my life and intelligence alert, and also provided me with the remaining necessities through the generosity of two emperors and the estates of my land, Upper Austria—that I have now, after discharging my astronomical duties ad satietatum, at long last brought to light. . . . Having perceived the first glimmer of dawn eighteen months ago, the light of day three months ago, but only a few days ago the plain sun of a most wonderful vision—nothing shall now hold me back. Yes, I give myself up to holy raving. I mockingly defy all mortals with this open confession: I have robbed the golden vessels of the Egyptians to make out of them a tabernacle for my God, far from the frontiers of Egypt. If you forgive me, I shall rejoice. If you are angry, I shall bear it. Behold, I have cast the dice, and I am writing a book either for my contemporaries, or for posterity. It is all the same to me. It may wait a hundred years for a reader, since God has also waited six thousand years for a witness. . . .

A comparison of Kepler's three laws with our algebraic deductions above shows us that we have, indeed, deduced these three laws from the two general laws of Newton—his law of motion and his law of gravity. These laws are more general than those of Kepler because, whereas the latter refer to planets, Newton's laws refer to the motion of any bodies acted upon by a central gravitational force. We can now see just how much more general and profound Newton's laws are than those of Kepler by comparing each of Kepler's laws with our results above.

As far as Kepler's first law is concerned, we have demonstrated that the orbit of a planet is characterized by two numbers: half the largest diameter a (the size of the orbit) which we now call the semimajor axis; and the shape e of the orbit which we call its eccentricity. The orbit is a symmetrical curve such that a is the mean distance of the planet from the sun, which is not at the center of the orbit but off to one side on the same major axis at a point called the focus of the ellipse. All of this information is contained in the previous paragraphs showing that one can, indeed, deduce Kepler's first law from Newton's laws by simple algebra. But we can demonstrate that Newton's laws give us a more general first law than Kepler's laws by predicting not only elliptical orbits but also circles, parabolas, and hyperbolas all of which, together with ellipses, constitute a family of curves called conic sections.

The equation for the orbit of a planet which can be derived directly from Newton's laws represents a whole class of curves (orbits) called conic sections so that according to Newton's laws the ellipse is but one of a number of orbits that a particle m can have in the gravitational field of another body. To show all the possible orbits we consider two cones formed by rotating the two intersecting lines AB and CD about their angle bisector EF. If we cut through either of these cones we obtain a curve called a conic section whose shape depends on the angle that the plane of the cut makes with the angle bisector.

When the cut is exactly at right angles to the angle bisector, we obtain a circle. This result shows us that the probability of getting a circle when we take a random cut of a cone is zero because only one angle (a right angle) out of a possible infinitude of angles gives a circle. This situation is to be compared with the exact conditions that must be imposed on the particle m if it is to be projected into a circular orbit: it must be projected exactly at right angles to r_1 and with exactly the circular speed $\sqrt{GM/r_1}$. Hence, the probability for a circular orbit for a planet is zero since there is an infinitude of angles and speeds of projection. As a result, no circular planetary orbits exist. Indeed, it is impossible to demonstrate that an orbit is a perfect circle since to do so requires infinite precision of measurement.

If the plane of the cut is not exactly perpendicular to the angle bisector, the curve is an ellipse as long as the plane of the cut is not parallel to any line like AB on the surface of the cone that passes through the vertex (the point) of the cone. We can thus get an infinite family of ellipses whose eccentricities lie between 0 and 1. If the plane of the cut is parallel to a line like AB, on the surface of the cone we obtain a parabola—a curve that never closes but goes off to infinity where the two branches ultimately become parallel to each other and to the line AB. Since a parabola can be obtained only if the plane of the cut is exactly parallel to AB, the chance of getting a parabola when we take a random cut out of the cone is zero. This is to be compared with the unique dynamical condition that must be fulfilled for the particle m to move in a parabolic orbit; it must move with exactly one and only one speed when it is projected into orbit—the parabolic speed (the speed of escape) at the point at which it is projected. The direction of its velocity at that point does not matter; it moves in a parabola as long as its speed at the distance r from M is $\sqrt{2GM/r}$.

If the cone is cut at a still greater angle, both the bottom and top cones are intersected by the cutting plane and a curve with two symmetrical branches is obtained. Such a two-branched curve is called a hyperbola. A random cut of a cone gives either an ellipse or a hyperbola but never a circle or a parabola. A similar conclusion can be drawn about the orbit of bodies projected at random at some point in the gravitational field of the sun. The orbits are either ellipses or hyperbolas, never circles or parabolas. Bodies with speeds less than $\sqrt{2GM/r}$ (the parabolic speed) travel in ellipses; all others (speeds greater than parabolic) travel in hyperbolas. From what we have just said, we see that Newton's law of gravity combined with his law of motion gives us a more general picture of possible planetary orbits than does Kepler's first law.

We now examine Kepler's second law more carefully and see how it is related to our discussion of the general properties of conic sections and the conservation principles. We have already noted that the angular momentum of m in its orbit relative to M is mvx where x is the perpendicular distance from the sun to the planet's velocity direction. But $vx/2$ is just the area of the triangle of which v is the base and x is the altitude. This region is also the area swept out (in a unit time) by a line from M to m (the radius vector from the sun to the planet). Kepler's second law therefore states that this area is a constant since it is the area swept out in each unit interval of time by the radius vector. But if this area $vx/2$ is a constant, so too is $(2m)(vx/2)$ which is just mvx the angular momentum of m relative to M. We see that Kepler's second law of planetary motion is equivalent to the statement of the conservation of angular momentum. If we call $vx/2$ the areal velocity $\dot{A}$ of the planetary motion (which we may, since it is the area swept out in a unit of time; the dot over the $\dot{A}$ represents rate), and J its orbital angular momentum, then $J = 2m\dot{A}$.

The angular momentum of a planet in its orbit is twice its mass times its areal velocity, which is a very useful result in astronomy. Since the period P of the planet in its orbit is the time it takes the planet to traverse its orbit once, then P is also the time it takes the radius vector from the sun to the planet to sweep out the complete area A bounded by the orbit of the planet. Hence, A/P is just the areal velocity A of the planet and we have $J = 2mA/P$.

If we know P at this point, we can use the expression we previously derived from J to obtain an expression for A, and this expression then tells us the shape of the orbit since the area and shape of a conic section are related. On the other hand, if we know A (its mathematical form), we can introduce it into the above expression and from this expression determine the period P.

We saw previously that the period P of a planet, the area A of its orbit, and its angular momentum J are related. Since we know J in terms of the size a and the eccentricity e of the orbit, we can express A in terms of a and e if we know P or we can express P in terms of a and e if we know A. In any case, we see that Kepler's third law is involved in this relationship since we are dealing with the period of the planet which is the subject of this law. To express Kepler's third law algebraically, we note that it tells us that the square of a planet's period P divided by the cube of its mean distance a from the sun is the same constant for all planets. Expressed algebraically, $P^2/a^3 =$ constant.

We can deduce this law directly from Newton's law of gravity (plus his law of motion) in a more general form than Kepler expressed it and we know that when we do this, the nature of the "constant" on the right-hand side is revealed. In fact, we see that Kepler's third law, in the form that Kepler stated it, is not quite correct because P^2/a^3 is not quite constant as one goes from planet to planet. It varies slightly, but by so small an amount, that it was not revealed in Tycho Brahe's observational data which Kepler used to establish his laws.

First, we derive Kepler's third law for circular orbits, which we can do very easily since we already have an expression for the circular speed of a body. In fact, we saw that $v_{\text{cir}} = \sqrt{GM/a}$ where a is the radius of the orbit. If we divide the circumference $2\pi a$ of the orbit by v_{cir} we obtain the period (the time for one revolution). Hence, $P = 2\pi a/\sqrt{GM/a}$. If we now square both sides, we obtain $P^2 = (4\pi^2/GM)/a^3$ and thus $P^2/a^3 = 4\pi^2/GM$, and this expression is just Kepler's third law since the right-hand side $4\pi^2/GM$ of this equation is the same for every planet; it consists of a pure number $4\pi^2$, the universal constant G, and the mass of the sun M. We have thus derived Kepler's third law from Newton's law.

But a careful analysis of the circular orbits shows that what we have just obtained applies only if the body M does not move; this would be so only if M were infinite. The sun is so much more massive than any one of the planets (it contains more than 99% of the entire mass of the solar system and is about 340,000 times more massive than the earth) that it is practically stationary so that Kepler's third law, as stated above, is quite accurate, but it is not exactly true. Since the sun's mass is not infinite, the sun does move owing to the gravitational pull of each planet on it, and these forces must be taken into account in calculating the period of a planet. On doing this we find that P^2/a^3 is not constant as we go from planet to planet because the denominator M on the right side of Kepler's law must be replaced by $(M+m)$ where m is the mass of the planet being considered. The reason for this procedure is that the equation that expresses Kepler's third law must involve M and m in exactly the same way since the pull of M on m exactly equals the pull of m on M, so that the formula for P must not distinguish between them. We thus obtain the correct third law of planetary motion in the form of $P^2/a^3 = 4\pi^2/G(M+m)$ and we see that the right-hand side of this formula varies slightly as we go from planet to planet because m varies since each planet's mass is different.

So far we have derived Kepler's third law for circular orbits only. Can we show that it holds for elliptical orbits? The answer is yes and we need only use the relationships $J = 2(mA/P)$ and $J = m\sqrt{GMa(1-e^2)}$ which we obtained previously for the angular momentum J. If we now equate the two right-hand sides (since they both equal J) and introduce for A the formula for the area of an ellipse $\pi a^2\sqrt{(1-e^2)}$, we obtain (with a little algebra) $P^2 = 4\pi^2 a^3/GM$ which is just Kepler's third law. This law shows us that the period of the body m depends only on the size a of its orbit and not on its shape e. The periods of all bodies of mass m moving in orbits of different shapes but the same size are equal.

We could reverse our procedure and deduce from the equation $J = 2(mA/P)$ and from the expressions for J and P that A must equal $\pi a^2\sqrt{(1-e^2)}$, showing that the orbits of the planets must be ellipses. Taking into account all that we have said above, we see that Newton's laws are more basic than Kepler's laws and thus give us a much deeper insight into the dynamics of a gravitationally bound structure like the solar system than do Kepler's laws.

Before leaving the two-body gravitational system we note the important role that the total energy of the system plays in determining its dynamical properties. Only if E (the total energy) is negative can such a two-body system form a bound gravitational structure. This condition means that if two bodies (like the sun and a planet) approach each other along a parabolic or hyperbolic path, they can never form a bound system (i.e., remain together with the planetlike body moving in an elliptical orbit around the massive body) as long as some of the mechanical energy is not dissipated in some manner or other. The more energy that is dissipated (enough must be dissipated in any case to make the total energy negative), the more tightly bound and compact the final two-body structure will be.

We saw previously that the force of gravity accounts for the dynamical binding of a given body of mass m to another one of mass M, so that a two-body structure is formed and one can predict the orbits of each of these bodies around the center of mass of the system. But the solar system is a collection of many bodies gravitationally bound to a central massive body, the sun. If all the bodies in the solar system were of about the same mass, the orbits of the bodies would be extremely complicated and one could not express them in simple mathematical forms (such as conic sections). Even if we had only three gravitationally interacting bodies to deal with, the general solution would be extremely complex because of the many different possible orbits that would be available to these bodies. One can therefore easily imagine how much more complex the solar system problem would be if all the bodies were about equally massive so that each one responded with the same acceleration to the forces acting on it. Fortunately, this state of affairs does not exist; the sun is so much more massive than the other bodies in the solar system that it dominates this structure and determines its overall dynamical properties. Owing to this gravitational dominance of the sun we may, to a first approximation, disregard the gravitational influence of the planets on each other and consider the motion (the orbit) of any one planet as being determined entirely by the sun. This assumption gives us the general, overall gravitational model of the solar system with each planet moving in its own elliptical orbit as though no other planet were present and we thus arrive at Kepler's three laws of planetary motion as the first approximation.

To obtain a more accurate picture of the motion and orbit of any one of the planets, one must calculate the gravitational action of all the other planets as well as that of the sun on this planet; this calculation can only be done by complex numerical methods. Fortunately, electronic computers have made such numerical calculations manageable and we can now check Newton's law of gravity to a very high degree of accuracy by comparing such calculations with the observed orbits of the planets. One then finds that the deductions from Newton's law agree with the observations amazingly well, but some discrepancies between theory and observation exist which cannot be accounted for by observational errors. The largest and most important of these discrepancies is present in a minor component of the motion of the planet Mercury around the sun which would be a closed ellipse if no other planets were in the solar system. But the actual orbit is not a closed curve but rather a figure that can be obtained by picturing Mercury as moving in an ellipse that is itself slowly rotating in the same direction as Mercury is moving. This motion (the rotation of the elliptical orbit) is called the advance of the perihelion of Mercury because the major axis of Mercury's ellipse revolves so that the perihelion, the point closest to the sun, revolves around the sun in the same direction as Mercury does. Most of this advance of the perihelion per century can be accounted for by Newtonian gravitational theory if the gravitational perturbing forces of all the other planets on Mercury are calculated. However, a small but significant amount still remains to be explained which indicates that the Newtonian theory of gravitation, though very good, cannot be entirely correct. One can see how good Newtonian theory is by noting that the disagreement between the observed motion of the perihelion of Mercury and the predicted motion (using Newtonian gravitational theory) is only 43″ of arc per century. But even though this is a very small quantity, it is a real discrepancy and hence enough to show that Newton's law of gravity is inadequate and must be replaced by a more general and more accurate law. This herculean task was completed in 1915 by Albert Einstein when he introduced his general theory of relativity which we discuss in Chapter 11.

CHAPTER 10

The Special Theory of Relativity

The views of space and time which I wish to lay before you have sprung from the soil of experimental physics, and therein lies their strength. They are radical. Henceforth space by itself and time by itself, are doomed to fade away into mere shadows, and only a kind of union of the two will preserve an independent reality.

—HERMANN MINKOWSKI, *Space and Time*

No single intellectual creation has so intrigued, excited, and challenged people in general as Albert Einstein's theory of relativity, surrounded as it is in the minds of most people by an aura of incomprehensibility, mathematical complexity, and strange but wonderful physical predictions that (it is generally believed) must be accepted on faith by the layman. That the theory, in spite of this almost mystical reaction to it, is still the one topic that most people first think of when modern science is discussed is an indication of the intellectual hold it has on people, so many years after its publication in 1905 when Einstein was an unknown young man working as a clerk and an examiner in the Swiss patent office. Since this theory is considered by many physicists to be the single greatest and most beautiful creation of the human mind, there is good reason for its preeminence among scientific theories in the mind of the layman, whose thinking about science is necessarily influenced and guided by the professional scientist.

But even if the theory of relativity were not the most publicized of all physical theories (the volume of literature about this theory in scientific, philosophical, and literary works far exceeds that about any other physical theory, and the name Einstein appears much more often in such writings and bibliographies than that of any other scientist), it would occupy a special place in humanity's thinking because of the nature of the theory of relativity itself. Whereas theories in general (other than the theory of relativity) deal with specific aspects of properties of the universe (space, time, and matter), the theory of relativity goes beyond that narrow scope. It does, indeed, give us a deeper insight into the nature of space, time, matter, and energy than any other physical theory, but it has the additional role of being the master theory of physics. We may refer to it as a theory of theories, for like a master template in the machine tool industry, which is the standard against which the precision tools are measured, relativity is a gauge of all physical theories in the sense that such theories, to be correct, must fulfill certain basic conditions, called relativistic invariance, set forth by the theory of relativity.

Before presenting the basic features of the theory of relativity, we contrast Einstein's usage of the word "relativity" and its common usage. As it is commonly used, the word "relative" means that two people observing the same "thing" or "phenomenon" see it differently owing to differences in their positions and perspectives with respect to the phenomenon and also owing to their different physiological, psychological, and philosophical makeup. The same tree appears much larger to a person standing a few feet from it than to one a few miles from it, and a nearby lamp appears brighter than an identical lamp far away. The same person ascribes different colors to a given object depending on the color of the light that illuminates it. As an example of the way our physiology affects the appearance of things, we note that the color of an object is purely subjective because it is a property of the observer's retina. If the rods and cones in the retinas of the eyes of two observers are not about the same (let us say one of the observers is color-blind), the two observers assign different colors to the same objects. We can go on giving many such examples of what is commonly understood by "relativity," but if this were all there is to the deductions of the true nature of phenomena from their descriptions by different observers, there would have been no need for a theory of relativity outside the accepted laws of geometry, perspective, and physiology as they were known before Einstein. But the relativity concept as it is used in the phrase "the theory of relativity" and as Einstein defined it refers to the description of events in the universe and the formulation of the laws of nature as deduced from these descriptions by observers who are moving relative to each other. The important thing here is the relative motion or velocity of the observers with respect to each other and not their positions or their physiological, psychological, or philosophical states.

To make the distinction between the classical (pre-Einsteinian) and the Einsteinian points of view as clear as possible, we again consider a physical measurement such as the height of a tree from the classical point of view, as measured by any two observers (regardless of how they may be moving) who, at some moment, are at different distances from the tree. Each observer, at the given moment, measures the angle which the tree subtends at his eye at that moment (the angle made by a line from the top of the tree to his eye with a line from the bottom of the tree to his eye). This angle is smaller for the distant observer than for the closer observer so that the tree appears larger for the latter than for the former. The tree's visual angle (the quantity that is actually measured by the observers) is thus relative. But classical science also ascribes an absolute physical quantity to the tree; namely its height, which is the number each observer obtains by multiplying the tangent of his visual angle by his distance from the foot of the tree. This number in classical science corresponds to the number one obtains by laying off a ruler from the foot of the tree to its top. This statement is nothing more than a statement of the validity of the theorem of Pythagoras for two right triangles with a common side (the tree).

As long as we accept the standard three-dimensional Euclidean geometry for the two observers, we must accept this relationship of the absolute height of the tree to the apparent (or relative) visual angle. Einstein's special theory of relativity does not deny this relationship as long as the two observers are at rest with respect to the tree. But it denies the concept of such an absolute height for moving observers and hence denies the validity of three-dimensional Euclidean geometry for such observers.

As another example of the difference between the classical and the Einsteinian concept of relativity, we consider the mass and the weight of a body. The weight of a body on the earth is different from its weight on the moon, but in classical physics its mass is absolutely the same regardless of where it is or how it is moving. This constancy of mass is not absolute in relativity physics. As we shall see, the mass of a body, as measured by different observers, depends on the state of motion of the observers relative to the body.

The general theory goes beyond the special theory in that it shows that such things as distances (lengths), time intervals, and masses associated with events being observed depend not only on the state of motion of the observer with respect to those events, but also on the distribution of masses (i.e., on the intensity of gravitational fields) in the neighborhood of the events.

Before leaving this general discussion of the meaning of the word "relativity," we caution the reader about thinking (erroneously) that the theory of relativity banishes all absolutes from the laws of nature. This is not true at all because the theory itself is really a theory of absolutes and is the most powerful intellectual tool that is available to humanity for separating the absolute entities in nature (which do, indeed, exist) from the relative ones. We shall see that the laws of nature are themselves absolute statements or absolute relationships (among measured quantities which themselves can be relative) that are the same for all observers regardless of how they are moving. Einstein arrived at his theory of relativity (or absolutes) by a profound analysis of the relationship of the temporal and spatial properties of the frame of reference of an observer to the events being observed and of how the measurement of the physical quantities associated with the event depends on the state of motion of the observer. To see why such a theory is necessary and why it departs so drastically from classical physics, we first consider the relativity and absolutism of Newtonian theory.

It is commonly believed that the concept of relativity was first introduced into science by Einstein, but this is not so since Newtonian theory also deals with relative and absolute concepts and sets up a logical criterion for distinguishing between these concepts. In Newtonian physics—like Einsteinian physics—the laws of nature are absolute relationships which are assumed to hold for all observers. The important (and crucial) difference between Newtonian and Einsteinian relativity lies in the criterion that is established for deducing or constructing the absolute quantities in nature from the observed or measured quantities. Since measurements are made by observers, we must analyze the role of the observer with great care if we are to see why the Newtonian absolute concepts had to be replaced by the Einsteinian concepts. Indeed, we must consider at least two different observers if we are to study the relationship between relative and absolute concepts since only by comparing the descriptions given by two different observers of the same events in nature can we pick out those features of the events that are absolute (the same for all observers) and hence are natural laws. We may in fact say that a law of nature is an absolute statement about events.

Since we are discussing events and the comparison of the descriptions of these events by different observers, we first define an "event" and the phrase "different observers" precisely. As commonly understood, an event is any kind of happening, but that is too general a definition for our purposes. Hence, we limit ourselves to the simplest kind of a physical event since all events in nature (however complicated) can be reduced to a series of such simple events and the laws of nature can be shown to be statements about such simple events. By a simple event we mean the presence of a particle at a particular spatial position at a particular time. The correlation of a series of such events by an observer leads to a law of nature. Thus, Newton's law of motion $\mathbf{F} = m\mathbf{a}$ is obtained by studying a particle as it moves from point to point describing an orbit (a series of simple events).

By the phrase "different observers" we mean observers who are moving with respect to each other in some manner. From this point of view all observers who are at rest with respect to each other but in different locations are identical. This definition of "different observers" is in line with the concept of relativity introduced above—relativity with respect to the states of motion of observers and not with respect to their particular positions. Our concern, then, in what follows, is with two observers who are moving with respect to each other. But to say that the two observers are moving with respect to each other is too broad and general a statement at this stage of our analysis. So we limit ourselves here to a very special kind of relative motion—uniform motion in a straight line—and call all observers moving with constant velocities with respect to each other (the motion of each observer is unaccelerated, but the speed and direction of motion may differ from observer to observer) inertial observers. Two inertial observers are the same if they are moving with the same speed in the same direction; otherwise they are different. The special or restricted theory of relativity (which was the first part of the theory of relativity to appear as proposed by Einstein in 1905) deals only with inertial observers; hence its designation as "special" or "restricted"—it deals with a "special" or "restricted" group of observers, namely those moving in inertial frames of reference. In 1915 Einstein generalized the theory of relativity (hence the name the general theory of relativity) by dropping all restrictions on the motions of the observers and thus considering the descriptions of events (actually comparing such descriptions) by observers moving in any arbitrary fashion. We discuss the general theory in the next chapter and study here the special theory as introduced by Einstein in 1905. But in this chapter we first consider the Newtonian relativity theory for inertial observers and then show why it breaks down and must be replaced by Einsteinian relativity.

The reason we restrict ourselves here to inertial observers is that we want to consider the laws of nature in their simplest form which we define as those statements about groups of events which are absolute in the sense that they are the same for all observers. A basic law in its simplest form is Newton's law of motion $\mathbf{F} = m\mathbf{a}$, and we know that this law does not apply in this simple form to objects and observers in accelerated frames. In accelerated frames of reference (coordinate systems), bodies left to themselves do not obey Newton's first law but move in complicated ways (rather than remaining at rest or moving in straight lines with constant speed) as though forces were acting on them although no sources of such forces are present. In fact, an inertial frame is defined as one in which a body left to itself obeys Newton's first law. One way, then, of determining whether or not we are in an inertial frame is to observe bodies that are free to move and see whether they are at rest or are moving with constant velocity when no force acts on them. In either case, we are in an inertial frame of reference. We see that the earth is not an inertial frame because the distant stars (which are freely moving bodies, or very nearly so, since they are so far away from one another that they exert hardly any gravitational pull on each other) do not appear to stand still, as they should if the earth's motion were entirely uniform translation (no acceleration or rotation). We may, in fact, define an inertial frame as one relative to which the distant stars appear to be fixed. If the earth were not spinning, it would almost be an inertial frame and we could then ascribe any departure from straight-line uniform motion of a freely moving body to the earth's gravitational field. It would still not be a perfect inertial frame because of its accelerated motion (its acceleration is quite small) around the sun.

We neglect here the rotation of the earth, its acceleration in its orbit around the sun, and its gravitational field and consider two different inertial observers on it. We take one of the observers to be fixed on the earth at the point O (on a railroad track) and the other to be at a second point O′ in a train that is moving to the right with constant speed v. Since our concern now is to discover the laws of nature by finding those intrinsic features of events which are the same for both observers, we ask each observer to describe the same set of events and we then compare these descriptions. To see what is involved in the description of any set of events it is sufficient to consider a single simple event. Since a simple event is just the presence of a particle at a given point of space at a given time, an observer describes this event properly by stating where the particle is relative to him and when it is at that specified point. Here we assume the particle to be pointlike so that we may picture it as occupying some particular point of space at a given moment. It is clear from this specification of the "description" of this "simple event" that each observer can best perform his task by introducing a coordinate system. The description, then, is given by four numbers: the three spatial coordinates of the point in the observer's coordinate system and the time (as given by a clock at rest in the observer's coordinate system) of the particle's coincidence with the point defined by the coordinates.

We assign a rectangular Cartesian coordinate system to each observer with the axes of the two coordinate systems parallel. The origin of the coordinate system of the fixed observer is to coincide with the position O of the fixed observer, and the origin of the coordinate system of the moving observer is to coincide at all times with the position O′ of the moving observer. We thus have a fixed and a moving coordinate system, with the axes of the fixed coordinate system labeled X, Y, Z and those of the moving system X′, Y′, Z′. The X and X′ axes coincide and are parallel to the track along which the train is moving, the Y axis is vertical, and the Z axis is horizontal and perpendicular to Y and X. Finally, the two observers are to have identical measuring rods and identical clocks, i.e., if the two sets of clocks and rods were at rest with respect to each other, they would be identical. The rod and the clock of the moving observer are to be moving right along with him.

We now arrange things so that both the fixed clock and the moving clock (which we may take to be stopwatches) are set going at the precise instant (the zero moment) that the two observers coincide on the railroad track. Moreover, let the event we are considering occur at a time t later than this coincidence of the two observers as measured by the fixed clock. At this later moment t, let x, y, z be the three spatial coordinates of the event (a particle's position) as measured by the fixed observer O and let x', y', z' be the spatial coordinates of this same event as measured by the moving observer in his co-moving coordinate system. Moreover, let t' be the reading on the clock of the moving observer when the event occurs.

At this stage of our analysis we say nothing about the equality or inequality of t and t'. We simply consider the readings on the two clocks as two bits of information given to us by two different observers and use the primed and unprimed designations merely as a means of differentiating the sources of these bits of information. We thus reduce the description of a simple event to the specification of four numbers (which are obtained by measurements) by each observer. The description of the event by the fixed observer O is given by the four numbers x, y, z, t while the description by the moving observer O′ is given by x', y', z', t'. Keep in mind that the words "fixed" and "moving" here are not meant in any absolute sense but only relative to the earth. We use the words "fixed" and "moving" simply to differentiate between the two observers, noting here that the relative motion of the two observers is the essential thing and that either one may be called fixed.

To discover the law of motion of the particle, each observer studies a whole series of these simple particle events, which we call an orbit of the particle, and deduces the law from this orbit. Our criterion for deciding whether these two observers have thus discovered the correct law of motion is that the statement of the law by both observers be the same. This idea is the content of Einstein's famous principle of invariance which states that the laws of nature are those statements about events or correlations of events which are valid in all inertial frames of reference. We restate this principle as follows: If the inertial observer O formulates what he thinks is a law in terms of his own spatial and temporal measurements x, y, z, t, and O′ formulates the same law in terms of his measurements x', y', z', t', then these two formulations must be the same if the two formulations are, indeed, statements of a law.

The mathematician expresses this principle by saying that the mathematical statement of a law must (if it is a correct statement) be invariant when we transform it from one inertial frame to another (i.e., translate it from the mathematical language of O into the mathematical language of O′). Since all formulations of laws by O are in terms of x, y, z, t and all formulations by O′ are in terms of x', y', z', t', we can translate any law expressed in the mathematical language of O into the mathematical language of O′ if we know how to translate x, y, z, t for any simple event into x', y', z', t' for that same event.

This translation from the description of O to that of O′ is called a transformation of coordinates from the fixed observer O to the moving observer O′. We may picture it as a kind of "mathematical dictionary" which enables us to compare what O says with what O′ says. If we have such a "dictionary," i.e., if we know how to transform from the "fixed" to the "moving" coordinate system (and vice versa), we can compare a description of any sequence of events given by O with the description of that same sequence of events given by O′. All we need to do is apply our "dictionary" to each event in the sequence.

Given such an overall prescription of what we must do to cull the laws of nature from the descriptions of physical events by different inertial observers, we come to the crucial question as to the nature of the transformation from the fixed to the moving frame:

$$x, y, z, t \xleftrightarrow[\text{Transformation?}]{\text{“Dictionary”?}} x', y', z', t'$$

To find this "dictionary" we follow three general rules that can be accepted without question by everyone: (1) The dictionary must conform to the accepted rules of logic; it must be self-consistent and it must not state that A equals B′ and at the same time that A equals C′ unless B′ equals C′; (2) it must conform to the accepted rules and laws of algebra and of geometry; (3) it must not violate or contradict the known laws (the accepted truths) of space and time. There is little to say about the first rule and the part of the second rule that applies to algebra. The laws of nature, dealing as they do with measurable quantities, are formulated algebraically, which is itself a very precise logical language. We are therefore forced to use algebra to express our transformations and must perforce accept all its rules and its logic. But the question of the laws of geometry is quite a different matter since we have a choice of different kinds of geometries. Since the question concerning the geometry that we must accept is related to the third rule stated above concerning the truths about space and time, we treat these together.

Since we are interested, at this point in our discussion, in Newtonian relativity, we go along with Newton and Galileo for the time being and accept their axioms about space, time, and geometry. This acceptance means that we assume space to be an absolute entity whose geometric relationships obey Euclidean geometry and that time is absolute. This assumption is equivalent to the statement that if observer O finds that two events are separated by a distance d and a time interval t, then all inertial observers find that these events are separated by the same distance d and the same time interval t. By accepting this statement we impose upon our dictionary the concept of absolute time and the concept of absolute distance or space.

With this point understood, we can now write down the Galilean–Newtonian transformations that take us from x, y, z, t to x', y', z', t'. To do this transformation we note that since time is now taken as absolute, the clock of observer O′ must read the same as that of O so we must have $t = t'$. Moreover, the distance of O′ from O at the time t as measured by O (which is just vt) must equal the distance of O from O′ as measured by O′. This means that x' differs from x just by this distance so that $x' = x - OO' = x - vt$. Since there is no motion along the Y or Z axes we also have $y' = y$ and $z' = z$. Collecting these statements into a group we obtain the complete set of Galilean–Newtonian transformations: $x' = x - vt$ ($x = x' + vt'$), $y' = y$ ($y = y'$), $Z' = z$ ($z = z'$), $t' = t$ ($t = t'$). The first series of transformations is written from the point of view of O who pictures O′ as moving to the right with the speed v. The second set of transformations, in parentheses (which are completely equivalent to those in the first series) gives the point of view of O′ who sees O moving in the opposite direction with the speed v. These transformations are the basis of what we call Newtonian relativity because, as we shall see, they conform completely to the laws of Newtonian mechanics. To see this congruency we consider the basic Newtonian law of mechanics $\mathbf{F} = m\mathbf{a}$. The content of the Galilean transformations, given above, is that two coincident, but different, inertial observers see the spatial and temporal arrangement of events in the universe in exactly the same way, the only difference being that the speeds of all the events as recorded by one of these observers differ by a constant amount in some particular direction (the direction of the relative velocity of the two inertial observers) from those recorded in the same direction by the other observer. Since a constant velocity has no effect on acceleration, force, mass, distance, or time in Newtonian mechanics, it is clear that Newton's laws of motion are the same for both inertial observers. Moreover, his law of gravitational force $F = Gm_1m_2/r^2$ is the same for both observers. From this law it follows that an inertial observer cannot deduce his velocity from the mechanical behavior of bodies because their orbits in a given inertial frame of reference, as seen by the observer fixed at the origin of such a frame (moving along with the frame), do not depend on how fast or in what direction this frame is moving with respect to the fixed stars. This statement is the content of Newtonian relativity. As far as the mechanical behavior of bodies is concerned (including gravitational interactions), there is no absolute velocity and one can only speak of the relative velocities of the inertial frames, but distance (space) and time are absolute.

We saw that the Newtonian laws of motion are not affected by the uniform velocity of an observer's frame of reference so that all such frames of reference (inertial frames) are equivalent insofar as such laws go. We can make no observations on moving bodies that tell us our absolute velocity if we are in an inertial frame. This statement is the content of the Newtonian–Galilean transformations we wrote down above and, hence, the content of what we call Newtonian relativity. If we limited ourselves to the laws that govern the motions of bodies, we would have every reason to accept the Newtonian concepts of absolute space and absolute time since these lead to the Galilean transformations which in turn leave the Newtonian laws of motion and the Newtonian law of gravity unaltered when we go from one inertial frame to another. This is precisely one of the properties that a correct set of transformations (dictionary) must have; it must leave all laws unchanged when it transforms them into the mathematical language of another inertial reference frame. Since scientists in the era between Newton and Einstein accepted the Newtonian laws as correct, they saw no reason to reject the Galilean transformations and, hence, the Newtonian concepts of absolute space and absolute time.

We now consider the relationship of Newtonian relativity to the laws that govern nonmechanical phenomena such as light. During the 19th century, when the great mathematicians of that era were expressing the laws of Newtonian mechanics in their most elegant mathematical form, scientists began studying the chemical, electrical, and optical properties of matter. From these studies the sciences of chemistry, electricity, magnetism, and optics evolved. Electricity, magnetism, and optics are different facets of a single branch of physics called electromagnetism. Optics is the study of light which, like all radiation, is an electromagnetic wave consisting of an electric field and a magnetic field which are rapidly oscillating at right angles to each other. In a vacuum these oscillating fields are propagated through space at a speed $c = 186{,}000$ miles/sec or 3×10^{10} cm/sec and in a direction which is perpendicular to the plane in which the electric and magnetic fields are oscillating. This description of light as an electromagnetic wave, which has been completely verified experimentally, was proposed by the great 19th century British physicist James Clerk Maxwell in his electromagnetic theory of light. This theory is contained in a series of equations that Maxwell derived from Michael Faraday's experiments on electricity and magnetism. One can deduce from these equations that an oscillating electromagnetic field of the sort described above must travel at the speed c through a vacuum and that this speed is independent of the motion of the source of the oscillating field relative to the observer. Maxwell's equations state that the speed c, as measured by an observer on the earth, is the same whether the source is an electric light bulb, a radio station on earth, or some vibrating atoms on the sun or in a distant star.

When Maxwell proposed his electromagnetic theory of light, he suggested that the oscillating electromagnetic field is not to be considered as a wave propagated through the vacuum (empty space) but rather as a wave propagated through an all-pervasive medium which he termed "an ethereal substance" and which since then has been called the "luminiferous ether." To Maxwell this ether was a real medium, the evidence for which he believed he had found in the behavior of light. He expressed his belief in the ether in the following paragraph which appears in his fundamental paper "A Dynamical Theory of the Electromagnetic Field" published in 1865: "We have therefore some reason to believe, from the phenomena of light and heat, that there is an aethereal medium filling space and permeating bodies, capable of being set in motion (vibrations) and of transmitting that motion from one part to another, and of communicating that motion to gross matter so as to heat it and affect it in various ways.'' Maxwell introduced the concept of the ether because he thought that such a medium is necessary for the transmission of electromagnetic waves as water is for the transmission of water waves. Since the propagation of acoustical waves and various types of matter waves (e.g., a wave along a taut cord) is always associated with a medium (air for an acoustical wave and a solid or a liquid for the matter waves, which are really also acoustical waves), scientists, in general, believed with Maxwell that a medium (the ether) is needed to transmit electromagnetic waves (light or radiation in general). According to Maxwell and his followers, this ether appeared to be necessary for another reason which stems from the presence in Maxwell's equation of a universal constant c, which can be determined entirely from electromagnetic measurements on electric charges and currents. This number is the same for all observers regardless of their state of motion since c is just the ratio of the unit of electric charge expressed in electrostatic units to the same unit expressed in electromagnetic units. On the other hand, Maxwell showed that c is also the velocity of electromagnetic waves and is the same regardless of how these waves are produced (regardless of how the source of the waves may be moving).

When one speaks of the motion of the source of an electromagnetic wave, one immediately faces an important question. Motion relative to what or whom? Maxwell and all physicists before the time of Einstein answered this question very simply and adequately, from their point of view (and in consonance with Newtonian ideas of space and time), as follows: Motion with respect to the all-pervasive ether which was considered to be at absolute rest in the universe and hence the frame relative to which absolute motion could be measured.

According to this point of view, the ether is a real substance without which electromagnetic waves (light or radio waves) cannot be propagated through space. Moreover, the speed of light in a vacuum (i.e., a vacuum except for the presence of the ether) has the value c only for observers at rest relative to the ether. Such observers (they were considered to be at absolute rest) are in a favored position vis-à-vis the laws of nature because only for them do Maxwell's equations assume their simplest form. For other observers (those moving with respect to the ether) Maxwell's equations are more complicated and the speed of en electromagnetic wave as measured by any observer depends on the speed of that observer relative to the ether. This argument would be valid if the Newtonian concepts of absolute space and absolute time were correct and we could use the Galilean transformations to transform Maxwell's equations from their form as expressed in terms of the coordinates of the observer whom we assume to be fixed relative to the ether (an observer for whom the speed of light is taken to be c) to their form as expressed in the coordinates of an observer moving with the velocity v with respect to the fixed observer (and hence with respect to the ether). We see that the laws of optics differ from the laws of mechanics insofar as Newtonian relativity is concerned. We saw that Newtonian relativity states that the laws of mechanics (including gravity) are the same for all inertial observers and that an inertial observer can make no observation of the motion of a body that can enable him to deduce his absolute velocity. But this is not true for optical phenomena because the Newtonian concepts of absolute space and absolute time, which lead to the Galilean transformations, necessarily require that optical phenomena vary from one inertial frame to another and that the basic optical equations (Maxwell's equations of the electromagnetic field in empty space) take on their simplest form (with the speed of electromagnetic waves equal to c) in a frame of reference that is at absolute rest (i.e., at rest with respect to the all-pervasive ether that was assumed to exist). We could then detect our state of absolute motion (if there were an ether) by measuring the speed of light in our own moving frame of reference and comparing it with c.

To analyze this point in detail we consider how the speed of light would depend on the motion of the observer's frame of reference if the Newtonian concepts of absolute space and absolute time and, hence, the Galilean coordinate transformations were correct. From the first transformation equation $x' = x - vt$ (where v is the speed of the moving frame of reference) we obtain the equation that tells us how velocities are transformed when we go from a fixed to a moving frame. We return to our two observers and imagine an object moving to the right with speed u as measured by the fixed observer O. What is its speed as measured by the moving observer O′? If x is the distance the body moves in a time t as seen by O and x' is the distance it moves during the same time relative to O′ (as seen by O′), then, from the definition of speed, we have $u = x/t$, and $u' = x'/t$. On applying the transformation equation $x' = x - vt$ to x' and x and dividing by t, we obtain $x'/t = (x/t) - v$ or $u' = u - v$.

The above equation gives us the law of the addition of speeds in Newtonian mechanics. As we see, it agrees with our daily experience because it tells us that if we are in an automobile moving in a given direction at 60 miles/hour, another automobile going in the same direction at 70 miles/hour has a speed relative to us (will pass us at a speed) of 70 − 60 or 10 miles/hour. That this result appears to be in complete agreement with our experience and with the accepted rules of arithmetic may predispose us to accept it as correct, but we must keep in mind that certain phenomena in nature are outside our direct experiences, and we must be prepared to find deviations from the expected classical behavior in such circumstances. The speed of light is such a phenomenon, as was already recognized by Einstein when, at the age of 16 (a year after he had dropped out of high school), he tried to picture what he would observe if he could travel as fast as light. This *gedanken* (thought) experiment which contained "the germ of the special theory of relativity" presented Einstein with what appeared to be a paradox which he describes in his *Autobiographical Notes* as follows: "If I pursue a beam of light with the velocity c, I should observe such a beam of light as a spatially oscillatory electromagnetic field at rest. However, there seems to be no such thing—neither on the basis of experience nor according to Maxwell's equations. From the very beginning it appeared to me intuitively that, judged from the standpoint of such an observer, everything would have to happen according to the same laws as for an observer who was at rest relative to the earth. For how is the first observer to know, i.e. be able to establish, that he is in a state of rapid uniform motion?" Einstein is saying here that if one could travel as fast as light and thus follow a beam of light, one would not see an electromagnetic wave (a moving electromagnetic field) as demanded by Maxwell's equations but instead a stationary electromagnetic field with electric and magnetic intensities varying from point to point in the frame of reference of the moving observer (rather than varying from moment to moment at any given point). To Einstein this conclusion meant that there is something peculiar about the speed of light insofar as one's usual understanding of speed is concerned because if the speed of light obeyed the ordinary rules of the addition and subtraction of speeds as defined by the Newtonian concepts of space and time, an observer traveling fast enough through the supposed ether would find a beam of light behaving in a most peculiar way.

Let us see how the speed of a beam of light would behave as seen by different inertial observers moving with respect to the "ether" if the Newtonian concepts were correct. If we call c the speed of the light along the X axis for an observer O at rest with respect to the "ether" which we take as the absolute standard of rest, then c' is the speed of the same beam as measured by an observer moving with speed v along the X axis with respect to the "ether." To find c' we just apply the formula (derived from the Galilean transformation) for u and u' given above since the beam of light can now be treated like a moving particle. We then have $c' = c - v$. We see from this equation that the speed of light as seen by the moving observer (moving with respect to the imagined "ether") is not the same in all directions but varies from $c + v$ to $c - v$ as the direction of the beam changes with respect to the direction of motion of the observer.

This statement means (if correct) that an observer on the earth should be able to measure (or at least detect) his velocity through space (relative to the "ether") by carefully measuring the speed of light in various directions. Any change in the speed of a beam of light as the direction of the beam is altered relative to the direction of the motion of the earth through the "ether" would then indicate that the earth is, indeed, moving and its absolute velocity could be ascertained in this way. Because Newtonian relativity does not apply to optical phenomena (the Maxwell equations change when we go from one inertial observer to another using the Galilean transformations) we should be able to determine our state of motion optically if the Newtonian concepts of space and time, which are the basis for the Galilean transformations, are correct. We shall see below that the earth's motion with respect to the ether is precisely what Michelson and Morley attempted to but failed to find in their famous experiment in 1887.

Since the earth is revolving around the sun, its direction of motion through the "ether" (if an ether exists) must change from day to day and this ether, according to Newtonian concepts, should manifest itself in the measured velocity of a beam of light that is always propagated in the same direction on the earth. In fact, the speed of a beam that is always propagated from west to east should change every twelve hours owing to the rotation of the earth which changes the east–west direction of an observer on the earth and hence the direction of the beam relative to the ether continuously. Since it is impractical to compare the results obtained by measuring the velocity of the same beam of light at different times of the day or year, Michelson proposed an experiment (which he performed with his colleague Morley) in which the velocities of two beams moving at right angles to each other are compared at the same time using a clever arrangement of mirrors with respect to a point source of light which is so stationed that a collimated beam of the light (a beam that moves along one direction without spreading) from it is directed toward a half-silvered mirror A that is tilted 45° to the direction of the beam. This mirror allows half the light to go through it (moving in its original direction) and to strike a mirror M′ at a distance L from A. The other half of the original beam is reflected 90° by the mirror A at a 45° angle to the original beam, and hits a mirror M at the same distance L from A. These two half beams are then reflected back by the mirrors M and M′ and then recombined at the point A′ along the beam from M, but their paths are different. The half beam that travels along the direction of the original beam (which we assume to be in the direction of the earth's motion through the ether) travels the path AM′A′ because the mirror A has shifted its position to A′ during the time the beam has traveled from A to M′ and back again. The half beam that travels from A to the mirror M and back describes the path AMA′, which is along diagonal directions.

We see from the above description of the two beams that we have a situation identical to that in the problem of the two swimmers that we analyzed previously. There we had one swimmer going downstream a given distance and back and the other going across the stream the same distance and back, and we saw that the cross-stream swimmer takes less time to complete his trip than the downstream swimmer and therefore returns first to the starting point. The same analysis applies to the two beams since the moving earth would create a stream of "ether" (if an ether existed) flowing opposite to the earth's motion which would affect the two beams just the way the stream of water affects the two swimmers. Michelson and Morley therefore expected the cross beam that travels to the mirror M to return first. This would produce interference between the two beams when they are reunited into a single beam after reflection. Since the distance of M and M′ from A could not be made exactly equal, the entire apparatus was mounted on a turntable so that the role of M and M′ could be interchanged. If the analysis based upon the Newtonian space and time concepts were correct, one would then expect to see a series of interference fringes at the recombination point drifting across the field of view as the apparatus was rotated through 90°. Michelson and Morley observed no such effect even though their apparatus was sensitive enough to detect the motion of the earth by comparing the observed speed of light in different directions, even if the earth were moving at a speed of 1 mile/sec instead of at $18\frac{1}{2}$ miles/sec in its orbit. From this null result they concluded that "[it] appears, from all that precedes, reasonably certain that if there be any relative motion between the earth and the luminiferous aether, it must be quite small. . . ."

Michelson and Morley did not conclude that an ether does not exist and that therefore one cannot detect one's absolute motion through space, but rather that there is no relative motion between earth and ether, or this relative motion is so small as to be unmeasurable with the apparatus that Michelson and Morley used. This conclusion, however, is untenable in light of the known revolutionary motion of the earth around the sun. Since the direction of the earth's motion is changing continuously, there must be some periods during the year when the linear motion of the entire solar system through space cannot compensate for the earth's motion so that Michelson and Morley should have observed some effect if the analysis based on the Newtonian concepts of space and time were correct. The negative result of the Michelson–Morley experiment therefore presented the late 19th and early 20th century physicists with a serious discrepancy between theory and observation.

A number of futile attempts were made to explain the negative result of the Michelson–Morley experiment within the framework of Newtonian space-time concepts, the most famous of which was the Fitzgerald–Lorentz contraction hypothesis. Both Fitzgerald and Lorentz (independently) proposed that the arm of the Michelson–Morley apparatus parallel to the direction of the earth's motion shrank slightly (owing to the earth's motion) thus permitting the beam of light moving parallel to the earth's motion to return to the half-silvered mirror at exactly the same time as the beam moving at right angles to the earth's motion. Fitzgerald had no physical reason for proposing his contraction hypothesis but Lorentz proposed it because of certain deductions about the change in the dimensions of charged particles when in motion that he had arrived at from his electron theory of matter. In any case, the Fitzgerald–Lorentz contraction hypothesis is not acceptable as an explanation of the results of the Michelson–Morley experiment because it was introduced as a kind of miraculous compensation just to explain this experiment with no justification other than that it does account for it. Such an ad hoc explanation is unsatisfactory because there is no way to determine whether this particular explanation or any one of a number of other such ad hoc explanations is the correct explanation. Instead of proposing that the arm of the apparatus parallel to the earth's motion contracts by just the right amount to give the observed result, one might propose that the arm of the apparatus perpendicular to the earth's motion expands by an appropriate amount. Or one might even propose a simultaneous expansion and contraction of the two arms. We see that there is an infinitude of possibilities. To escape such arbitrary possibilities one must seek an explanation which is based on or can be deduced from a set of general principles or laws, just the way the orbits of the planets can be deduced from Newton's law of motion and his law of gravity. Einstein was the first to do this and he showed that in doing so one must discard the Galilean–Newtonian principle of relativity and replace it by a more general principle of relativity that takes into account optical phenomena as well as mechanical phenomena. This replacement, as we shall see, also leads to a denial of the Newtonian concepts of absolute space and absolute time and the Galilean transformations, as given above, that stem from these absolute concepts.

Einstein, in introducing his theory of relativity, was guided by deep philosophical reasons which had their origin in his belief that all natural phenomena (all the laws of nature) are on the same footing; if a principle of relativity holds for mechanical phenomena (the Newtonian–Galilean principle), it should also hold for optical phenomena. If this consistency did not exist, optical laws would hold a special place in nature and the symmetry vis-à-vis the laws of nature that Einstein felt must be present in the universe would be destroyed. To follow Einstein's procedure, we first recall that the Galilean transformations differentiate between optical and mechanical phenomena in the following sense: Although the Galilean transformations state that no mechanical experiments (the observations of the motions of physical bodies) can be devised to enable an inertial observer to determine his absolute velocity through space (the Newtonian relativity principle), these transformations state that optical experiments (the observations of the propagation of light beams; for example, the Michelson–Morley experiment) can be devised for such a purpose. That the Michelson–Morley experiment failed to detect the earth's motion shows us that there is something wrong with the Galilean transformations and hence with the Newtonian principle of relativity. But since this principle and the Galilean transformations rest on the Newtonian concepts of absolute space and absolute time, the breakdown of Newtonian relativity necessarily leads to the breakdown of people's long-held (and cherished) belief in absolute space and absolute time.

There is good reason to believe that Einstein, when he published his revolutionary paper on the special theory of relativity in 1905 (one of the epochal events in the history of human thought), was unaware of the Michelson–Morley experiment; he certainly did not write his paper to explain the experiment. He was well aware of the Galilean transformations and of Newtonian relativity and was disturbed by and unhappy about the asymmetrical relationship of this principle to mechanical and optical phenomena. He felt strongly that if the principle of relativity is true for mechanical phenomena, it must also hold for optical phenomena, and if this fact required the denial of the absolute concepts of space and time, then one must accept such a denial and replace the Newtonian concepts of absolute space and absolute time by a new space-time concept that would lead to a correct, all-embracing principle of relativity.

In his synthesis of these ideas into a new theory, Einstein relied heavily on a series of what he called *gedanken* (thought) experiments: experiments that are designed, carried out, and analyzed entirely in one's thoughts. These *gedanken* experiments can lead to conclusions that either support or contradict previously held conclusions and can thus support an old theory or lead to a new one. Einstein was profoundly influenced by two such experiments, one of which we have already discussed and which dealt with the speed of light, and the other of which dealt with the measurement of time. As we shall see, the second of these experiments (which convinced Einstein that the concept of absolute time is untenable) stemmed from the first *gedanken* experiment which first occurred to Einstein when he was 16 years old.

In his first experiment, Einstein considered the following question: "If I can travel as fast as a beam of light, what will this beam look like?" He concluded that the beam would look like something that is never observed in nature (a stationary electromagnetic field whose intensity varies from point to point) and which contradicts Maxwell's electromagnetic theory. To Einstein this image meant that the speed of light plays a very special and important role in the laws of nature and in our concepts of space and time. To see how Einstein used the speed of light first to alter the traditional (Newtonian) concepts of absolute space and absolute time and then to extend (actually, to change) the Newtonian principle of relativity to include optical phenomena, we note that Einstein's *gedanken* experiment dealing with the impossible optical and electromagnetic phenomena an observer would see if he could travel as fast as light means that no observer (in fact, no physical body) can ever acquire a speed equal to that of light. No matter how fast an inertial observer is moving (always slower than light) he always finds that a beam of light passes him at the same constant speed c. Einstein made this idea a starting point of his theory stating that the speed of light in empty space is the same in all directions for all inertial observers, regardless of how fast they are moving with respect to each other.

This belief enabled Einstein to extend the principle of relativity so that it applied to optical and electromagnetic phenomena in general as well as to mechanical phenomena. That this extension follows from the constancy of the speed of light is obvious if we recall the Michelson–Morley experiment, which was designed to measure or detect the motion of the earth by measuring the differences in the speeds of beams of light moving in various directions relative to the earth's velocity. This experiment shows that the speed of the earth (and hence of an observer attached to the earth) has no effect on the speed of light as measured by such an observer; it thus supports Einstein's contention that an inertial observer cannot determine his velocity by means of optical experiments.

We may now state Einstein's principle of special relativity as follows: No experiment or observation (whether mechanical, optical, or electromagnetic) that an inertial observer can make within his own frame of reference enables him to determine whether he is at rest in space or in a state of uniform translational motion. The concept of absolute velocity is thus banished from physics since there is no way for one to determine such a velocity. If our earth were at rest instead of moving in its orbit, we would observe no differences in the behavior of physical bodies, beams of light, or electrical circuits.

In Einstein's hands, the special theory of relativity acquired a much more profound meaning than Newtonian relativity had, for Einstein introduced, as an adjunct of the theory (or a consequence of it), the general concept of the invariance of all the laws of nature. This principle of invariance, which states that the laws of nature must have the same algebraic form for all inertial observers, is another way of stating Einstein's special theory of relativity.

The principle of invariance becomes a powerful tool in the hands of physicists because it imposes fairly strict conditions on a law of nature. Only those statements about events in nature are laws which retain the same form when we transform from one inertial frame of reference to another. This principle of invariance thus permits us to check a statement to see whether it is or is not a law by noting whether it is altered when it is transformed.

In line with this principle we see that the Galilean transformations cannot be correct because when we use them to transform the propagation of a beam of light from one inertial frame to another, we obtain different speeds for the beam in the two frames, which denies that the speed of light is the same for all inertial observers. With this thought in mind, Einstein showed how the Galilean transformations must be changed to give the same value for the speed of light in all inertial frames. In his first step in this direction he showed that the Newtonian concept of absolute time must be discarded and replaced by a concept of relative time; he did this by describing and analyzing the second of the two *gedanken* experiments mentioned above.

That the Newtonian concepts of absolute space and absolute time are inconsistent with the constancy of the speed of light for all inertial observers can be deduced easily enough if we consider the speed of the same beam of light as measured by two observers, one of whom is at rest on the earth, and the other of whom is moving on a train in the same direction as the light. We supply each observer (when they are both at rest) with identical clocks and two equal and weightless measuring rods. To make our analysis simple, let each rod be 186,000 miles long and let the beam of light (and hence the train) be moving from east to west. Each observer obtains the speed of the beam by placing his 186,000-mile-long rod along the east–west direction (parallel to the light beam) and noting how long the beam takes to go from the eastern end to the western end of the rod. The constancy of this speed means that each observer finds that exactly 1 sec elapses (as measured by his own clock) while the beam moves from one end of his measuring rod to the other.

To see why these measurements conflict with the Newtonian ideas of absolute space and time, we consider the fixed observer and the moving observer at the very moment when the eastern ends of their two rods coincide and let that be the moment when the front end of the moving beam of light touches the eastern ends of their two rods. The fixed observer notes that the beam moves westward along his rod, reaching the western end in 1 sec, as shown by his own clock. He also notes that this same beam is advancing along the moving rod, but since this moving rod is itself moving westward (in the direction of the beam) he concludes (knowing that his fixed rod and the moving rod are identical and were exactly equal when they were both at rest) that the beam of light cannot possibly reach the western end of the moving rod in 1 sec because this western end of the moving rod (according to the Newtonian concept of absolute space) will have advanced in 1 sec some distance beyond the western end of his own fixed rod. To be specific, let the moving rod have a westward speed (relative to the fixed observer) of 1000 miles/sec. The fixed observer will then (if he adopts the absolute concepts of Newton) conclude that after 1 sec has elapsed on his clock (at the moment that the front end of the advancing beam has reached the western end of his rod), the western end of the moving rod will have advanced 1000 miles beyond the western end of his own fixed rod. The beam of light will therefore (according to his reasoning) have more than an additional 1000 miles to go before reaching the western end of the moving rod, so that more than 1 sec will have elapsed from the time the front end of the beam coincides with the eastern end of the moving rod and the time it reaches the western end. The Newtonian point of view therefore tells him that the speed of light relative to the moving observer is less than 186,000 miles/sec.

What is the fixed observer's surprise then on learning that the moving observer did not obtain a lower speed but found the speed of light in his frame of reference to be exactly 186,000 miles/sec? The moving observer will tell the fixed observer that he clocked the beam of light as it advanced from the eastern to the western end of his 186,000-mile-long rod (just as the fixed observer did) and found that it took the beam just 1 sec, thus giving the same result as that found by the fixed observer. If we think about this idea for a moment, we see that this can be so only because distances and times in the moving frame are different from those in the fixed frame. Suppose (as measured by the fixed observer) that the rod of the moving observer shrinks by a given amount (owing to the relative motion of this rod) and the clock of the moving observer slows down somewhat compared to that of the fixed clock. We see then that the moving observer can get the same answer for the speed of light as does the fixed observer if the shrinkage of his rod and the slowing down of his clock are just right. Einstein contended that the distance and time between two events for a moving observer are different from the distance and time between these same two events as measured by a fixed observer (or, for that matter, by any other inertial observer moving with respect to the first moving observer). These differences between the measurements of the two inertial observers must be just sufficient to lead to the same value for the speed of light in both inertial frames.

Einstein demonstrated the nonabsolute character of time (which, as we shall see, implies the nonabsoluteness of space) most dramatically by describing and analyzing the second of his two *gedanken* experiments. This second experiment deals with two inertial observers who are to determine whether two events that they both observed happened simultaneously or not. In this experiment, Einstein considered two inertial observers—the observer O fixed on the earth next to a railroad track, and the other observer O′ in a train moving with constant speed to the left, who observe two events at two points A and B on the railroad track. We arrange things in such a way that O is exactly midway between A and B. Moreover, let A′ and B′ be two markers on the train which coincide with A and B respectively when the train is at rest, and let O′ be midway between A′ and B′ in the train. If the train were standing still and A′ and B′ coincided with A and B respectively, then O′ would coincide with O. In other words, if O′ and O were both stationary, the length AB would exactly equal the length A′B′, and both observers would be at the midpoints of these lengths. Let the event at A be a bolt of lightning which strikes the track at A at the moment that A′ coincides with A. The event at B is also to be a bolt of lightning striking the track at B. We say nothing about the origin of the two bolts of lightning since all we are concerned with here are the observations by the two observers of when the bolts struck the ground at A and B. One bolt may have originated from a distant star and the other from Jupiter for all we care.

When the bolt of lightning strikes the ground at A and B, two light signals (one from A and one from B) immediately start moving toward the fixed observer O. Suppose that O receives these two signals at exactly the same time. He then concludes (because he is midway between A and B) that the two bolts struck A and B simultaneously. To O the events A and B were simultaneous.

Consider now observer O′ who is moving to the left (toward A) relative to O. He also notes the bolts of lightning striking points A and B but since he is moving toward A (and away from B) the light signal from A, having a smaller distance to travel than does the signal from B, reaches him before the signal from B does; he therefore concludes that the event at A occurred before the event at B did. To O′ the events at A and B were not simultaneous. Since neither observer can deduce anything at all about his absolute state of motion, it is impossible for either one to say that the statement of the other is incorrect. They are both right. The concept of simultaneity, and hence of time in general, is relative; the span of time between two given events has no absolute meaning but differs from observer to observer.

Now that we have seen that the time between two events has no absolute meaning, we deduce from it the relativity of distances. In the experiment we have just described, we have arranged things so that the mark A′ on the train coincides with the point A on the track at the moment that the lightening bolt strikes A. Both observers O and O′ therefore see A and A′ lined up when they receive the signal telling them that the lightning has struck. If the train were brought to rest at that moment, the two observers would see that B and B′ are also coincident and they would both agree that the length AB equals the length A′B′. But as long as the train is moving the two observers do not find the two lengths equal. If we look at things from the point of view of the moving observer O′, we note that he sees the lightning strike A before he sees it strike B. He therefore notes that A and A′ coincide before B′ coincides with B, because the latter will occur as seen by O′ only when O′ sees the lightning strike B. This means, insofar as O′ is concerned, that when the point A′ on the train has reached A, the point B′ has not yet reached B. Thus, to O′ the length A′B′ is larger than the length AB. We may put it somewhat differently as follows: O′ sees the lightning strike A and notes that A′ and A coincide. A moment later he sees the lightning strike B and notes that B′ is then lined up with B. But then A′ has advanced beyond A so that AB (with B coinciding with B′ but with A to the right of A′) is shorter than A′B′. Thus, the distance AB between the two events has no absolute meaning.

This relativity of space and time stems from the finiteness of the speed of light and the fact that this speed has the same value for all inertial observers. If the speed of light were infinite, all observers, regardless of their state of motion, would agree about the simultaneity of events because the information about such events would be propagated instantaneously to all observers.

We have seen that the constancy of the speed of light for all inertial observers (which we must now accept as a law of nature) destroys the Newtonian concepts of absolute space and absolute time and therefore invalidates the Galilean transformations which are based on these absolute concepts. We must therefore replace these Galilean transformations by a set which gives the same speed of light for all inertial observers. Since the speed of light c (a universal constant) plays a dominant role in the determination of distances and time intervals between events, this constant must appear in the transformations that are to replace the Galilean transformations if the Einsteinian concepts of space and time are to replace the Newtonian concepts.

Before Einstein proposed his revolutionary ideas of space and time, H. A. Lorentz, the great Dutch physicist of the early 20th century, had extended Maxwell's electromagnetic theory of light to take account of the interaction of light and matter. Maxwell's theory, which deals with the propagation of electromagnetic waves (light and all other forms of radiation) in a vacuum, cannot, in its original form, account for the way light behaves when it is propagated in a medium such as glass (the speed of light through such a medium is not the same for all colors), or account for its absorption in some substances (opaque substances) and its reflection from others (metals). To account for such phenomena,

Lorentz, in the early 1890s, proposed his electron theory of matter. He pictured matter as consisting of electrically charged particles which he called electrons. When an electromagnetic wave such as light enters a material medium, these electrons, according to the Lorentz theory, are set into oscillations by the oscillating electric field of the wave, and these oscillations in turn alter the character of the wave. Using this electron pictured, Lorentz explained, to a good approximation, many of the phenomena that Maxwell's theory alone cannot explain. Moreover, he showed that a spherical electron, moving through the ether, contracts in the direction of its motion (thus becoming an ellipsoid) and he reasoned that this contraction accounts for the negative result of the Michelson–Morley experiment. He also showed that if one introduces a special way of transforming space and time coordinates when dealing with electromagnetic phenomena (different from the Galilean transformations), Maxwell's equations are the same for all inertial observers and, in particular, the speed of light is the same for all observers. These transformations, the famous Lorentz transformations, which Einstein derived independently from a more general and universal point of view, were introduced by Lorentz just to treat light. He did not consider these transformations as universal and applicable to all phenomena, but rather as a special and rather artificial way of treating space and time when one deals with electromagnetic phenomena (light). Lorentz accepted the Galilean transformations as correct for all mechanical events (motions of bodies), which he pictured as governed by absolute space and time, but not adequate for the description of electromagnetic events, and so he introduced a variable time and space for the latter. He did not consider this variable time and space (varying from one inertial observer to the next) as the real time and space in which the observer lives, but as a mathematical invention or artifice which leads to correct results.

Einstein's approach was quite different because he drew no distinction between his treatment of mechanical and electromagnetic phenomena; he discarded the Newtonian concepts of absolute space and absolute time entirely, and replaced them by his own concepts of relative time and relative space that apply to all events in nature whether mechanical or electromagnetic. The only requirement he imposed on his concepts of relative space and relative time was that they be such that when one transforms from one inertial frame to another, the value for the speed of light remain the same. This requirement alone enabled Einstein to deduce in a very simply way the transformations that Lorentz had introduced to deal with electromagnetic phenomena.

To deduce the Einstein–Lorentz transformation for inertial observers in as simple a manner as possible (using only elementary algebra), we again consider the same beam of light (moving along the coincident X and X′ axes) as viewed by our two inertial observers O and O′. Let this beam start its X-ward journey from the origin of observer O at the moment time $t = 0$, and let the origin of observer O′ coincide with that of O at this same zero moment. Further, let the clocks of our two observers also coincide at this starting moment, so that t and $t' = 0$ when the series of events we are considering begins. We now ask each observer to specify the position of the front of the beam in his own frame of reference after a time t has elapsed as given by the clock of the fixed observer O and a time t' has elapsed as given by the clock of the moving observer O′. Let x be the distance from the origin of O that the beam has traveled in the time t as given by the clock of O, and let x' be the distance from the origin of O′ that this same beam has traveled in the time t' as given by the clock of O′. Since we are discarding absolute time and absolute space (and the Galilean transformations), we no longer place $t = t'$ since the clocks of the two observers give different times. How are we to proceed then if we reject the equality of these two times? We use the observed fact that the speed of light is the same for both observers which means that we must write $x = ct$ and $x' = ct'$.

These relations (or equations) simply tell us that c, the speed of light, is the same for both observers since they state that the distance the beam travels divided by the time it takes to travel that distance, as measured by both observers, is the same: $c = x/t = x'/t'$.

These relationships apply only to the propagation of a beam of light, but from them we deduce the Einstein–Lorentz transformations that apply to the coordinates of all kinds of events. To do this, we start from the Galilean transformations, which we know to be approximately correct (they are extremely accurate if we are dealing with inertial observers whose relative speed v is very much smaller than the speed of light), and alter these transformations just sufficiently to be consistent with the relations written above. This is not an arbitrary procedure. One can use various procedures to deduce the Einstein–Lorentz transformations, and although these methods seem to differ from each other considerably, they all lead to the same final result. If we start from the

Galilean transformations and then change them in some manner or other, but insist that this change be consistent with the relations $x = ct$ and $x' = ct'$, we inevitably obtain the same final result—the Einstein–Lorentz transformations.

Taking the Galilean transformations previously given as our starting point, we alter them as follows: First we multiply the right-hand side of the first equation by the constant factor f (which we determine later) to obtain $x' = f(x - vt)$. This is the simplest alteration we can make to obtain a set of transformations that give us relative distances between events rather than absolute ones. The constant f can depend only on the relative speed v of the two observers and on c, the speed of light. Thus, the constant f brings the speed of light into our transformations.

The equations for y' and z' remain the same, since we are still dealing with relative motion along the x direction only, but we must change the equation for t'. We first add a term containing x to the right-hand side. The reason we do this is that we want t' to depend on x and t just as x' does. In doing this we treat space and time symmetrically so that time is placed on the same footing as space insofar as the transformation equations go. We then add $-\alpha x$ to the right-hand side of the equation for t', where α is another constant to be determined and which depends on v and c.

If we change the expression for t' in only this way, we have $t' = t - \alpha x$, and the expression for t' still differs in its appearance from that for x'. To make the expression for t' look as much as possible like that for x', we multiply the entire right-hand side by f thus obtaining $t' = f(t - \alpha x)$. This equation and the one for x' written above are the basic equations of the Einstein–Lorentz transformations. As they stand now, we cannot use them since we have not yet given explicit expressions for f and α in terms of v and c. Once we have done that, we shall have completed our task of finding the Einstein–Lorentz transformations which must replace the Galilean transformations.

Collecting all our equations at this point (with α and f still to be determined) we have $x' = f(x - vt)$ and $t' = f(t - \alpha x)$ as well as $x = f(x' + vt')$ and $t = f(t' + \alpha x')$. To find α and f we now apply these equations to the beam of light we discussed previously. We use these equations as they stand, with α and f undetermined to transform from the fixed observer's description of the propagation of the beam to the moving observer's description. This transformation leads to a set of relations involving only f, α, v, and c, from which we deduce the correct dependence of f and α on v and c. We shall then have the correct transformations for the space-time description of an electromagnetic wave. But since according to Einstein's principle of relativity all phenomena (mechanical as well as electromagnetic) must obey the same transformations, the results we obtain for the electromagnetic wave as outlined above must be universally true and apply to all events.

To determine α we divide x' by t' to obtain $x'/t' = (x - vt)/(t - \alpha x)$ which gives us the equation $\alpha = v/c^2$ and the transformations become on introducing this value of α $x' = f(x - vt)$ and $t' = f[t - (v/c^2)x]$ and $x = f(x' + vt')$ and $t = f[t' + (v/c^2)x']$. Again we apply these equations to a beam of light so that $x = ct$ and $x' = ct'$. Using some algebra we also obtain the equation $f = 1/\sqrt{1 - (v^2/c^2)}$. We have thus deduced the value of f and can now write down the complete set of Einstein–Lorentz transformations $x' = (x - vt)/\sqrt{1 - (v^2/c^2)}$ and $t' = [t - (v/c^2)x]/\sqrt{1 - (v^2/c^2)}$ as well as $x = (x' + vt')/\sqrt{1 - (v^2/c^2)}$ and $t = [t' + (v/c^2)x']/\sqrt{1 - (v^2/c^2)}$.

A brief consideration of these equations shows that they differ very slightly from the Galilean transformations if the relative velocity, v, of the two observers is small compared to the speed of light c, for then $(v/c)^2$ and v/c^2 are very much less than 1, and we may neglect these terms. The equations then reduce to the Galilean transformations since the denominators on the right-hand sides of the equations for x', t', x, and t reduce to 1, and $(v/c^2)x$ is negligible compared to t or t'. The Newtonian concepts of space and time and the Galilean transformations were accepted without question for hundreds of years because ordinary events are associated with very small speeds so that there was no reason to question Newton's concepts. Even the speeds of the planets around the sun are too low to give easily observable discrepancies between observations and classical theory. Not until scientists began investigating high-speed phenomena very accurately (as in the Michelson–Morley experiment) were such discrepancies detected.

An examination of the Einstein–Lorentz transformations shows how important the constant c (speed of light in a vacuum) is in the special theory of relativity. These transformations show that v (the speed of any material body as measured by any inertial observer) must always be less than c. If v equaled c, the denominators in the Einstein–Lorentz transformations would vanish and the expressions for x' and t' would become infinite, a result which is physically meaningless. This already tells us that the Galilean–Newtonian law for the addition of speeds is wrong, because if it were not, we could bring the speed of a body up to any value desired just by accelerating it slowly. If a body moved with the same acceleration it has when it is falling freely on the surface of the earth (about 1000 cm/sec^2) and if Newtonian laws held, it would acquire a speed of 3×10^{10} cm/sec (the speed of light in a vacuum) after about a year, simply because a speed of 1000 cm/sec would be added to its instantaneous speed each second. This result is denied by the special theory of relativity which says that the law of the addition of speeds must be such as to make c an unattainable value. In a sense, c must behave like an infinite speed as compared to all other speeds since it can never be reached although, in principle, one can approach it as closely as may be desired.

This unattainable property of the speed of light and the correct law of addition of speeds should be deducible from the Einstein–Lorentz transformations. To do this we consider a particle traveling with a constant speed u as measured by the fixed observer O, and constant speed u' as measured by the moving observer O$'$. How are u and u' related? In Newtonian mechanics we know that $u' = u - v$ (the usual arithmetic law of addition of speeds) but this expression is only an approximation which breaks down when u is large. To find the correct law of addition of speeds, we use the relationships $u' = x'/t'$, $u = x/t$ which are the definitions of u' and u, and introduce these expressions into the Einstein–Lorentz transformations.

If we divide the expression for x' by the expression for t' as given by the transformations, we obtain $u' = (u - v)/[1 - (uv/c^2)]$. This formula is the Einstein law of addition of speeds that must replace the simpler but incorrect Newtonian law. The denominator in the above expression for u' very nearly equals 1 for ordinary speeds u and v, but differs considerably from 1 when u and v separately or together approach the speed of light. This formula for the addition of speeds shows us at once that the speed of light must be the same for all inertial observers. To see this, consider a beam of light traveling along the X axis to the right so that $u = c$ (its speed as measured by O). To find u' (the speed of the beam as measured by O$'$) we apply the above addition formula and obtain $u' = (c - v)/[1 - (cv/c^2)] = c(c - v)/(c - v) = c$. Thus, u' also equals c regardless of how fast O$'$ is moving.

Suppose now that a beam of light is traveling to the left along the X axis so that $u = -c$. In the classical case, using the Newtonian law of addition of speeds, we would obtain $u' = u + c$ (a speed greater than c), but we now have $u = (-c - \mathrm{v})/[1 + (cv/\mathrm{c}^2)]$, again obtaining c. No matter how we arrange things, we can never get a speed greater than c.

We saw that the Galilean transformations are based on the Newtonian concept that the distance between two given events is the same for all inertial observers so that lengths, in general, are absolute in Newtonian mechanics. We have already demonstrated, however, in our discussion of Einstein's concept of simultaneity, that this idea of absolute lengths is not tenable and must be discarded with the Newtonian concept of absolute time. We can now use the Einstein–Lorentz transformations to calculate just how the measurement of the distance between two events varies as we go from one inertial observer to another.

Consider two identical rods, each of length L, and let the fixed observer O hold one of the rods parallel to his X axis and let the moving observer hold the other rod parallel to his X′ axis, with its left end coincident with the origin O′ and its right end at $x' = L$. We now ask the fixed observer to compare the length of the moving rod with the length of the fixed rod he is holding. To measure the length of the moving rod the fixed observer O must look at both ends of this moving rod simultaneously and note the coordinates of these two ends in his frame of reference at that moment. To make things specific, let the moment that O chooses to observe the two ends of the moving rod be the zero moment ($t = 0$) as given by his clock. At this moment let the left end of the moving rod (which coincides with the origin of the moving observer) be coincident with the origin of the fixed observer (the origins of the two inertial coordinate systems coincide). What is the coordinate in the fixed coordinate system of the right end of the moving rod? In other words, if at the moment $t = 0$ the left ends of the two rods coincide, where is the right end of the moving rod along the X axis as seen by O?

To answer this question we use the first equation of the Einstein–Lorentz set of transformations. Since $t = 0$, we have from this equation (applied to the right end of the moving rod) $x' = x/\sqrt{1 - (v^2/c^2)}$ or $x = x'\sqrt{1 - (v^2/c^2)}$. But $x' = L$ since x' is the coordinate of the right end of the moving rod in the moving frame of reference. Hence, $x = L\sqrt{1 - (v^2/c^2)}$. Since $\sqrt{1 - (v^2/c^2)}$ is smaller than 1, we see that the right end of the moving rod does not coincide with that of the fixed rod whose fixed coordinate is $x = L$. Thus, the moving rod as measured by the fixed observer is smaller than the fixed rod by the factor $\sqrt{1 - (v^2/c^2)}$. As the speed of the moving rod increases, it shrinks more and more, approaching zero as v approaches c.

There is complete reciprocity between the two observers. Each one finds that the measuring rod of the other (measured along the line of their relative motion) has shrunk by exactly the same factor. The distance between two given events is thus not a fixed, invariant quantity but depends on the state of motion of the observer. We can see now why an observer cannot travel at the speed c because if he could, all distances along the line of his motion would shrink to zero and he would be everywhere at once.

Having discussed the relativity of length, we now consider the relativity of time. To do this, we take two identical clocks (one attached to O and the other to O′) and determine by means of the Einstein–Lorentz transformations how the time between the two events as measured by one of the clocks (let us say the clock O) compares with that measured by the other clock. To do this we use either one or both of the equations that relate t' to t in the Einstein–Lorentz set. Since the relationship between the readings of the two clocks depends on where each clock is in its coordinate system, we place each one at the origin of its own coordinate system so that $x' = 0$ at all times for the moving clock (the clock of observer O′) whereas the x coordinate of the moving clock (the position of the moving clock in the fixed coordinate system as seen by the fixed observer O) changes continuously.

We consider now the following two events: the initial coincidence of the origins of the two observers (so that the fixed and moving clocks are also coincident) when both clocks read zero and the coincidence of the moving clock with the point x on the track at some time later as given by the clock of the fixed observer so that $x = vt$. These two events are separated by the time interval t according to the clock of observer O. What is the time interval t' as given by the clock moving along with observer O′? To compare t and t' we use the last of the Einstein–Lorentz transformation equations. This equation relates t to t' and x', the coordinate of the moving clock in the moving coordinate system and reads $t = [t' + (v/c^2)x']/\sqrt{1 - (v^2/c^2)}$. But $x' = 0$ (the moving clock is at the origin of the moving system) so that we can write $t = t'/\sqrt{1 - (v^2/c^2)}$ and $t' = t\sqrt{1 - (v^2/c^2)}$. These equations show us that the clock in the moving system slows down by the factor $\sqrt{1 - (v^2/c^2)}$ compared to the fixed clock as seen by the fixed observer O. By the same reasoning and using the same equations we deduce that O′ finds that his clock runs faster than the clock attached to O. We see that there is complete reciprocity here; each observer concludes from the Einstein–Lorentz transformations that his clock runs faster than that of the other observer. That this must be so is obvious since neither observer can know anything at all about his absolute state of motion; all each observer knows is that the other observer is moving past him at speed v.

If two observers are moving with constant speed with respect to each other, each observer finds that his own clock is going at a faster rate than the clock of the other observer even though the two clocks are identical and were carefully synchronized when they were at rest with respect to each other. It may seem that this retardation is only an apparent effect since the same clock goes slow or fast depending on its state of motion relative to the observer but this is not so since this retardation can be verified experimentally. This effect was pointed out in Einstein's first paper on relativity in which he noted that such a retardation gives rise to what, at first sight, appears to be a paradox and hence an inconsistency in the theory. A careful analysis of the retardation of a moving clock shows, however, that the theory is not at all inconsistent and no paradox really exists.

To see how the retardation of a moving clock can actually be measured, we consider two identical synchronized clocks which are both initially at rest at the origin of some inertial coordinate system. We now set one of the clocks in uniform motion with constant relative speed v along the X axis of the inertial coordinate system. Let this moving clock continue along the X axis until it has reached some point P on this axis and a time t has elapsed on the fixed clock. The time recorded by the moving clock will then be $t\sqrt{1 - (v^2/c^2)}$. At this moment t the velocity of the moving clock is reversed, and it is sent back to the origin of the coordinate system at the constant speed v. When the moving clock returns to this origin, the fixed clock will show that the total time $2t$ has elapsed for the entire trip of the moving clock. But the time shown on the moving clock when it returns will be $2t\sqrt{1 - (v^2/c^2)}$ since it is slowed down by the same amount going and coming. In other words, when the moving clock has returned, it will show that the amount of time that has elapsed for it is less than the amount of time that has elapsed for the moving clock by the factor $\sqrt{1 - (v^2/c^2)}$. This result appears quite inconsistent with the whole principle of relativity since it tells us that even though two inertial observers who are passing each other have no way of telling which one is "really" moving, yet there is a difference between the two, insofar as the passage of time goes; if one of them reverses his direction and catches up with the other, and then continues to move along with the other, the two clocks are no longer synchronized. The clock of the observer who reversed his velocity shows a shorter passage of time. The paradoxical quality of this situation stems from the following consideration: If observer O sees observer O′ move past him in a given direction, and then reverse his direction of motion and return, then O′ (not knowing his absolute state of motion) also sees O move past him (in the opposite direction) and return. Yet the two clocks do not behave the same way so that there appears to be a paradox. There really is no paradox because only one of the observers has actually suffered an acceleration and therefore altered his inertial frame of reference so that the two observers are not really equivalent.

We cannot analyze the above experiment in the framework of special relativity because special relativity deals with inertial frames of reference at all times. In the above experiment, one of the observers (the one who undergoes an acceleration and returns) changes from one inertial frame to a different one four times during his trip. Initially the two observers (or clocks) are in the same inertial frame, then one of them, say O′, is accelerated relative to the fixed stars until he acquires a velocity v relative to the other observer O. After a time t (as shown by the clock of O and a time $t\sqrt{1 - (v^2/c^2)}$ as shown by the clock of O′), O′ is slowed down (negative acceleration) to a stop and then accelerated again in the opposite direction until he again acquires a speed v. He is then again slowed down to zero relative speed when he returns to the observer O and comes to rest. After each acceleration during his journey, observer O′ changes to a new inertial frame so that the two observers do not undergo the same experience and there is therefore no reason to think that the flow of time is the same for both.

A dramatic and striking example of this dilation of time for a moving observer is the twin "paradox." We consider two identical twins, one of whom sets out on a trip to a distant star (let us say Sirius which is about 9 light-years away) at very nearly the speed of light while the other remains on the earth. We now ask each twin to measure the time that has elapsed (as he measures it) by counting the number of times that his heart has beaten during the trip. If N is the number of heartbeats per year of each twin (the hearts of the two twins are exactly synchronized so that N is the same for both) the twin on the earth finds that his heart has beat $(2L/v)N$ times, where L is the distance to Sirius and v is the distance traveled by the moving twin in 1 year. On the other hand, the traveling twin finds (owing to his speed v toward Sirius) that the distance to Sirius and back is just $2L\sqrt{1 - (v^2/c^2)}$. He therefore finds that his heart has beat $(2L/v)\sqrt{[1 - (v^2/c^2)]N}$ times. If v differs from c by a small amount, the twin left back on the earth counts about 18 years' worth of heartbeats whereas the traveling twin counts a few days' worth of heartbeats or less. In fact, by making v large enough (but always less than c) we can reduce the number of heartbeats of the traveling twin to as small a number as we please. Thus, the stay-at-home twin ages by at least 18 years while the traveling twin hardly ages at all.

This is such an astonishing deduction and one so contrary to our everyday concept of aging and time that many people (including some scientists) refuse to accept it, arguing that it is contrary to the very idea of relativity, but, as we have seen, this opposition is unjustified. We must simply readjust our thinking and accept the consequences of the theory. We find ourselves in a similar situation over and over again as we probe more deeply into modern theories. We must then adopt the following attitude toward any deduction from a theory: If a theory is known to be correct (it is in agreement, within experimental error, with all known observations), then all deductions that logically follow from the theory (all mathematical deductions) must be accepted as correct. If one rejects any part of a theory, one must reject the entire theory. The special theory of relativity has so far agreed to an incredible degree of accuracy with all phenomena that can be described in inertial frames. We must therefore accept the time dilation as deduced from the theory. But we need no longer accept it as a pure theoretical deduction from the theory because we now have two kinds of direct experimental evidence that there is a dilation of time for moving clocks.

The first reliable experimental evidence for the slowing down of clocks comes from the lifetimes of certain fundamental massive particles than can be created in the laboratory (they are also found in cosmic rays) by bombarding protons and neutrons with beams of high-energy (very rapidly moving) protons. In such collisions new particles, more massive than the original protons and neutrons, are created. The additional mass required to create these so-called "strange" particles comes from the high kinetic energy of the protons in the bombarding beam. These strange particles have a very short lifetime which ranges (depending on the kind of strange particle that is created) from much less than a trillionth of a second (10^{-12} sec) to a billionth of a second (10^{-9} sec). They then decay into protons and other particles. Consider a strange particle that is at rest relative to the laboratory when it is born. The span of time such a particle exists before it decays is called the lifetime of the particle; each species of strange particle has its own well-defined lifetime. When strange particles of a particular species are created, they are born with a variety of speeds relative to the laboratory and one finds that the lifetimes of these particles are not all the same. The particles that are moving fastest have the longest lifetimes. The lifetime of a particle increases with speed in exact accordance with Einstein's formula for the dilation of time.

Since some people may not be inclined to accept the increase of lifetime of strange particles with their speeds as direct evidence for the slowing down of time, we now cite another, more direct experiment that proves conclusively that moving clocks slow down compared to fixed clocks. This experiment was performed during October 1971 with special cesium beam atomic clocks having an accuracy of one part in billions. Four such clocks were flown twice around the world (once eastward and once westward) on regularly scheduled commercial jet flights. The theory predicts that the eastward-flown clock should slow down compared to the clock fixed on the earth whereas a westward-flown clock should run faster than the earthbound clock.

The analysis here is fairly simple and one can go through it with simple algebra. We consider an inertial observer fixed in space at the north celestial pole looking down at the earth. Relative to him a clock on the earth's equator at any moment is moving with a speed $R\omega$ where R is the earth's radius and ω is the angular speed of the earth. Such a clock therefore runs slow by the factor $\sqrt{1-(R^2/\omega^2)/c^2}$ or approximately $1 - (R^2\omega^2)/2c^2$ compared to the inertial clock fixed at the north celestial pole. If a clock is now moving eastwardly around the equator at a speed v (relative to the earth), its speed relative to the inertial observer is $R\omega + v$ and it runs slow by the factor $\sqrt{1-(R\omega+v)^2/c^2}$ or approximately $1 - (R\omega + v)^2/2c^2$ compared to the clock at the pole. Thus, the clock traveling eastward at the speed v relative to a clock fixed on the equator records a smaller passage of time than the clock fixed on the equator after one circumnavigation. In fact, the time difference is $(2R\omega v + v^2/2c^2)(T)$ where T is the time registered by the fixed clock. If, on the other hand, a clock is sent around the equator westward at a ground speed v, its speed relative to the inertial clock is $R\omega - v$ and it runs slow by the factor $\sqrt{1 - (R\omega - v)^2/c^2}$ or approximately $1 - (R\omega - v)^2/2c^2$. Hence, such a clock actually runs ahead (fast) with respect to the clock fixed at the equator if v is very nearly equal to $R\omega$ (about 1000 miles/hour).

This analysis, applied to the clocks on the eastward and westward circumnavigating jets, shows that compared to the fixed clock on the equator, the clocks moving eastward should lose about 40 nanoseconds (1 nsec equals one billionth of a second) during a single round-trip whereas the westward-moving clocks should gain about 275 nsec in a single trip (compared to the fixed clock). Put differently, the theory predicts that if two clocks are sent in opposite directions around the equator with each clock moving at a speed of about 1000 miles/hour relative to the ground, then when they return to the starting point, the clock that traveled eastward will have lost about 315 nsec compared to the clock that moved westward. When the actual experiment was performed, these predictions were completely verified. The observed results agree with the predicted results with an accuracy that was well within the experimental errors, so that we have conclusive proof of the slowing down of moving clocks as predicted by Einstein. There is no real clock paradox.

We now consider an inertial observer moving along with a clock at a constant speed v relative to another inertial observer whom we may take as "fixed." The time t' as measured by the moving clock is called the proper time of the moving observer. If t is the time as measured by the clock attached to the "fixed" observer, then the moving observer's proper time equals $\sqrt{[1 - (v^2/c^2)]}t$. Each inertial observer has his own proper time and inertial observers fixed with respect to one another have the same proper time. The proper time is important because it gives (or is a measure of) a quantity that has an absolute significance for all observers.

Before Einstein introduced his theory of relativity, to say that all observers measure the same distance and the same time interval between two specific events was meaningful; but the theory of relativity shows that such "absolutes" do not exist. This fact leads to the question as to whether any really absolute quantity is associated with such events; we now show that an absolute interval between any two events can be defined by combining, in a certain specified and well-defined way, the distance and the time interval between the two events. To do this we again consider the same two inertial observers we dealt with before and have them both observe the same two events that occur along the X axis. Let x and t be distance and time between them as measured by O′ and x' and t' the distance and time interval as measured by O′. We know that x does not equal x' and t does not equal t', but we can easily show that the quantity $\sqrt{(x^2 - c^2t^2)}$ equals $\sqrt{(x'^2 - c^2t'^2)}$. In other words, a quantity which we may call the space-time interval and which is the same for all observers exists. The square of this space-time interval for any observer is obtained by squaring the distance between the two events (as measured by this observer) and subtracting from this quantity the square of the product of the speed of light and the time interval between the two events.

The special theory of relativity does not say that no absolutes exist in nature, but rather that space and time combine to give a more general absolute (the space-time interval) than space alone or time alone gives in Newtonian theory. The space-time interval $\sqrt{(x^2 - c^2t^2)}$ is the same for all observers; it is an absolute.

The negative of the space-time interval between any two events equals the product of c^2 and the square of the proper time interval shown on a clock that coincides with the first event at the moment that event happens and is moving fast enough to reach (and to coincide with) the second event just when it happens. To see that this is so we may picture the two events as happening at the origin of the moving observer O′ so that the distance x' between these two events is always zero. Hence, the square of the space-time interval is just $-c^2t'^2$ since $x'^2 = 0$. But t' is the proper time of the moving clock so that we have demonstrated the truth of our statement. We can also see this from the point of view of O by noting that the square of the space-time interval for O is $x^2 - c^2t^2$ or $v^2t^2 - c^2t^2$. But this is just $-c^2(1 - v^2/c^2)t^2$ and $(1 - v^2/c^2)t^2$ is the square of the proper time of the moving clock.

The square of the space-time interval between two events may be positive, zero, or negative, and it is interesting and important to see when these three different possibilities occur. Since the square of the space-time interval, which we call s^2, is $x^2 - c^2t^2$, we see that $s^2 = 0$ when $x^2 = c^2t^2$ or $x = ct$. But this equation is exactly the condition for the propagation of a light signal in a vacuum. If two events are separated by a distance x which is just large enough so that a beam of light starting from the first event at the moment that it occurs reaches the position of the second event just when that event happens, then the space-time interval between the two events is zero. We then speak of a null (zero) interval between the events. If a beam of light triggers off a series of events along its path, these events are all separated by a null interval.

If the distance x between the two events is such that the light signal from the first event reaches the position of the second event after the second event has happened, then $x > ct$ and $x^2 - c^2t^2 > 0$ and the space-time interval between the two events is positive. We then call the space-time interval space-like. Two events that are separated by a space-like interval cannot be causally related because there is no way for one event to influence the other since no signal starting from the first event can travel fast enough to reach the position of the second event (and hence influence it) before that second event occurs. If two events E_1 and E_2 are separated by a space-like interval, it is always possible to arrange three different inertial observers at E_1 who are moving relative to each other in such a way that E_1 happens before E_2 for one of the three observers, E_1 is simultaneous with E_2 for another one of the observers, and E_1 occurs after E_2 for the third. Events that are separated by space-like intervals have no absolute time sequence; their chronological order depends on the state of motion of the inertial observer. The reason such an interval is called space-like is that an inertial observer always exists (the one for whom the events are simultaneous) for whom the events are separated only in space (no time separation).

If the distance x between the two events is such that $x^2 < c^2t^2$ the square of the space-time interval $x^2 - c^2t^2$ is time-like. This means that an observer starting at event E_1 at the moment it happens can always reach the position of event E_2 before E_2 occurs. This observer can thus influence E_2. We see that two events that are separated by a time-like interval can be causally related. The temporal sequence of events separated by time-like intervals is the same for all observers. If E_2 is separated from E_1 by a time-like interval and it occurs after E_1 as seen by an observer fixed at E_1, it occurs after E_1 as seen by all inertial observers. The reason such an interval is called time-like is that we can always find an inertial observer at E_1 (at the moment E_1 happens) who coincides with E_2 at the moment it happens so that for this observer there is no spatial separation between E_1 and E_2; they are separated only in time. Time-like and space-like intervals do not change their character when we go from one frame of reference to another; a time-like interval is time-like for all observers as is a space-like interval.

Taking into account the dependence of distances and time intervals on the velocity of the observer and on the constant c (the speed of light in a vacuum), we see that the Newtonian laws of motion are not valid because time and distance in these laws are treated as absolutes, and the constant c is absent from these laws. We must therefore reformulate the laws of motion to bring them into conformity with Einstein's theory. We first note that the Newtonian laws deal with two kinds of quantities separately—vectors and scalars. Newton's second law of motion relates a vector (force) to another vector (acceleration) with inertial mass (a scalar) appearing as a proportionality factor. Another law (which can be deduced from the second law) states that the time rate of change of a certain vector (momentum) is zero under certain conditions (conservation of momentum). Still another law deals with the time rate of change of a scalar called mechanical energy (conservation of energy). Another classical law dealing with a scalar states that mass is conserved. Since vectors are spatial quantities that are defined in terms of three-dimensional space and spatial relationships are not absolutes in the theory of relativity, we see that classical vector relationships are not invariant and hence are not laws in relativity mechanics. In the same way, statements about the time rate of change of scalars (e.g., mass, energy) are not laws in relativity since time is not an invariant. To obtain laws of mechanics that are relativistically correct, we must replace the Newtonian vector concept by a vector concept that does not refer to space in any absolute sense. Since space and time are mixed together in an inseparable way in the theory of relativity, we get correct laws only if we replace the usual three-dimensional vectors by vector entities which have time parts as well as spatial components. Such vectors, which are called four-vectors (four-dimensional vectors), can, indeed, be introduced, and the correct laws of mechanics can be expressed in terms of such four-vectors.

To see how four-vectors are defined we consider the simple four-vector that is related to the space-time interval; this four-vector combines space and time into a single space-time vector consisting of four components, but before doing this we note how we go from a vector viewed in two-dimensional space to this same vector in three-dimensional space. We can picture a three-dimensional orthogonal coordinate system and a vector **r**, which has the components x, y, and z along the three mutually perpendicular axes X, Y, and Z. Suppose now that two-dimensional beings live in the horizontal X–Y plane with no vertical perception at all. To such flat creatures the idea of a third spatial dimension at right angles to their plane is nonsensical and completely at variance with their everyday experience. They do not perceive the vector **r** in its totality but only its horizontal component **h**, which is the shadow of the vector **r** projected onto the X–Y plane by vertical rays of light L coming from a very distant source. Their picture of the actual three-dimensional world is thus a shadow picture and their geometry is only a two-dimensional portrayal of the real world. Instead of ascribing the length r to the vector **r** they ascribe the length h to it and write for its square $h^2 = x^2 + y^2$ (the theorem of Pythagoras). Two different flat observers in the horizontal plane, using different horizontal axes (x, y) and (x', y'), still assign the same length h to the vector **r** even though for one it is $h^2 = x^2 + y^2$ and for the other $h^2 = x'^2 + y'^2$. The individual values of the X and Y components are different but the sum of the squares of the components is the same for all observers. Thus, h^2 is the invariant for all observers. We see from our three-dimensional vantage point that the flat horizontal observers see a foreshortened picture of the actual world.

Suppose now that there were another set of flat, two-dimensional creatures living on a plane that is tilted at some angle with respect to the horizontal plane of our first set of flat beings (let us call it the E–N plane where E and N are the Cartesian axes for these other flat beings). They cannot perceive the full vector **r** either, but only the two-dimensional component of it projected onto their plane. The length of this component is different from h since it equals $\sqrt{(E^2 + N^2)}$ where E and N are the components of r along the E and N axes, respectively. We see that each set of two-dimensional observers gets a different picture of the length of the vector **r** (and hence of real space) and the observers in each set think their picture is the correct and only picture. If these different sets of two-dimensional observers could communicate with each other, they would see at once that their individual pictures of the universe are separately incomplete because each such set gives a different picture. The quantity $x^2 + y^2$ or $E^2 + N^2$ is not a three-dimensional absolute since it varies from one set of flat observers to another. To obtain a true picture of the vector **r** each set of two-dimensional observers must introduce a third dimension (a third orthogonal axis) at right angles to his plane. The length of **r** is then given by $\sqrt{(x^2 + y^2 + z^2)}$ and this length is the same for all sets of flat observers who enlarge their geometry from two to three dimensions. All observers thus ascribe an absolute character to three-dimensional spatial vectors and these vectors must be used to express the laws of nature. Although the individual components x, y, z of any particular vector vary from one observer to another, depending on the choice of coordinate system, the quantity $x^2 + y^2 + z^2$ for a given vector equals $x'^2 + y'^2 + z'^2$ or $E^2 + N^2 + m^2$ for the same vector representing the spatial separation of two events. This absolute three-dimensional geometry is the basis of Newtonian mechanics.

We now draw an analogy between the passage from the two-dimensional picture to the three-dimensional picture of the geometry of the universe and the passage from the Newtonian absolute space and absolute time picture of the universe to the Einstein space-time picture. We are three-dimensional creatures and if we were ignorant of Einstein's discovery, we would quite naturally describe events in our universe in terms of a three-dimensional geometry with the vectors **r** as real geometric absolute entities. We would also introduce separately the time t between events as a scalar invariant. Thus, we would break up the description of events in our universe into a separate absolute three-dimensional space description plus a one-dimensional time description. There would be no problem in accepting such a division of our universe into absolute space and absolute time if all inertial observers obtained the same spatial description and the same temporal description. But this is not so; we have seen that the separate spatial description r and temporal description t vary from one inertial observer to another but the quantity $\sqrt{(r^2 - c^2t^2)}$ (the space-time interval) remains the same. Thus, $x^2 + y^2 + z^2$ and t^2 vary from one inertial observer to another but $x^2 + y^2 + z^2 - c^2t^2$ does not. But this situation is analogous to the situation of the flat observers who see only the two-dimensional part of each spatial vector and thus obtain a wrong picture of space. To obtain a correct picture they must replace $x^2 + y^2$ by $x^2 + y^2 + z^2$. In the relativistic world we obtain the correct picture by replacing the three-dimensional $x^2 + y^2 + z^2$ by a four-dimensional invariant $x^2 + y^2 + z^2 - c^2t^2$. Just as we look upon $x^2 + y^2 + z^2$ as the square of the length of a three-dimensional vector, we must look upon $x^2 + y^2 + z^2 - c^2t^2$ as the square of the length of a four-dimensional vector or a four-vector.

The three-dimensional vector **r** has the three components **x, y, z** and the square of its length is the sum of the squares of these components. We must therefore look upon our four-vector as consisting of four components, the sum of whose squares gives us the square of the length of the four-vector. From these considerations and from the first three terms in the expression $x^2 + y^2 + z^2 - c^2t^2$ for the square of the space-time interval we can at once pick out three of the four components of our space-time four-vectors. They are **x, y, z,** the same as the components of the vector **r.** The fourth component of this four-vector must then be a term which when squared gives us $-c^2t^2$ and the only possibility is ict where $i = \sqrt{-1}$. The appearance of the unit imaginary i is remarkable here since it shows us that the fourth component has a physical reality that is different from that of the other three components and lies along an imaginary axis which is orthogonal to the three real spatial axes. The multiplication of a quantity by i rotates it into an orthogonal direction. We may refer to the three components x, y, z as the spatial components of our space-time four-vector s and to ict as the time component and write s: (**r**, ict) or (**x, y, z,** ict). If we square each component of this four-vector and add them, we obtain $s^2 = x^2 + y^2 + z^2 - c^2t^2$.

In addition to s, other natural four-vectors exist, each of which has three space components and one time component. The time component of a four-vector, in general, contains i and c as factors in some manner or other. Just as the square of a four-vector (obtained by adding the sum of the squares of the four components) is an invariant for all inertial observers, so too is the scalar product of two four-vectors which is obtained by multiplying together the respective components of each four-vector and adding these four products. If one four-vector has the components $\mathbf{x}_1, \mathbf{y}_1, \mathbf{z}_1, ict_1$, and the other has the components $\mathbf{x}_2, \mathbf{y}_2, \mathbf{z}_2$, ict_2, then the scalar product of these two four-vectors is $x_1x_2 + y_1y_2 + z_1z_2 - c^2t_1t_2$ and this product is the same for all inertial observers.

Although the basic laws in Newtonian mechanics must be expressed in terms of three-dimensional spatial vectors, the laws in Einsteinian mechanics must be expressed in terms of four-vectors, a special example of which is the space-time four-vector s: (**r**, ict). We now deduce another four-vector—the momentum–energy four-vector—which leads to a single conservation principle involving momentum, mass, and energy. In our discussion of Newtonian mechanics we introduced three separate conservation principles—the conservation of momentum, the conservation of energy, and the conservation of inertial mass. The first of these principles deals with a three-dimensional vector (momentum) and the other two deal with two scalars: energy and mass. These conservation principles as stated within the context of Newtonian mechanics are incompatible with the special theory of relativity. We can construct a single four-vector which incorporates the three components of momentum and energy of a particle in such a way that the square of the length of this four-vector gives us at once, in a single statement, the conservation of momentum–energy, and mass.

We consider a particle of rest mass m_0, moving with constant speed v with respect to some inertial observer, whom we may take to be at rest. By the "rest mass" or "proper mass" of a particle we mean the mass the particle has when it is not moving relative to the inertial observer when $v = 0$. We differentiate here between the rest mass m_0 of a particle and its mass m when in motion because in the theory of relativity the mass of a particle, as measured by a given observer, may depend on its velocity relative to that observer. We construct the quantity $v^2 - c^2$ and write down the identity $(v^2 - c^2)/(v^2 - c^2) = 1$ or $v^2/(v^2 - c^2) - c^2/(v^2 - c^2) = 1$ where c is the speed of light. Since $v^2 = v_x{}^2 + v_y{}^2 + v_z{}^2$ where $\mathbf{v}_x$, $\mathbf{v}_y$, $\mathbf{v}_z$ are the x, y, z components of the velocity v of the particle in the observer's coordinate system, we can rewrite the previous equation as $v_x{}^2/c^2(1 - v^2/c^2) + \cdots + c^2/-c^2(1 - v^2/c^2) = 1$, where the dots stand for the y and z terms which have the same form as the x term where we have factored $-c^2$ out of the denominator of each term. If we multiply through by $-c^2$, we obtain $[v_x/\sqrt{1 - (v^2/c^2)}]^2 + \cdots + [ic/\sqrt{1 - (v^2/c^2)}]^2 = -c^2$ where $i = \sqrt{-1}$, the imaginary unit so that $i^2 = -1$.

If we examine the right-hand side of the equation, we see that it consists of the negative of the square of the speed of light, which is the same for all observers and is thus an invariant. Hence, the left-hand side must be an invariant (the same for all inertial observers). But we note that the left-hand side consists of the sum of the squares of four components, three of which are related to the x, y, z spatial coordinates of an inertial observer and the fourth of which involves i and c. These four terms must then be the squares of the four components of a four-vector since their sum is an invariant and there are just three space components and a fourth component that has all the properties of a time component.

The three space components, as we see, are just the velocity components of the particle divided by $\sqrt{1 - (v^2/c^2)}$, and for very small values of v they are practically equal to these velocity components. We may therefore consider the four components above as constituting a four-vector which we may call the velocity four-vector of a particle.

Returning now to the previous equation that shows that the square of the velocity four-vector equals $-c^2$, we multiply each term by m_0^2 thus obtaining a four-vector whose square is $-m_0^2c^2$. Since the rest mass m_0 of our particle is the same for all inertial observers, the right-hand side of this equation is still an invariant (the same) for inertial observers. Hence, the left-hand side must be the square of the length of a four-vector whose three space components are $m_0v_x/\sqrt{1 - (v^2/c^2)}$, $m_0v_y/\sqrt{1 - (v^2/c^2)}$, $m_0v_z/\sqrt{1 - (v^2/c^2)}$ and whose time component is $im_0c^2/c\sqrt{1 - (v^2/c^2)}$. How are we to interpret this four-vector? We note that for very small values of v [which means that $(v/c)^2 \ll 1$] the three space components are very nearly m_0v_x, m_0v_y, m_0v_z. But these are just the three components of the classical or Newtonian momentum of a particle. Hence we see that in relativity theory the components of the momentum are $m_0v_x/\sqrt{1 - (v^2/c^2)}$, etc., which are just the space components of the four-vector we constructed. If we call these components $\mathbf{p}_x$, $\mathbf{p}_y$, $\mathbf{p}_z$ and place $m_0c^2/\sqrt{1 - (v^2/c^2)} = E$, we see that we have a four-vector with the three space components $\mathbf{p}_x$, $\mathbf{p}_y$, $\mathbf{p}_z$ and a time component iE/c. We may write this four-vector as $(\mathbf{p}, iE/c)$, where $\mathbf{p}$ has the components p_x, p_y, p_z. If we consider the x component of this four-vector (or the y or z component) we have $p_x = m_0v_x/\sqrt{1 - (v^2/c^2)}$. If we now define the moving mass m of the particle as $m = m_0/\sqrt{1 - (v^2/c^2)}$, we can write $p_x = mv_x$ for the x component of our four-vector. But this expression is just like the Newtonian expression for the x component of the momentum of a particle with m taking the place of m_0. In other words, we can still write mv for the momentum of a moving particle in Einsteinian mechanics, but m is now not a fixed invariable quantity which is the same for all observers, but rather one that varies with the velocity v of the particle relative to the observer. We have in fact $m = m_0/\sqrt{1 - (v^2/c^2)}$ which is the famous Einstein formula for the variation of the mass of a particle with its velocity. This formula is in complete agreement with the measured masses of cosmic ray particles that are detected on the earth. Such particles travel with speeds that are close to the speed of light and their measured masses vary with their speeds according to the Einstein formula.

We have seen that the three space components of the four-vector $(\mathbf{p}, iE/c)$ are just the Einsteinian or relativistic components of the momentum of the particle, but what about the fourth or time component, iE/c? It has the dimensions of a momentum, but the quantity E that appears in this fourth component is an energy. We can see what E means by examining it for small values of v. Since $E = m_0c^2/\sqrt{1 - (v^2/c^2)}$, we can show that for $v \ll c$ $E \approx m_0c^2 + (\frac{1}{2})m_0v^2$, since all higher powers of v/c such as $(v/c)^4$ are negligible. We see from this approximate expression for E, that is valid for small values of the speed v, that, except for the term m_0c^2, E is just the Newtonian kinetic energy. In relativity theory, E is the kinetic energy plus an additional term m_0c^2 which is the same for all observers. We may therefore call E the total energy of the particle.

We deduce from this statement that a freely moving particle in relativity theory has a total amount of energy, as measured by a given inertial observer, which consists of a part that is invariant (the same for all observers) and a part that depends on its speed relative to this observer. The total energy of a free particle (a particle that is not in a force field) can never be less than m_0c^2, which is a result that is quite unexpected and is entirely outside the domain of Newtonian mechanics.

That the fourth component of the four-vector $(\mathbf{p}, iE/c)$ is related to the energy shows us that the momentum p of a particle and its energy E cannot be treated separately. They are related to a single four-vector (the momentum–energy four-vector) whose space part reduces to the Newtonian momentum for small values of v and whose time part is related to the total energy of the particle. We have already seen that the square of the momentum–energy four-vector equals the invariant quantity $-m_0c^2$. In fact, we have from above the expression $p_x^2 + p_y^2 + p_z^2 - E^2/c^2 = -m_0^2c^2$ or $p^2 - E^2/c^2 = -m_0^2c^2$. This relationship, which was first derived by Einstein, shows us how the energy, the momentum (as measured by a given inertial observer), and the rest mass m_0 of a freely moving particle are related. If we multiply through by c^2 and solve for E^2 we obtain $E^2 = c^2p^2 + m_0^2c^4$. Thus, the total energy of a freely moving particle depends on its momentum p (which changes from observer to observer) and on its rest mass m_0 (which is the same for all observers). To obtain E from this expression, we take the square root of both sides, so that $E = \pm\sqrt{c^2p^2 + m_0^2c^4}$. We call special attention to the negative square root which plays an important role in the theory of antimatter.

This expression, relating E to the momentum p of a particle, is to be compared to the Newtonian expression $E = p^2/2m_0$; we see two important differences: (1) in Newtonian mechanics E is zero if p is zero, but in Einsteinian mechanics $E = m_0c^2$ (the famous Einstein equation between mass and energy) if $p = 0$; (2) the energy E has only positive values in Newtonian mechanics whereas it can be both positive and negative in relativistic mechanics. The equation $E = m_0c^2$ which expresses the equivalence between mass and energy is one of the most important consequences of the theory of relativity without which we could not understand such things as nuclear transformations, nuclear energy, and the vast luminosities of the stars.

We close this chapter by introducing one other important four-vector that is associated with the propagation of light in a vacuum, as described by Maxwell's electromagnetic theory of light, which we have already mentioned. According to this theory, light is a wave which, in its simplest form, we may picture as a train of electric and magnetic oscillations propagated at the speed c. We may associate with these oscillations a spatial aspect and a temporal aspect. The spatial aspect is given by a three-dimensional vector pointing in the direction of the propagation of the wave. The magnitude of this vector is proportional to the reciprocal of the wavelength λ of the wave (the distance between two successive crests of the wave). The temporal aspect of this four-vector is proportional to the frequency ν of the wave (the number of vibrations per second). Since $1/\lambda$ is proportional to the space part of the wave vector and momentum is also the space part of a four-vector, it follows that $1/\lambda$ is proportional to momentum. We may therefore assign momentum p to such a wave and write $p = h/\lambda$ where h is a proportionality factor. By the same reasoning, we see that the energy E of the wave must be proportional to its frequency ν because ν and E are both the time parts of a four-vector. We therefore write $E = h\nu$ where h is the same proportionality factor as that in the relationship between p and $1/\lambda$. These two relationships are the basis for the quantum theory of radiation and the wave theory of matter (quantum mechanics).

CHAPTER 11

The General Theory of Relativity

A theory is the more impressive the greater the simplicity of its principles is, the more different kinds of things it relates, and the more extended is its area of applicability.

—ALBERT EINSTEIN, *Autobiographical Notes*

In the previous chapter on Einstein's special theory of relativity, we saw that this special theory is necessary because the relativity of Galileo and Newton places the Newtonian laws of motion in a preferred position with respect to the laws of optics. As we have seen, only the Newtonian laws of mechanics remain the same in all inertial frames if we adopt the Galilean–Newtonian concepts of absolute space and absolute time; the laws of optics (in particular, the law concerning the speed of light in a vacuum as deduced from Maxwell's electromagnetic theory of light) take on different forms in different inertial frames. To obtain a description of nature which treats all the laws on the same footing so that neither mechanical nor optical phenomena can be used to differentiate between any two inertial observers, we must use Einstein's four-dimensional space-time concept. Thus, the special theory of relativity is required if we are to consider all inertial frames of reference to be equal or equivalent in the eyes of nature. If the special theory were not true, the laws (in particular the laws of optics) would have to be so formulated that they are valid in only one frame of reference—the frame which is at absolute rest in the universe. But since the concept of absolute rest has no meaning from an experimental point of view (no mechanical or optical experiment can ever tell us whether we are or are not at rest), we must discard it and, with it, the idea of a single preferred inertial frame of reference (the one at absolute rest), as the only one in which the laws of physics are valid. In other words, the special theory of relativity was introduced to eliminate the restriction imposed upon the laws of nature by the concepts of absolute space and absolute time and the notion of a preferred coordinate system.

But this theory does not remove all the restrictions on coordinate systems since it says nothing about the validity of laws in noninertial frames of reference (accelerated frames, rotating frames, or frames in gravitational fields). Indeed, it says that only in inertial frames of reference do the laws of nature take on their simplest and valid forms. To Einstein this restriction of observers (who are seeking the correct laws) to inertial frames of reference is a very undesirable feature of the special theory which indicated to him that something essential is missing in the theory as it stands. He felt that to be complete the theory of relativity should be general enough to include all observers regardless of how they are moving, and the laws of nature should be the same for all such observers. This conclusion prompted Einstein to reformulate the theory in such a way that all frames of reference are equivalent in the eyes of nature. When Einstein had completed this task, he obtained what we now call the general theory of relativity.

When we consider that it took Einstein (the greatest mind of the 20th century, and possibly of all time) ten years to achieve this generalization, we can understand why the general theory of relativity is described as the single greatest intellectual synthesis of all time and as the most beautiful of all physical theories. We shall see that its beauty lies in the magnificent manner in which it unites nature and geometry so that natural laws become a branch of geometry and geometry becomes a part of natural laws. We saw that the special theory is the first step in this direction in replacing the three-dimensional Euclidean geometry that is the basis of Newtonian mechanics by a four-dimensional space-time Euclidean geometry. The step that Einstein took in going from the special to the general theory of relativity replaced the four-dimensional Euclidean space-time geometry of special relativity by a four-dimensional non-Euclidean space-time geometry. The need for this substitution becomes apparent as we analyze the physical reasoning that guided Einstein.

One may perhaps feel that Einstein was laboring a point needlessly in insisting that all frames of reference (inertial or not) must be acceptable frames in which to express the laws of nature, for what actual difference can it make in the end (in our understanding of nature) whether we accept the equal status of all frames of reference or not? This question is not merely a philosophical matter, without any real physical content, because some very practical consequences flow from the equivalence of all coordinate systems. If we deny this equivalence and try to formulate the laws of nature within the framework of Einstein's restricted (special) principle of relativity, we find it impossible in practice to compare our theoretical deductions (laws) with the experimental evidence; for to do so, we must perform all our experiments and measurements in an inertial frame of reference since the special theory states that only in such frames do the laws of nature take on their correct forms. But where is such a frame of reference? The earth is certainly not such a frame since an inertial frame is one in which the Galilean–Newtonian law of inertia is valid. The frame must be so far away from all other bodies (no gravitational field) and moving in such a way (uniform straight-line motion) that a body left to itself obeys the law of inertia (Newton's first law that an object moves with uniform speed in a straight line). To pursue such science we must have a frame of reference that is undergoing no acceleration. But since the bodies all around us affect us gravitationally, we can never find an inertial frame of reference. We can never apply the special theory of relativity to the real world (except approximately) because we can never find an inertial frame. Einstein was fully aware of this difficulty inherent in the special theory and felt that it is thus flawed in its description of nature. To eliminate this flaw, Einstein introduced the general principle that the laws of nature must be formulated in such a way that they are valid in all frames of reference. This concept is known as the general principle of covariance which may be interpreted as follows: Only those statements about the universe which have the same mathematical form in all frames of reference are laws of nature. This idea is to be compared with the special principle of relativity which states that a law of nature must have the same form in all inertial frames but can change its form in a noninertial frame.

Einstein was prompted by broad arguments concerning the invariance of the laws of nature in all frames of reference to seek a generalization of his theory of relativity but he began this search for another strong reason. The special theory cannot deal with gravitational phenomena since a frame of reference is not an inertial frame when it is in a gravitational field. Moreover, Newton's law of gravity is not the same in all inertial frames if one uses the Einstein–Lorentz transformations to pass from one inertial frame to another. To convince ourselves of this we note that the quantities (such as the inertial masses of two gravitationally interacting bodies and the distance between them), in terms of which the law is expressed, have no absolute meanings. If one calculates the force of gravity between two bodies using Newton's law, the result depends on the inertial frame of reference used. If we were traveling at a very great speed (along the line connecting the two bodies), their masses would be very large and the distance between them would be small. We would observe a very large gravitational force, whereas an observer traveling slowly with respect to the two interacting bodies detects a much smaller force. In any case, Newton's law of gravity cannot be correct, as it stands, because it disagrees with the invariance requirements of the special theory of relativity. Before Einstein introduced his general theory of relativity, which led to the geometrization of gravity, various attempts were made to alter Newton's theory and bring it into conformity with the special theory. But such attempts were fruitless. The principal obstacle to such attempts is the action-at-a-distance character of Newton's law of force which requires that gravitational action be transmitted instantaneously from one body to another. This requirement is contrary to the basic principle of the special theory which states that no physical effects can be transmitted at speeds greater than the speed of light c.

From Einstein's point of view, the special theory of relativity is flawed because it gives preference to a special category of coordinate systems (inertial frames) for transmitting nature's laws; this flaw can be eliminated only if we treat all coordinate systems as equivalent. This notion was the starting point of Einstein's search for a correct and general formulation of the theory of relativity, and we shall see that, in spite of what seemed to be insurmountable difficulties, Einstein succeeded in his quest. In doing so he arrived at a law of gravity which flows naturally from the non-Euclidean four-dimensional space-time geometry in terms of which the general theory is formulated. Gravity then manifests itself not as a force acting instantaneously at a distance but rather as a distortion of the space-time geometry in the neighborhood of a point arising from the surrounding masses.

Before we go on to Einstein's geometric generalization of the theory of relativity and his geometric theory of gravitation, we consider the difficulties that he encountered in proving that all motion (accelerated as well as uniform) is relative. This proof was clearly essential if his principle of general covariance (the principle that the laws of nature are those relationships among events that have the same mathematical form in all coordinate systems) was to be accepted because if (contrary to Einstein's strong intuition) acceleration were an absolute kind of motion (as distinct from uniform motion) the laws of nature could not be generally covariant for they would have to assume a special form in an accelerated frame to permit an observer in such a frame to detect his absolute state of acceleration. At first sight, in terms of our daily experience, it appears that Einstein's intuition had led him astray and that accelerated motion is, indeed, absolute and quite different from uniform motion. We are all aware that when we are in an unaccelerated vehicle (such as a smoothly flying plane or a uniformly moving ship in still water) we are completely unaware of our motion and we can do everything as easily as we can when we are at rest on the earth. We can walk about, pour coffee from a pot into a cup, play billiards, write on a surface, and so on without experiencing the slightest disturbance. But let there be the slightest interruption in the motion of the vehicle (a speeding up, a slowing down, or a change in its direction of motion) and we become aware of it immediately; we lurch to one side of another as we walk; the coffee we pour misses the cup; the billiard ball we hit no longer moves in a straight line; our pen ceases to follow its accustomed course. These easily observable phenomena seem to tell us that we are being accelerated and hence appear to give credence to the concept of absolute acceleration. How did Einstein overcome this apparent denial of his hypothesis that all motion is relative?

To understand Einstein's motivation and his procedure we consider in some detail the relationship of Newton's second law of motion to the frame of reference in which it is expressed. We recall that Newton's second law, expressed in the form $\mathbf{F} = m\mathbf{a}$, tells us that if we observe that a body of inertial mass m has an acceleration $\mathbf{a}$, then a force $\mathbf{F}$ must be acting on it, where $\mathbf{F}$ equals $m\mathbf{a}$. The second law relates any observed acceleration of a body to a force acting on the body, but according to Newtonian mechanics, such a relationship can be assumed to exist only in an inertial frame of reference. If we are in an accelerated frame of reference, all bodies in the frame (ourselves included) suffer easily observable accelerations (opposite to the acceleration of the frame) which arise from the inertial properties of the material bodies, but since these observable accelerations occur without any apparent forces acting on the bodies, we conclude (again in accordance with Newtonian principles) that they arise not from some mysterious external forces acting on the bodies but rather from the acceleration of the entire frame. In other words, if we find that Newton's second law is not obeyed (accelerations of bodies without applied forces) we do not discard this law but simply assume that we are in a noninertial frame. It appears, then, that Newton's second law is an insurmountable barrier against Einstein's concept of the relativity of all motion.

But we now examine the situation of the observer in the accelerated frame of reference more carefully and see how Einstein prevailed. If we study the accelerated motions of the bodies in the noninertial (accelerated) frame without asking how these accelerations arose, we note that the bodies behave just as though a force were acting on them and the accelerations induced by this apparent force are the same for all the bodies regardless of their inertial masses. From this point of view we might, in fact, say that all the bodies in our accelerated frame of reference behave as though there were a force (which we may call an "inertial force") acting on them and which imparts to each one the same acceleration. From the Newtonian point of view, such inertial forces are fictitious and are introduced simply to enable us to apply Newton's second law to bodies in accelerated frames. By introducing such inertial (but fictitious) forces in Newtonian mechanics we can still apply the law $\mathbf{F} = m\mathbf{a}$ to accelerated frames if $\mathbf{F}$ consists not only of all external forces that are applied to the bodies but also all the "inertial forces" such as centrifugal and Coriolis forces arising from rotations and other kinds of accelerated motions. Einstein stepped in at this point and used these very inertial forces, that seemed to operate against his concept of the relativity of all motions, to establish this general relativity concept.

Einstein first noted that inertial forces have precisely the same property as do gravitational forces in that they impart the same acceleration to all bodies regardless of their masses. This last fact was very suggestive to Einstein and pointed the way to achieving his goal. The gravitational field has the very interesting property (first discovered by Galileo) that all bodies, regardless of their masses, have the same acceleration when placed at the same point in a given gravitational field. All bodies in a vacuum at the same point on the earth fall with the same acceleration. This constancy is also true for all falling bodies on the moon as was demonstrated by the astronauts who landed on the moon. In other words, a gravitational force imparts the same kind of acceleration to all bodies as an "inertial force" does so that there is no way to distinguish between a gravitational and an inertial force. This conclusion led Einstein to enunciate one of the most famous principles in physics—the principle of equivalence, which Einstein used as the basis of the general theory of relativity. That bodies respond to "inertial forces" in exactly the same way as they do to gravitational forces led Einstein to the assertion that "inertial forces" are as real as gravitational forces and that there is no way for an observer in a noninertial (accelerated) frame to distinguish between gravitational forces and "inertial forces" acting on bodies in that frame. The two kinds of forces are completely equivalent—hence the "principle of equivalence."

This principle goes beyond the mere statement that gravitational and inertial forces are equivalent because it asserts that every effect produced (on any physical system) by acceleration, or observed by an observer in an accelerated frame can be produced by appropriate gravitational forces and observed by an observer at rest in the gravitational field. This statement of the principle of equivalence shows us how Einstein used it to proclaim the equivalence (the relativity) of all frames of reference. To say that all motions are relative and that therefore inertial and noninertial frames are equivalent means that an observer in an accelerated frame of reference can perform no experiment (e.g., mechanical, optical) nor make any observation within his frame of reference on any phenomena that can tell him whether he is at rest, moving with constant speed in the same straight line, or is accelerated. This result follows from the equivalence of inertial and gravitational forces; the accelerated observer cannot know whether he is really accelerated or whether he is at rest or moving uniformly in a gravitational field of force because it produces the same effects as the accelerated motion does. We see that in spite of our feeling that we can detect accelerated motion, in actual practice it is impossible to do so because of the validity of the principle of equivalence. The validity of the principle of equivalence rests on the exact equality of the gravitational and inertial mass which has been established experimentally by a number of different investigators with incredible accuracy. Newtonian mechanics accepted this exact equality as a purely phenomenological accident and ascribed no theoretical significance to it. But Einstein could not accept such a property of matter as being a mere accident of nature and therefore assigned to it a profound theoretical significance.

In our discussion of the special theory of relativity, we saw that Einstein performed imaginary (*gedanken*) experiments which led to questions that had to be answered in such a way as to reveal important features of nature. These experiments were designed to answer a few specific questions but these questions were of such a nature and so penetrating that some particular theory survived or fell according to the answers given by the experiment. Einstein's imaginary elevator experiment is perhaps the most famous of his thoughts experiments leading, as it did, to the concept of the curvature of four-dimensional space-time and to a new gravitational theory based on non-Euclidean geometry.

Since Einstein's elevator experiment was designed to show that an observer cannot distinguish between accelerated motion and uniform motion or rest, we can best understand and analyze this experiment by considering two observers in two identical elevators, one of which is at rest on the earth and the other of which is somewhere in empty space and so far from any large masses that it experiences no gravitational force. If the second elevator were a freely moving body, it would be an inertial frame of reference moving in accordance with Newton's first law, but in Einstein's experiment a rope is attached to this elevator and the rope is being pulled by a constant force that is just large enough to impart to this elevator a constant acceleration of 32.2 ft/sec^2. Owing to this acceleration, all bodies (including the observer) in this distant elevator experience "inertial forces" just as all bodies in the first elevator, at rest on the surface of the earth, experience gravitational forces. We can picture one elevator at rest on the earth and the other at a great distance from the earth and accelerated. Each observer now describes events in his elevator and draws whatever conclusions he can from them.

The observer in the fixed elevator on the earth (whom we may take as a 17th century scientist) notes that an apple that has just dropped from a tree (which we may picture as being enclosed in the elevator) descends with a constant acceleration $g = 32.2 \text{ ft/sec}^2$. He also notes that this acceleration is true of all bodies (regardless of their masses or chemical natures) that are in free fall (bodies in a vacuum that are in contact with nothing and are not being acted upon by any observable external force) and that a body thrown horizontally falls in an arc. He also notes that when an object of mass m is attached to a spring suspended from the ceiling of his elevator, the spring is stretched as though it were being pulled by a force exactly equal to mg. Finally, he feels the pressure of the ground against his feet as though he were being pulled downward by some force and he observes that a scale on which he happens to be standing records his weight as Mg where M is his mass. To explain all these phenomena, he assumes first that he is at rest (or in a state of uniform motion) and to explain his observations (the accelerations), he invokes a force, emanating from the earth, which he calls gravity, and notes that this force has the remarkable property of imparting the same acceleration to all freely falling bodies, regardless of their weight.

The observer (a 20th century scientist) in the distant elevator also notes that an object thrown horizontally falls in an arc; that an apple torn free of the apple tree in his elevator "falls" to the floor of the elevator (as do all other free bodies in his elevator) with an acceleration g; and that if an object of mass m is attached to a spring suspended from the ceiling of his elevator, the spring stretches as though it were being pulled with a force mg. Finally, he feels the bottom of his feet being pushed up by the floor of the elevator and notes that the scale on which he is standing registers a weight Mg for him. This 20th century scientist, however, is more circumspect in his conclusions than the 17th century scientist because he reasons that there are two possible (and completely equivalent) ways of explaining his observations. On the one hand, he may conclude that his elevator is at rest on some surface (as did the 17th century scientist) and that all the observed phenomena are caused by a gravitational force emanating from the surface on which the elevator rests. On the other hand, he may conclude that no force is acting on the elevator or the objects in it and that all the observed phenomena arise because his elevator is undergoing a constant acceleration. He then points out that since there is no way to tell which conclusion is correct or to choose one in preference to the other, it is meaningless to distinguish between the two and, indeed, even to assert that there is any difference. If, as dictated by the facts, we accept this position (which is essentially a restatement of the principle of equivalence), we must accept Einstein's assertion that there is no way for an observer in a given frame of reference to distinguish between an accelerated state of motion of his frame and nonaccelerated motion. This absence of distinction between the two cases abolishes the idea that accelerated motion is absolute and that it can therefore be detected by observing the behavior of bodies in the accelerated frame.

In stating his principle of equivalence, Einstein went beyond the idea that this principle applies only to mechanical phenomena (the motion of particles). He stated that no phenomena (e.g., mechanical, optical, electrical) can show the absolute state of motion of a frame of reference because whatever occurs in accelerated frames also occurs in a frame that is at rest in an appropriate gravitational field. If we think about this idea for a moment, we see that this statement has a profound physical significance because it tells us that we can deduce what happens to matter, light, and electrical phenomena in a gravitational field by studying these same phenomena in an appropriately accelerated frame of reference. This aspect of the principle of equivalence led Einstein to the immediate prediction that light must be attracted by a gravitational field as are all material bodies.

In our analysis of Einstein's elevator experiment we saw that we can create a force field inside an elevator, by accelerating the elevator, that is completely equivalent to the gravitational field. But we can also reverse things and eliminate a gravitational force field in a given frame of reference (our elevator) by allowing the frame to move appropriately. We consider the elevator with the 20th century scientist in it, but this time it is at rest, suspended by a rope from the ceiling of a room on the earth. The observer now finds that everything behaves just as it did before when the elevator was accelerated: The apple is accelerated, the spring is stretched, and the scale registers his weight. These phenomena, however, now arise from the force of gravity instead of from a constant acceleration. If the rope by which the elevator is suspended is suddenly cut so that it begins to fall freely, with an acceleration g, everything in the elevator begins to fall with the same acceleration. The apple, instead of continuing to fall to the floor of the elevator with increasing speed, falls with constant speed (the speed it had at the moment the rope was cut); the spring with the mass m attached to it does not remain stretched but contracts to its normal state; the observer on the scale no longer feels the platform of the scale pressing up against his feet so that the scale reads zero (registers no weight for him); and the horizontally thrown object does not continue to move in a parabolic arc as it falls, but moves horizontally across the elevator in a straight line. In other words, everything inside the elevator behaves (insofar as the observer in the elevator is concerned) just as they would in an inertial frame or frame of reference. We can thus create the equivalent of a gravitational force field or eliminate a gravitational field by appropriately accelerating a frame of reference. In free fall objects become weightless.

We have seen that Einstein introduced the general theory of relativity to remove the restrictions imposed on the laws of physics by the special theory which states that only in inertial frames of reference can the laws take on their simplest and correct form. With the general theory guiding us we must now formulate our laws so that they have the same form in all frames of reference. To see what this means and how we are to use this idea to express the laws of nature we consider Newton's first law of motion. Newton stated this law in a form which makes it correct only in inertial frames of reference and which would enable us to differentiate between inertial and noninertial frames. Newton's first law states that a body left to itself (no forces acting on it) continues in a state of rest, if initially so, or moves with constant speed in a straight line. But we see from Einstein's elevator experiment that the above statement of Newton's first law is not invariant since the concept of a straight line can have no absolute meaning. The behavior of the same horizontally thrown body as viewed by the two observers is different; to the observer in the elevator on the ground (the elevator at rest in the gravitational field), the body is not moving in a straight line and hence, according to Newton's first law, is being acted on by a force, whereas, to the observer in the freely falling elevator, it is moving in a straight line and therefore has no force acting on it.

Which statement is correct? According to the principle of equivalence, they are both correct but that means that the form in which Newton's first law is stated is incorrect. The reason it is incorrect is that it uses such concepts as ''force'' and ''straight line'' which have no invariant or absolute meaning. We now reformulate Newton's first law to express it in terms that do have absolute meaning so that in its new form it is invariant when we transform it from one coordinate system to another. To do this we must express the first law without invoking the terms ''force'' and ''straight line'' as used by Newton, which have no absolute, invariant meaning. If we limit our discussion to bodies in accelerated frames of reference and in gravitational fields, we can see how the first law is to be brought into conformity with the general principle of relativity; we must describe the orbit of the horizontally thrown particle in a way which has the same absolute meaning for both observers. Obviously we fail to do this if we use such expressions as ''force'' and ''straight line''; ''straight'' for one observer is not ''straight'' for the other. But there is a way of solving this problem which is related to the general concept of a ''straight'' line but which does not have the undesirable features of the term ''straight.''

To see just what is to be done and how we are to get rid of the unacceptable concept of ''straight'' we consider two bodies, one of which is moving on a perfectly flat surface and the other on the surface of a perfect sphere. In accordance with Newton's first law of motion, the particle (which we may assume to be a small ball bearing) on the flat surface moves in a straight line, but the particle on the sphere cannot move in a straight line since there are no straight lines on a sphere. How then does the particle on the sphere move if it is always constrained to move along the surface of the sphere by a force that always keeps the particle in contact with the sphere (a force that is always directed toward the center of the sphere)? By noting that the surface of a very large sphere is very nearly flat we conclude that the path of the particle on such a surface is a curve that is equivalent to a straight line. Now it can be shown both theoretically and experimentally that the path of the particle on the sphere is the arc of a great circle (a circle whose radius equals the radius of the sphere) and we know that the arc of a great circle on a sphere is equivalent to a straight line on a plane because each is the shortest distance between any two points they connect. We find, in fact, that a particle moving freely on any smooth surface, moves along a curve that is the shortest distance between any two points on the curve.

From this example of a particle moving on a surface we see how we can express Newton's first law without any reference to forces or to an absolute straight line. A free particle moving in any frame of reference moves along a path that is the shortest distance between any two points on the path. The beauty of this way of stating Newton's first law of motion is that it is invariant in this form since the shortest distance between two points (a curve that is called a geodesic) can be expressed in the same mathematical form in all coordinate systems. Einstein took this statement as the natural extension of Newton's laws of motion to the general theory of relativity. But at first sight this statement, that a free particle must always move along the shortest path, seems to contradict our direct experience for we know that an object thrown from the earth at some angle with the vertical moves along the arc of a parabola which is certainly not the shortest geometrical distance between the point from which the particle starts and the point where it lands on the earth. The shortest geometrical distance between these two points is a straight line. If Einstein's statement of the law of motion is correct, why does the particle not move along a geometrical straight line (which is the geometrical shortest distance) rather than along an arc? To answer this question we must consider the path of the particle in the four-dimensional space-time continuum rather than in three-dimensional space alone, and we then find that the path it takes (the arc) is, indeed, the shortest space-time distance (a geodesic). We recall that the special theory of relativity merged the separate Newtonian concepts of three-dimensional space and one-dimensional time into a single four-dimensional space-time continuum and the laws of optics and mechanics in the special theory are theorems in four-dimensional Euclidean space-time. The general theory still retains the four-dimensional space-time continuum as the geometrical framework in which the laws of nature are to be represented, but this continuum is a non-Euclidean continuum in which the shortest distance between two space-time points is not the usual Euclidean straight line. Einstein's great contribution to the theory of gravity was to show that what we call a gravitational field in Newtonian physics is a distortion of the geometry of space-time from a Euclidean to a non-Euclidean type.

Since Einstein's theory of gravitation is a geometric theory in which the gravitational effects of masses are viewed as the geometric properties of the non-Euclidean geometry imposed on space-time by these masses, we must discuss the general features of non-Euclidean geometries before we can get a clear understanding of Einstein's general theory.

In our previous discussion of geometry and reality we saw that geometry as a mathematical discipline plays a special role in its relationship to the laws of nature (to physics) because the universe (entirety of space-time, which is subject to these laws) is itself a geometrical structure. The laws must be expressed within the framework of some kind of geometry but until Einstein introduced the general theory of relativity it was accepted as a matter of fact that the universe is governed by Euclidean geometry. We have seen that various kinds of geometries are possible in the mathematicians' world, each of which is based on a set of axioms which are assumed to be true or are taken as "self-evident" truths, and Euclidean geometry is a very special highly restricted kind of geometry that separates one group of geometries from another. The word "truth" as it is used by mathematicians in the above sentence means that all the axioms are self-consistent and that no inner contradiction exists among them, but it does not mean that the universe is governed by the geometry based on those axioms. The only way we can determine whether one or another kind of geometry applies to the universe is by observation and comparison of the actual geometrical features of the universe with the theoretical features as deduced from the particular geometry we are considering.

The non-Euclidean geometries arose in the middle of the 19th century when geometers at that time began to question what is known as Euclid's "parallel axiom." It seemed to mathematicians like Gauss, Lobachevsky, and Bolyai that this famous axiom is not really an axiom but rather a theorem that should be deducible from the other Euclidean axioms. Gauss therefore set out to prove this axiom, but he quickly convinced himself that this cannot be done and showed that if this axiom is replaced by another one, a new kind of (non-Euclidean) geometry results. Gauss showed that two alternatives to the parallel axiom exist, each of which leads to a different kind of non-Euclidean geometry, but he never publicized his work so that the honor of discovering the first of the two non-Euclidean geometries goes to Lobachevsky.

To understand the nature of the Lobachevskian (or hyperbolic) non-Euclidean geometry we consider the parallel postulate of Euclid which may be stated as follows: Given a straight line and a point P outside this line, then one and only one straight line can be passed through P parallel to the given line. Any other line through P must cut the given line either to the right or to the left of P. This statement means that if the parallel line through P is tilted ever so slightly to either side, it cuts the given line.

We now replace this axiom by another kind of parallel axiom which leads to a non-Euclidean geometry. In addition to passing a line through P that does not cut the given line, we take one which does cut the given line in some point q to the right of P. As we tilt the line Pq so that it is brought closer to the first line (the noncutting line), the point q moves off to the right, finally reaching infinity. The parallel postulate of Euclid is based on the assumption that if the line Pq is now tilted still more by the minutest amount, it immediately cuts the given line to the left of P. Now this seems to be a "self-evident truth" since it appears that a moment arrives when the line Pq coincides with the noncutting line and q is off at infinity. It seems therefore that any further tilting causes Pq to cut the given line on the left.

Lobachevsky (as did Gauss and Bolyai) rejected this assumption and proposed in its stead the idea that when q is infinitely far to the right an additional slight tilt of the line Pq need not cause it to cut the given line to the left. He assumed that the given line is cut on the left only after Pq has been tilted by the additional finite angle Ω after q has moved off to infinity on the right. In other words, Lobachevsky assumed that two parallel lines exist—a right one (the line Pq with q at infinity), and a left one (the line $q'P$ with q' at infinity to the left) making an angle Ω with each other at the point P. This is a perfectly acceptable axiom which can be used in place of the parallel axiom of Euclid, and a complete self-consistent geometry can be built around it, as was demonstrated by Lobachevsky. Although the lines $q'P$ and Pq are called the right and the left parallel lines through P, it is clear that in this geometry any line through P lying in the sectors $q'P$ and Pq does not cut the given line. We may say that an infinite number of lines through P parallel to the given line exists if we define parallel lines as lines that do not cut. According to this point of view, an infinite number of lines can be parallel to a given line and still pass through a point outside the given line in Lobachevskian (or hyperbolic) geometry.

One can show that in this hyperbolic geometry the sum of the three angles of a triangle is less than 180° and the circumference of a circle is larger than 2π times its radius. The actual value of the sum of the angles of a triangle varies with the size of the triangle. The larger the triangle, the greater is the difference between the sum of its angles and 180°; indeed, this sum can even be zero. For a very small triangle the sum is almost 180°. That the sum of the three angles of a triangle must be less than 180° can be seen from the lines Pq and q′P. These lines meet at P to form the triangle q′Pq where q′ and q are the points off at infinity on either side of the given straight line. Hence, the angles of the triangle at q′ and at q are practically zero. The sum of the angles of this triangle is therefore essentially equal to the angle at P, which is clearly not a straight angle and hence less than 180°. Obviously, we cannot use this geometry on the surface of a plane (which is Euclidean geometry) to illustrate hyperbolic geometry, but we can see that such a geometry is a possible geometry on certain curved surfaces. In fact, this geometry is the kind of non-Euclidean geometry that applies to a saddle surface. This surface consists of a part ABC, which is concave up, and a part CDE, which is concave down. The entire surface is open, extending to infinity upward and down. The shortest distance between the two points AC on this surface is the curved line (a hyperbola) ABC. Consider now a point P on the bottom part of the surface. We can picture two "straight" lines qP and q′P (two geodesics) passing through P, but neither of them cuts the "straight" line (the geodesic) ABC. The geometry on this saddle surface is hyperbolic geometry. Notice that in this geometry no similar triangles of different sizes exist (the triangles of different sizes cannot be similar since the larger the triangle, the smaller is the sum of its angles).

Following the discovery of hyperbolic geometry by Lobachevsky, the great German mathematician Bernhard Riemann (a student of Gauss) discovered and developed an *n*-dimensional non-Euclidean geometry in which the parallel axiom of Euclid is replaced by the axiom that no parallel lines exist (all "straight" lines in this geometry intersect each other). With this axiom in place of the parallel axiom, the sum of the angles of a triangle is always greater than 180°; the larger the triangle, the greater is the sum of its angles so that triangles of different sizes cannot be similar. In this geometry, Riemannian or elliptical geometry, the circumference of a circle is always smaller than 2π times its radius. In fact, the circumference goes to zero as the radius of the circle increases.

A good example of elliptical geometry that is easy to understand is the geometry on the surface of a sphere. The "straight" line on this surface is the arc of a great circle which is the shortest distance between two points. We see that in any triangle such as ABC, the sum of the angles is greater than 180° since the angles at B and at C (which intersect the equatorial circle at right angles) are both equal to 90° so that the sum of just two of the angles is 180°. In fact, as we make the triangle larger and larger, the sum of the angles approaches twice 360°. The circumference of a circle around either pole such as DEF (which we may picture as a sort of arctic circle on a sphere like the earth) is smaller than 2π times its radius AE where A is the polar point, and this discrepancy between the circumference and $2\pi r$ increases as the radius r of the circle increases.

Knowing that Euclidean geometry is a very restricted kind of geometry which separates two very broad geometries (hyperbolic and elliptical) we must determine which geometry governs our four-dimensional space-time universe if, following Einstein, we agree that the gravitational field of a body is not to be viewed as a force field but rather as a region of non-Euclidean geometry. This statement means that where masses are present, the geometry of space in the neighborhood of these masses becomes non-Euclidean, but as we move away from these masses, the geometry becomes less and less non-Euclidean and, finally, in a region of space that is very far from all masses, the geometry becomes Euclidean. With this thought in mind we can now follow Einstein in the formulation of his general theory of relativity which places all frames of reference on an equal footing and gives a geometrical interpretation of gravity. To do this we start from the four-dimensional space-time geometry of the special theory of relativity.

The special theory, which deals with inertial observers (no accelerated frames of reference and no gravitational fields), is based on the invariance of the four-dimensional space-time interval between two events. The square of this interval is written as $\Delta S^2 = \Delta x^2 + \Delta y^2 + \Delta z^2 - c^2\Delta t^2$ where the Δ in front of S, x, y, z, and t means that the interval is tiny. This expression is just the Pythagorean theorem in four-dimensional Euclidean geometry expressed in a four-dimensional orthogonal Cartesian coordinate system. We can make this point more obvious by relabeling our four space-time coordinates as $x = x_1$, $y = x_2$, $z = x_3$, $ict = x_4$, where $i = \sqrt{-1}$. We then have (noting that $i^2 = -1$) $\Delta S^2 = \Delta x_1{}^2 + \Delta x_2{}^2 + \Delta x_3{}^2 + \Delta x_4{}^2$. This invariant expression, from which all the physical consequences of the special theory flow and which is the natural extension of the Pythagorean theorem into four dimensions, shows us that the geometry of the special theory is Euclidean and that only nonaccelerated coordinate systems are allowed. If, as Einstein did, we now generalize the theory so that all kinds of coordinate systems are permitted and gravitational fields are considered, we must introduce the most general kind of space-time interval to encompass them instead of the very special Euclidean interval. We first note that we need not limit ourselves to an orthogonal coordinate system (axes all perpendicular to each other). If we allow the axes x_1, x_2, x_3, x_4 to cut each other at any angle, we obtain terms such as $\Delta x_1 \Delta x_2$, $\Delta x_2 \Delta x_4$ (all possible mixed products) in the expression for ΔS^2. All together then, ΔS^2 is a sum of 16 terms, 4 of which are squares such as $\Delta x_1{}^2$, Δx_2^2, etc. and 12 of which are mixed products such as $\Delta x_1 \Delta x_2$, $\Delta x_3 \Delta x_4$, etc.

This enlargement of the space-time interval merely takes account of the permissible use of nonorthogonal coordinate systems, but it does not take care of the possible non-Euclideanism of space-time or the use of curvilinear (Gaussian or non-Cartesian) coordinates. To do that and thus to obtain the most general expression for the four-dimensional space-time interval we multiply each term in the expression for ΔS^2 by a different quantity that may, in general, differ from point to point of our space-time continuum. Thus, we multiply Δx_1^2 by a quantity g_{11}, Δx_2^2 by g_{21}, etc. When we are all finished we obtain the sum $\Delta S^2 = g_{ik}\Delta x_i \Delta x_k$, where it is understood that the right-hand side stands for a sum of terms obtained by choosing for the subscripts i, k all possible integers from 1 to 4. This procedure yields just 16 terms, but since $g_{ik} = g_{ki}$ (e.g., $g_{12} = g_{21}$, $g_{23} = g_{32}$, etc.) and $\Delta x_i \Delta x_k = \Delta x_k \Delta x_i$, only 10 distinct terms remain in the sum.

The coefficients g_{ik}, taken as a group, define what mathematicians call a second-order tensor which is a quantity that can be obtained by multiplying all the components of one vector with all those of another in pairs; the quantities g_{ik} themselves are the components of this tensor, which is called the metric space-time tensor because it determines the geometry of space-time. Because $g_{ik} = g_{ki}$, the metric tensor is said to be symmetric. If we knew the values of all the g_{ik}'s at all points of space-time, we would know the nature of our space-time geometry. Each of the g_{ik}'s changes as we move from point to point in our space-time continuum and, in general, they all change if we alter our coordinate system. If we try all possible coordinate systems and find that in no case (i.e., for no coordinate system) do we obtain a set of g_{ik}'s such that $g_{11} = g_{22} = g_{33} = g_{44} = 1$ everywhere while $g_{12} = g_{23} = \ldots = 0$, then we know that our space-time is non-Euclidean (curved, not flat). We can always find a frame of reference in a given small-enough region of space such that $g_{11} = g_{22} = g_{33} = g_{44} = 1$ with all other g_{ik}'s equal to zero in that frame of reference. This expression simply means that we can always find an observer who is moving in such a way that the geometry of the space-time in his own immediate neighborhood is Euclidean or very nearly so. This statement is true for an observer in an elevator that is falling freely in a given gravitational field since for him it appears that there is no gravitational field (the weightlessness of the astronauts demonstrates this fact). This notion is similar to the impression one gets that the earth is flat because it appears flat in a small region around a given point.

If we could determine whether space-time is Euclidean or not only by trying all possible coordinate systems to see if one exists for which the g_{ik}'s are everywhere 1 or 0, as described above, our task would be hopeless, but we do not have to engage in such a futile task. Riemann, who discovered elliptical geometry, proved that one can construct a fourth-order tensor from the components g_{ik} of the metric tensor and their rate of change with changing position which tells us whether space-time is Euclidean or not. If the components of this fourth-order tensor, which is known as the Riemann–Christoffel curvature tensor, are known at all points of space-time in any coordinate system, the geometry of space-time is completely known. The condition for space-time to be flat (Euclidean) is that the Riemann–Christoffel tensor (actually, all its components) be zero everywhere. Actually, we do not have to know all the 256 components (the components are labeled with four subscripts i, j, k, l, each of which can assume values from 1 to 4) of this tensor to know the geometry of space-time. Some of the 256 components always vanish and others are repetitious so that only 20 distinct (or independent) components are left. Although these 20 components (in a given region of space-time) vary individually as we alter our coordinate system, they determine (in their aggregate) a quantity (the tensor itself) which is the same in all coordinate systems and which defines the geometry of space-time. If these 20 components are known in any one coordinate system (for any frame of reference or coordinate mesh), they can be calculated in any other coordinate system, so that all we need to do to determine our space-time geometry is determine the 20 components of the Riemann–Christoffel tensor in any convenient coordinate system. If, for example, all these components vanish everywhere for one coordinate system, they vanish everywhere in all coordinate systems and the geometry is Euclidean.

If we are to construct the components of the Riemann–Christoffel tensor from the 10 components g_{ik} of the metric tensor, we must first have some way (a law) to determine these components. Since the values of the g_{ik}'s depend on the particular frame of reference that we use and on the gravitational fields that may be present, the law (actually a set of equations) that tells us how to find the g_{ik}'s becomes the law of gravity that must replace Newton's law. This law is contained in what are known as Einstein's field equations for the g_{ik}'s.

We have seen that the g_{ik}'s (the components of the metric tensor of the space-time continuum) depend both on the frame of reference that is used and on the actual departure of the geometry of space-time from flatness (the curvature of space-time), induced by gravitational fields arising from masses. An observer in an accelerated frame of reference in a region of space-time where no gravitational fields are present finds that the g_{ik}'s are not constant but change from point to point. But these changes arise not from gravity but from the curved (Gaussian) coordinate axes (a non-Cartesian mesh) imposed by the acceleration and is similar to mapping points on a plane in a polar coordinate system or a coordinate system constructed from parabolas and hyperbolas rather than in an orthogonal Cartesian coordinate system. Although the g_{ik}'s as calculated by the accelerated observer vary from point to point, this does not mean that the geometry of his space-time is non-Euclidean. It may, indeed, appear to be non-Euclidean for the accelerated observer, but for an inertial observer it appears Euclidean. The question as to whether the geometry is Euclidean or not must be settled by the evaluation of the Riemann–Christoffel curvature tensor. If the components of this tensor vanish everywhere in any coordinate system, these tensor components vanish in all coordinate systems and the space-time geometry is Euclidean even though it may appear to be non-Euclidean for an observer in an accelerated (e.g., a rotating) frame of reference.

If masses are present in a given neighborhood, the space-time in that region is intrinsically curved and the components of the Riemann–Christoffel tensor are nonzero at all points in every frame of reference. In other words, no frame of reference exists in which the 20 components of the Riemann–Christoffel tensor are zero everywhere, although frames can be found in which these components are zero (or very nearly so) in a very limited region of space-time (in the neighborhood of the origin of such a coordinate system). The freely falling elevator in Einstein's *gedanken* experiment are frames of this sort for in them, the gravitational field vanishes. This is equivalent to the assertion that it is always possible to set up a Cartesian coordinate system in a small neighborhood of any point on a curved surface (e.g., a sphere). We know this assertion to be true from our own experience which tells us that a small enough piece of the earth's surface looks flat to us.

Before we write down Einstein's field equations for the g_{ik}'s, we consider their physical significance. We know that these 10 quantities determine the geometry (the metric properties) of space-time, but what is their physical meaning? What do they correspond to in Newtonian theory? We recall that the gravitational field at various points in space can be defined in two different ways: either in terms of the force that the field exerts on unit masses placed at these points (the acceleration of gravity at these points), or in terms of the work we must do on each of these masses to move it off to infinity. Since it is more difficult to work with vectors than with scalars, the latter way of defining the field is the more convenient and simpler one because work is a scalar whereas force (field intensity) is a vector.

Consider now the gravitational field in the neighborhood of a mass such as the sun. If a particle of unit mass is placed at some point in this field, it is pulled toward the sun and we must do work to move it off to infinity. The negative of this work is called the potential of the gravitational field at the point occupied by the particle of unit mass. Since this work depends on how far from the massive body the particle is, to begin with, the Newtonian potential of a gravitational field varies from point to point; it is a scalar function of the spatial coordinates of the field. In the Newtonian gravitational theory the gravitational field is defined by a single potential function, which varies from point to point in the field.

The physical significance of the Newtonian potential of a gravitational field is that its variations determine how a unit mass placed at any point in a field moves; if left to itself, such a particle always moves from points of higher potential to those of lower potential. Thus, rocks roll down a steep hill because the higher up a rock is on this hill, the higher is its potential.

From this discussion of the role of the potential in Newtonian theory, we see that the g_{ik}'s of the general theory of relativity are related to the Newtonian potential. We recall that the g_{ik}'s determine the curvature (the geometry) of space-time and hence determine the path of a free particle. If the g_{ik}'s are all constants (either 1 or 0), space-time is flat (Euclidean geometry) and the natural path of a particle is the ordinary Euclidean straight line and a particle has no tendency to move when placed at a point. This path corresponds to the absence of gravitational fields (or accelerated frames of reference) in Newtonian theory and therefore to a constant Newtonian potential which shows us why the g_{ik}'s of Einstein correspond to Newton's potential. Recognizing this correspondence we see how much richer Einstein's gravitational theory is than Newton's because whereas Newton's theory is associated with a single potential function, Einstein's theory introduces the 10 potentials given by the 10 distinct components of the metric tensor g_{ik}. The Newtonian gravitational potential corresponds to the g_{44} component (the one that multiplies c^2t^2 in the expression for ΔS^2) of the metric tensor. The Newtonian theory is thus a theory in which only time is curved (non-Euclidean) whereas in Einstein's theory the three spatial directions are also curved.

Einstein, knowing that the g_{ik}'s of his theory correspond to the potential function of Newton's theory, used the equation that governs the Newtonian potential as a guide for finding the equations that govern the g_{ik}'s. Einstein's task was more complex than that of the Newtonian physicists because whereas the latter had to set up only one equation for the Newtonian potential (only one Newtonian potential) Einstein had to find 10 equations because there are 10 independent g_{ik}'s and we need one equation for each g_{ik} if these are to be determined. In spite of this complexity, Einstein found the 10 equations for the g_{ik}'s.

To see how Einstein did this and to follow his reasoning and procedure, we go back to the Newtonian potential and see what kind of equation it obeys. This equation, first written down in 1811 by Poisson, the great French mathematician and physicist, equates the density of matter at any point in a gravitational field to the rate at which the intensity of the gravitational field changes as one moves in any direction away from the point. Since this equation is, in turn, related to the spatial rate at which the Newtonian potential changes with change of position, Poisson's equation equates the spatial rate of change of the rate of change (the second spatial rate of change) of the Newtonian potential to the density of matter at a point. If we know the density of matter at the points in the region of space we are dealing with, we can then solve Poisson's equation and find the Newtonian potential. Einstein saw (in analogy with Poisson's procedure) that he had to find 10 equations (each similar to the single Poisson equation), each of which equates a quantity constructed from the spatial rate of change of the rate of change of the g_{ik}'s at each point of space-time, to some quantity that is related to the density of matter or to some quantity that is the equivalent of matter. Since there are 10 potentials g_{ik} in the general theory, the 10 equations that they (or their various rates of changes) obey cannot just involve the density of matter alone since that does not allow us the mathematical freedom to construct 10 distinct equations. What physical quantities in addition to matter can we call upon to play a role in determining the gravitational field (the curvature of space-time)? The answer is contained in the special theory which tells us that energy and mass are related by the famous equation $E = mc^2$. To describe the dynamics of particles in the special theory we must replace the separate three-dimensional momentum vector $\mathbf{p}$ and the energy scalar E by a single space-time four-vector $(\mathbf{p}, iE/c)$. Einstein used this energy–momentum four-vector to construct a symmetric second-order energy–momentum–mass tensor that plays the same role in his theory of gravitation as the density of matter does in Newtonian theory.

Since energy and momentum as well as mass contribute to the second-order energy–momentum–mass tensor, we see that in addition to mass, energy in any form (kinetic, potential, thermal, radiation) contributes to the gravitational field. This second-order energy–mass tensor can be constructed from the four momentum–energy components p_x, p_y, p_z, iE/c by forming all possible products of these terms taken two at a time and multiplying each such product by c^2. The resulting symmetric tensor which contains just 10 distinct components is written as T_{ik}.

Since the tensor T_{ik} in the general theory of relativity plays the role played by the density of matter in Newtonian theory, the tensor T_{ik} must determine the geometry of space-time just as the density of matter determines the Newtonian gravitational field. If there were no matter or energy present anywhere, the components of T_{ik} would be zero everywhere and space would be Euclidean everywhere. But the presence of matter and energy distorts the space-time geometry so that instead of being flat, space-time is curved. With this thought in mind, Einstein saw that he could construct a system of 10 equations to describe the geometry of space-time (equivalent to the single Poisson equation) if he could relate the tensor T_{ik} to the curvature tensor in some way. But Einstein could not do this directly because whereas T_{ik} is a second-order tensor, the Riemann–Christoffel curvature tensor is a fourth-order tensor, which we may write as $R^i{}_{jkl}$, where each of the symbols i, j, k, l can take any one of the four values of 1, 2, 3, 4, thus giving rise to 256 components ($4 \times 4 \times 4 \times 4$). But as we have seen, only 20 of these components are distinct and nonzero because of certain symmetry conditions.

Since one cannot equate a second-order tensor to a fourth-order tensor, Einstein had to construct from these 20 components of the Riemann–Christoffel tensor $R^i{}_{jkl}$ a symmetric second-order tensor with just 10 components. Fortunately, the great Italian mathematician and geometer G. Ricci had already done this: if we combine four components of the Riemann–Christoffel tensor at a time into 10 sums in a certain prescribed way, we obtain the symmetric, second-order Ricci tensor R_{jk} which has just the properties to set up the field equations that one wants.

Since the Ricci tensor is related to the curvature of space-time and the matter–energy tensor T_{ik} is a measure of the curvature of space-time, it is natural to equate R_{ik} to some universal constant times T_{ik}. One thus obtains the 10 equations $R_{ik} = AT_{ik}$ where A is a universal constant but, as Einstein was quick to see, this leads to an inconsistency for the following reason: The tensor T_{ik} obeys a conservation principle; for a given distribution of matter, energy, and momentum, it remains constant in time. This constancy is not true of R_{ik} because the Ricci tensor is not conserved. We may not equate one of these tensors to some multiple of the other since one side of the equation would be conserved whereas the other side would not.

To obtain a consistent set of 10 equations, Einstein enlarged the tensor R_{ik} by adding to it another tensor that can be obtained from R_{ik}. Although R_{ik} is related to the curvature of four-dimensional space-time, it does not give the curvature itself, which, however, can be obtained by summing certain components of R_{ik}. If we call this curvature R (it is the Gaussian curvature), we can obtain a new tensor by multiplying R by $(\frac{1}{2})g_{ik}$. It can be shown that if the density of matter at any given point in space is ρ and the matter is at rest there, then $R = -8\pi\rho G/c^2$ where c is the speed of light and G is the universal constant of gravity. When Einstein subtracted the tensor $(\frac{1}{2})Rg_{ik}$ from R_{ik}, he obtained a tensor that he could equate to some multiple of T_{ik} and thus obtained his 10 famous gravitational field equations, $R_{ik} - (\frac{1}{2})Rg_{ik} = -(8\pi/c^4)GT_{ik}$, which Einstein first wrote down in his 1915 paper on the general theory of relativity.

We saw how Einstein set up his gravitational field equations that replace Newton's law of gravity, but the field equations alone cannot tell us how a body placed at some point in a gravitational field moves. For this we need equations of motion as well as the field equations. We recall that Newtonian gravitational theory is based on two sets of equations: one set is just the equations for the gravitational force between two bodies and the other set consists of the equations of motion $\mathbf{F} = m\mathbf{a}$. To determine the orbit of a particle, we need both sets. In Einstein's gravitational theory we also need two sets of equations, one of which we have in Einstein's field equations and the other of which (the equation of motion) we now consider.

To see how Einstein obtained his equations of motion, we recall that the principle of equivalence tells us that a particle, when left to itself (a freely moving particle), must move along the shortest space-time path between two world events. Consider then a particle moving freely from the event A in space-time to the event B. We can connect these two space-time events by an infinite number of distinct space-time paths. Along which one of these paths does a free particle move? The answer may be determined by dividing any particular path into a large number n of small intervals where $\Delta S_1, \Delta S_2, \ldots, \Delta S_n$ are the lengths of these successive intervals. The total length S of the given space-time path from A to B is obtained as the sum of all of these separate intervals. The particle moves along the one path for which S is a minimum; such a path is called a geodesic.

Since the manner in which any interval ΔS depends on the g_{ik}'s is a known expression (we have seen that $\Delta S^2 = g_{ik}\Delta x_i \Delta x_k$) the equation for the geodesic can be written down in terms of the g_{ik}'s and this equation (actually a set of four equations) is Einstein's equation of motion that replaces $\mathbf{F} = m\mathbf{a}$. The importance of the equation of the geodesic is that it has the same form in all coordinate systems so that it really expresses a law of nature. Although Einstein adopted the geodesic as his equation of motion for a particle as a consequence of the principle of equivalence, he later deduced this equation directly from his field equations. In the general theory of relativity the equations of motion need not be introduced separately; they follow from the field equations themselves. This is a remarkable feature of the general theory of relativity which is not present in Newtonian theory. In Newtonian theory we cannot deduce the law $\mathbf{F} = m\mathbf{a}$ from the law of force $F = Gm_1m_2/r^2$.

To find the path of a particle in a gravitational field in the general theory we proceed as follows: We first set up the 10 Einstein field equations for the given distribution of matter in some suitable coordinate system (frame of reference) and solve these equations for the 10 g_{ik}'s. Since we may choose any frame of reference, we choose that particular frame which makes the field equations as simple as possible. Having obtained the 10 g_{ik}'s (some of which may be zero; the best coordinate system to use is the one in which the maximum number of g_{ik}'s is zero) we then substitute these into the geodesic equation (four in number) and obtain the orbit of the particle.

Depending on the complexity of the distribution of matter in space-time, the field equations (and hence the geodesic equations) are more or less complex. The simplest case is that of the gravitational field surrounding a single mass-particle concentrated in a point (a point particle). In this case the T_{ik}'s on the right-hand side of Einstein's field equations are all zero because there is no matter in the space surrounding the particle that generates the gravitational field. The reason this problem is so simple is that everything is spherically symmetric so that in the space surrounding this particle only one space coordinate enters. We can call this coordinate r (the distance of any point in the field from the point particle which is the source of the field). The other coordinate that enters into the problem is *ict*, the time coordinate. We see that only two of the ten g_{ik}'s need be considered: the one that multiplies Δr^2 and the one that multiplies $c^2\Delta t^2$ in the expression for ΔS^2. These two g_{ik}'s are the only ones that vary from point to point. Two of the other g_{ik}'s equal 1 and the other six g_{ik}'s are zero. The rigorous mathematical solution of Einstein's field equations for this simple case was first obtained by the German astronomer and physicist Karl Schwarzschild in 1917. One can also obtain the path of a beam of light in the given gravitational field by placing $\Delta S = 0$ since a beam of light moves along a space-time path whose length is zero (a null interval).

Although Schwarzschild was the first to obtain the g_{ik}'s for a gravitational field like that of the sun by solving the field equations using rigorous mathematics, Einstein, using the principle of equivalence, had already deduced all the results obtained later by Schwarzschild. We recall that the principle of equivalence tells us that we can deduce gravitational effects by studying phenomena in appropriately accelerated frames of reference. Einstein followed this principle in his analysis of the behavior of physical bodies (e.g., planets and atoms) and beams of light in a gravitational field. Using appropriately accelerated reference frames and the principle of equivalence we can, in the manner of Einstein, not only deduce the orbits of particles and beams of light but also the behavior of clocks and measuring rods, in general, in gravitational fields; we can thus deduce the expression for the square of the line element ΔS^2 and obtain Schwarzschild's solution of Einstein's field equations.

To obtain this solution we consider a particle of mass M (the sun, for example) fixed at the origin O of the observer's coordinate system. Let A and B represent two events that are separated by the small distance Δr and by the small time interval Δt as measured by rods and clocks that are at rest in the gravitational field at the distance r from M. If we know how these rods and clocks are affected by the gravitational field at the point r, we can see how the expression for ΔS^2 (the square of the space-time interval) at r when M is at O differs from the expression ΔS^2 when no mass M is present (which corresponds to Euclidean space-time or special relativity). If there were no mass present at O (no gravitational fields), we would have $\Delta S^2 = \Delta r^2 - c^2\Delta t^2$, where we consider only the single space coordinate r and the time coordinate $ic\Delta t$. In other words, if an observer (whom we call E) were fixed at the event A (or B) and there were no material bodies such as M (or fields of energy) within a finite distance r from him, he would find the quantity $\Delta r^2 - c^2\Delta t^2$ for the square of the space-time interval between the two neighboring events A and B. But with M at O this is not so and we must write $\Delta S^2 = g_{11}\Delta r^2 - g_{44}c^2\Delta t^2$ where g_{11} and g_{44} (as stated above) are the only surviving components of the metric tensor; they tell us how the presence of M alters the space-time geometry (the rods and clocks) at the point r. Noting that ΔS^2 must be the same for all observers at r, regardless of their frames of reference, we introduce an observer at r whose frame of reference is such that for him there is no observable gravitational field. If this observer (whom we call E) compares his rods and clocks with those of E, he can determine g_{11} and g_{44} by noting that the quantity ΔS^2 that he obtains for the two events A and B must be identical with that obtained by E.

To set the stage properly so that we can apply the principle of equivalence to the description of the two events as given by the two observers, we picture the observer E as being in an elevator at r that is suspended vertically (along the radial direction) and is at rest relative to M. This observer is to have a clock in his elevator and a rod of length l which is aligned parallel to r. We now introduce the observer E′ and assign to him a frame of reference in which for him there is no gravitational field. We can do this using the principle of equivalence if we recall that it is applicable to an observer who is falling freely. We therefore place E′ in an elevator at r that is falling freely toward M, having fallen from an infinite distance. The freely falling observer E′ has in his elevator measuring rods and clocks that are identical with those in the fixed elevator of observer E. As far as E′ is concerned, the events A and B occur in field-free space, and, using his rod and clock, he expresses the square of the space-time interval as $\Delta S^2 = \Delta r'^2 -$

$c^2\Delta t'^2$ where $\Delta r'$ is the spatial distance between A and B as measured by E′ and $\Delta t'$ is the time between the two events as measured by E′. To obtain g_{11} and g_{44} E′ must compare $\Delta r'$ with Δr and $\Delta t'$ with Δt keeping in mind that ΔS^2 is the same for E as it is for him.

We note that since the observer E′ has fallen to the point r from an infinite distance, his speed toward M is just the speed of escape from M at distance r from M. But this speed is just $v = \sqrt{2GM/r}$ so that this is the speed at which E′ is falling past the two events A and B and past E. It follows from the Einstein–Lorentz contraction that E′ finds that the measuring rods of E held radially (parallel to r) are shorter than his rods by the factor $\sqrt{1-(v^2/c^2)}$ or $\sqrt{1 - (2GM/c^2r)}$. This means that his ΔS^2 can be the same as that of E only if he replaces Δr^2 in the expression for ΔS^2 by $\Delta r^2/[1 - (2GM/c^2r)]$. E′ now finds the value of g_{44} by noting that E's clock is slower than his clock (the time intervals are longer on E's clock) owing to the motion of E's clock past E at the speed $v = \sqrt{2GM/r}$. E must multiply the interval Δt by $\sqrt{1 - (v^2/c^2)}$ or $\sqrt{1 - (2GM/c^2r)}$. Hence, the quantity $-c^2\Delta t^2$ must be multiplied by $[1 - (2GM/c^2r)]$. E′ thus obtains for ΔS^2 for a point at a distance r from the gravitational source (the particle of mass M) the expression $\Delta S^2 = \Delta r^2/[1 - (2GM/c^2r)] - [1 - (2\mathrm{GM}/c^2r)]c^2\Delta t^2$ for the square of the space-time interval between two events separated by the space interval Δr and the time interval Δt as measured by an observer stationed (motionless) at the distance r from M. The factor in front of Δt^2 in the above equation may be interpreted as the reduction of the speed of light in a gravitational field.

To understand the physics of the general relativistic expression for ΔS^2 we note that since E′ is falling freely, he experiences no gravitational field so that his observations are equivalent to those of another observer at an infinite distance from M who also detects no gravitational field. To the distant observer, then, the situation is the following: he observes that a measuring rod aligned with the lines of force of a gravitational field shrinks as it approaches the source of the field whereas the lengths of rods placed at right angles (transversely) to the field are unaltered. This shows that the geometry of space-time in the neighborhood of a massive particle (e.g., the sun) is not Euclidean.

To determine the nature of the geometry in such a gravitational field we compare the circumference of a circle, whose center is at M, with its radius. Since the circumference is measured with a rod tangent to the circumference at each point, and hence transverse to the gravitational field, its (the rod's) length is unaltered by the gravitational field so that the length of the circumference is the same as it would be if no gravitational field were present. On the other hand, since the radius of this circle must be measured by a rod laid along the radius, and hence along the lines of the field, the rod's length is shortened so that the radius, as measured by this rod, is longer than it would be if no field were present. It follows that 2π times the radius does not equal the circle's circumference, as in Euclidean geometry, but is larger than the circumference. Hence, the geometry in the neighborhood of a massive body is elliptical (Riemannian) non-Euclidean geometry. This is verified by the path of a beam of light which is slightly curved toward the sun when it grazes the sun. Two such beams from two different sources, coming to a focus after grazing the sun on opposite sides, form an angle, and if the rays themselves are the two legs of a triangle, the third side of which is the line connecting the sources, the sum of the three angles of this triangle exceeds 180°, which, again, means elliptical geometry. This discovery was made during the famous 1919 total eclipse of the sun, which completely confirmed Einstein's prediction that the path of a ray of light is bent in a gravitational field.

Two other important physical results flow from the change in the space-time geometry from Euclidean to elliptical in going from field-free space into a gravitational field. The first concerns the slowing down of clocks and, therefore, of the vibrations of atoms, which are measured by the frequency of the light they emit. Since the rate of a clock is diminished by the factor $\sqrt{1 - (2GM/r)}$ if the clock is at the distance r from the source of a gravitational field of mass M, the light emitted by atoms on the surface of a small massive sphere must be measurably different from that emitted by atoms in field-free space. White dwarfs, stars no larger than the earth but as massive as the sun, are such spheres so that the light from such a star should show this phenomenon and it does. If the frequency of the light emitted by an atom on the earth is ν, the frequency of the light emitted by the same atom on a white dwarf of mass M and radius R is $\nu\sqrt{1 - (2GM/r)}$. In other words, the frequency is reduced so that the wavelength is increased, which means that the light is reddened. This so-called Einstein redshift has been completely confirmed by the spectral lines in the light from the white dwarf companion of the star Sirius.

Another important consequence of the non-Euclidean character of space-time in the neighborhood of a massive body is that Keplerian elliptical orbits cannot correctly describe the paths of planets around the sun. This departure of the true orbits of the planets from the closed elliptical orbits deduced from Newtonian theory is most pronounced for a rapidly moving planet close to the sun such as Mercury. Einstein's gravitational theory predicts that the closed elliptical orbit Newtonian theory predicts for such a planet should be replaced by an ellipse which itself is rotating in the same direction in which the planet is revolving around the sun. This phenomenon, called the advance of the perihelion (point of closest approach) of the planet, had been known to be present in Mercury's motion and had been measured very accurately before Einstein had proposed his general theory. Most of the advance arises from the gravitational interaction of Mercury with the other planets (primarily with Jupiter) but a small residue stems from the Einsteinian curvature of space-time.

To see in a rough way how the non-Euclidean geometry of Mercury's space-time causes its perihelion to advance, we apply Kepler's third law of planetary motion to Mercury which tells us that the square of the planet's period (the time for one revolution) is proportional to the cube of the planet's mean distance from the sun. These two quantities have a different character as far as measuring them goes since we measure one (the period) with a clock in our neighborhood that is not affected by the gravitational field and the other (the mean distance) by a rod that must be placed along the gravitational field near Mercury. Since this rod is shortened it tells us that the mean distance is larger than the mean distance determined from the period. In other words, if we observe Mercury and the sun lined up with a particular star at some moment and wait until it is lined up with that star again (after one period or one revolution of Mercury) it will not be back in the initial point of its orbit—the planet will have to go a bit farther. This conclusion is equivalent to saying that the orbit has rotated slightly. If the orbit were perfectly circular, this rotation could not be observed because the orbit's rotation is then indicated by the change in the position of the perihelion.

CHAPTER 12

The Perfect Gas Law

It has long been an axiom of mine that the little things are infinitely the most important.

—SIR ARTHUR CONAN DOYLE, *"A Case of Identity"*

Up to this point in the development of our subject we have been discussing the structures that are controlled by and, in a sense, stem from the force of gravity. We noted that if there were no forces at all in our universe, there could be no differentiation of the matter into the great variety of objects (from galaxies down to atoms) all about us. The universe would consist of noninteracting particles moving around in total randomness and completely filling all regions of space—a completely unordered and symmetrical universe. The first step in the evolution of structures was taken when the force of gravity, early in the life of the universe, began to tame the disorderly motions of particles and to bring these particles together to form gaseous spheres (the stars) and the more solid structures (the planets). But the force of gravity under ordinary conditions is far too weak to account for the structure of the matter that we see all about us. The great variety of chemical and physical properties associated with the different kinds of matter is clear enough evidence that the forces which produce these material structures are quite different from the force of gravity. Moreover, from the strengths of solids such as the metals, we deduce that the material structural forces are much stronger than the ordinary gravitational force.

The physical and chemical properties of a structure indicate the nature and strength of the forces that govern this structure. We know that matter exists in three different states—solid, liquid, and gaseous—and it may appear that the most direct way to see the relationship between forces inside matter and the structure of this matter is to study matter in its solid state. In principle this suggestion is correct but it is very difficult to carry it out in practice because it is hard to penetrate a solid and to disentangle the complex skein of forces associated with such closely packed particles. A more feasible approach to this problem is to begin by studying the behavior of particles when they are not squeezed together but rather when they are moving about independently of each other. This state is analogous to the condition of the particles in a gas. If we begin with the gaseous state, we can see what happens when the freely moving particles of the gas are subjected to various influences or when they are brought closer together.

We begin our study of forces among the particles that constitute matter by studying these particles when they are exerting no forces on each other as when they are in the gaseous state. We first note how a gas differs from other states of matter, examples of which are all around us. But since we see the same substance in its three possible states only occasionally, we might mistakenly assume that only certain substances can exist as solids, liquids, and gases. We are all familiar with ice, water, and water vapor, but how many of us have seen solid hydrogen, liquid helium, or iron vapor? With the possible exception of helium (for which a solid state has never been observed), all substances can, under the proper conditions, exist as solids, liquids, and gases. Usually, we associate the solid state of any substance with what we refer to as "coldness" and the liquid and vapor phases with "hotness," but the concepts of "hotness" and "coldness" have not yet been defined here. In fact, from the basic quantities discussed up to now—length, time, force—we cannot construct or derive the concept of "hotness." We therefore introduce hotness (or something related to it) as a new basic entity, like length and time, rather than as a derived quantity.

The three states of gross matter are related to some physical entity that we associate with hotness and coldness, but these terms are not precisely defined and therefore cannot be used in our analysis of the states of matter. We can, however, describe a measuring procedure, just as we did for length and time, which gives a number that is related to the hotness or coldness of a body. This number, called the temperature of the body, then enters into the basic laws of nature just as length and time do. The temperature concept must be treated as one of the basic undefinable physical entities in nature for the purpose of our present discussion of the gross properties of matter. We shall see later that it can be expressed in terms of length, time, and force, but only if we first introduce a new basic natural constant (like the gravitational constant) which itself must then be treated as a fundamental, undefinable entity. To proceed then with an analysis of matter, we must introduce some new fundamental entity, and we can choose this entity to be either a basic constant or a measurable quantity. We choose the measurable quantity rather than the constant because we lay down a precise procedure for measuring it. The reader should note that we do not, at any point, define the basic measurable entity—in this case, the temperature—but merely provide a method for measuring it.

Since we associate temperature with hotness or coldness, we obtain a method for measuring temperature by observing how a body responds to hotness and coldness. We know that bodies tend to expand when they grow hotter and to contract when they get colder. This behavior is precisely the property of matter we use to introduce a temperature scale. First we place a body in a very cold environment and note its size or dimensions. We then place the same body in a hot environment and again measure its dimensions. We use the change in the body's dimensions when its environment changes from hot to cold and vice versa to define a temperature scale.

The most convenient material for this purpose is a liquid since the change in its dimension can be most easily controlled and observed. A solid expands, as do all substances, in all directions, and the amount of expansion in any one direction for small changes in temperature is quite small and hence difficult to measure accurately. This difficulty can be overcome in a liquid or gas because the expansion of a body in the three spatial dimensions we have just described increases the volume of the body, which is the essential physical aspect of the phenomenon. This volume expansion is the same regardless of the shape of the body or the amount of expansion in any one direction. If we suppress or greatly reduce the expansion in one direction, the expansion in the other two directions increases to keep the volume expansion the same. It is difficult to do this sort of thing with a solid but fairly easy with a liquid. If the liquid is placed in a very narrow cylinder made of a transparent substance (e.g., Pyrex glass) which expands or contracts negligibly with thermal changes, then the liquid in the cylinder can only rise or fall with the change in its height making up for the restriction in any changes in its lateral dimensions. Such a device—a liquid in a closed, evacuated cylinder with a reservoir of the liquid in a bulb at the bottom of the cylinder—is called a thermometer.

The bulb of a narrow glass tube containing a liquid (such as mercury) is immersed in a homogeneous mixture of ice and water; then place a scratch on the tube at the height of the column of liquid above the bulb and mark this scratch zero. This point is the zero point of our temperature scale, which means that the temperature of a homogeneous mixture of ice and water is zero degrees (0°) on the temperature scale we are about to introduce.

To establish the complete scale, we now define another point on it, which we obtain by immersing the bulb of the thermometer in a homogeneous mixture of steam and boiling water. The liquid in the thermometer now rises to a different height at which we place another mark and label it 100. This mark is the 100° point of the temperature scale. On this scale, the boiling point of water is 100°. We now divide the section of the cylinder between the 0° and 100° marks into 100 equal sections. Each of these sections represents one degree change of temperature on the centigrade temperature scale.

The material that we use in our thermometer (in this case, mercury) is called the thermometric substance, and the way it expands or contracts with the change of temperature is determined by the size of the thermometer and its geometrical features in general. The important physical quantity involved here is the coefficient of thermal expansion, which we define as the expansion of the thermometric substance per degree rise of temperature on the given temperature scale—the centigrade scale in this instance.

If we have a thermometric substance whose coefficient of expansion is a on some temperature scale t and if the height of this substance is h_0 for $t = 0$ and if its height is h for any other temperature t, then $h = h_0 + at$. This relation establishes the temperature scale t for the given thermometric substance according to the relation $t = (h - h_0)/a$. Using this relationship for various temperature scales, we can always change from one scale to another and express the temperature in any scale we wish. As an example of how two different scales are related to each other, we consider the centigrade and Fahrenheit scales. The zero point of the centigrade scale is taken as the temperature of a mixture of ice and water, but this same point is assigned the temperature 32° on the Fahrenheit scale. The temperature of boiling water is 100° on the centigrade scale but 212° on the Fahrenheit scale. We use these data to relate the two scales: if t_f is the Fahrenheit temperature and t_c the centigrade temperature, we have (since both scales are based upon the same thermal expansion of the thermometric substance) $t_c = 5/9(t_f - 32)$ or $t_f = (9/5)t_c + 32$. These two expressions permit us to go from one scale to the other with ease.

We now consider another temperature scale which plays a very important role in the laws of nature. In our procedure for introducing a temperature scale we began by picking a zero point, but this selection was made quite arbitrarily which permits us to have temperatures that are lower than the chosen zero point. On the centigrade scale, we have temperatures that lie below the freezing point of water. These temperatures must be taken as negative on the centigrade scale since the zero point of the scale is itself the temperature of freezing water. The question which naturally arises here is whether an absolute lowest temperature exists in nature and just how far below the freezing point of water we must go to reach this absolute zero point. If such a lowest point does exist, we can obtain an absolute temperature scale by placing the zero point of this scale at this lowest realizable temperature. As we shall see later, such a lowest point does exist and it is at 273.15°C below the zero point of the centigrade scale (the freezing point of water). We can thus introduce an absolute temperature scale by placing its zero point at −273.15°C. If we call this temperature scale T, we then have $T = t_c + 273.15$. On this scale (which has the same size temperature intervals as the centigrade scale) the freezing point of water is 273.15° and the boiling point of water is 373.15°. For the sake of simplicity we define the absolute scale as the centigrade scale plus 273, dropping the 0.15. On this scale, absolute zero is at −0.15°K where K stands for Kelvin (after the great 19th century British physicist Lord Kelvin) so that $T = (t_c + 273)$°K.

Before leaving the subject of temperature scales, we caution the reader about a possible uncertainty that can arise in applying the procedure outlined above. In choosing the zero point for a scale (except the absolute scale) we use some natural phenomenon—the freezing of water, for example—that can depend on various physical conditions such as atmospheric pressure. We must therefore specify these conditions precisely if our scale is to be exact and self-consistent.

Now that we have introduced the temperature scale and given a procedure for measuring temperature, we can give a precise meaning to the solid, liquid, and gaseous phases of a substance. Suppose we heat a solid at a definite temperature by placing a flame under it or putting it into an oven, arranging things in such a way that the solid heats up uniformly so that a thermometer in contact with the solid rises steadily. Does the thermometer rise steadily as the solid is heated? The answer is no because the temperature reaches a constant value at which the solid begins to change into a liquid. The temperature then remains at this constant value until the solid has changed entirely to liquid. This constant temperature is called the melting point of the solid.

We may view this phenomenon differently by starting from the liquid phase, allowing the liquid to cool steadily and uniformly. The temperature then drops steadily until it reaches the same constant value—the melting point of the solid. Again, the temperature of the liquid remains constant and the liquid begins to solidify. This constant temperature is called the freezing point of the liquid, which is equal to and is the counterpart of the melting point of a solid.

The temperature of a uniform mixture of a solid and its liquid phase remains constant whether we cool or heat it, as long as the mixture is stirred uniformly. If we heat it, some of the solid changes to liquid, and if we cool it, some of the liquid freezes, with no temperature change occurring in either case. By slightly changing the thermal conditions of the environment of the mixture we can make the process solid⇔liquid go in either direction.

Using the concept of the melting point (or freezing point) we now define a solid precisely as a substance which has a definite melting point. According to this definition, certain substances which appear solid to our touch and sight are, strictly speaking, not true solids. For this reason glass is not a true solid since its temperature rises steadily as it is heated and it softens gradually rather than changing abruptly from a solid to a liquid at a given temperature. This is also true of substances such as butter, honey, waxes, and fats in general. These substances are generally referred to as amorphous solids, and those that have definite melting points are called crystalline solids. Metals (crystalline solids) have melting points that range from values of about 3400°C for tungsten to −38.9°C for mercury.

The temperature of a true solid remains constant while it is melting, but once the substance has changed into its liquid state completely, its temperature rises as it is heated, until it begins to boil, at which point the temperature again remains constant. This state of the liquid is characterized by the continual and spontaneous formation of bubbles throughout its volume which rise and escape through the surface of the liquid into the surrounding region. The constant temperature at which this occurs is called the boiling point of the liquid. All liquids have a definite boiling point under the same set of environmental conditions, such as atmospheric pressure, and this temperature remains constant during the process of boiling until all the liquid has boiled away and the substance had changed into a gas (or vapor). The transformation of a liquid into a gas is quite obvious during boiling, because we can see bubbles of gas form in the liquid. But the transformation of a liquid into a gas goes on continuously at all temperatures by a process of evaporation, even though we cannot see it directly. A solid can also pass directly into the gaseous phase (without first becoming a liquid) by the process known as sublimation. The transformation of snow directly to water vapor is a good example of this process.

Now that we have discussed the three phases (or states) of matter in a general way, we consider the properties of a gas in detail. Some people find the concept of a gas difficult to grasp because a gas, in general, is not visible, and when it can be seen, as in the case of the atmosphere (the blue sky), it does not give the impression of being a palpable substance. A solid has a definite shape and a fixed volume, and a liquid has a definite volume but no fixed shape as it assumes the shape of the vessel containing it. A gas has neither fixed shape nor fixed volume, but spreads out uniformly to fill any volume in which it is placed. How are we to specify a gas or describe it in a way which is physically meaningful? To do this we must introduce a set of measurable physical characteristics.

To find these characteristics, we consider a gas placed in a cylinder with a movable piston at one end. We use this kind of container so that we can change the volume of the gas at will, which we do by simply raising or lowering the piston. Since a gas fills any region of accessible space uniformly, the volume of the gas depends on the position of the piston. If A is the surface area of the piston and h is its height above the bottom of the cylinder, then the volume V of the gas is Ah. This volume is a measurable quantity and, therefore, is one of the characteristics that describe a gas.

We now introduce two other measurable quantities which, together with the volume V, describe a gas fully. One is the temperature t on some scale (let us say the centigrade scale), which we can measure quite accurately by inserting a thermometer into the cylinder. The third gas characteristic is not as obvious a property of the gas as the volume and temperature, but it can be expressed and measured in terms of the basic space, time, and force concepts we have already introduced. In our discussion of the earth we introduced and defined the concept of atmospheric pressure. But the concept of pressure is not limited to the earth's atmosphere; it can be applied to any gas. As we noted in a previous chapter, we define the magnitude of the pressure exerted on a surface as the force applied to the surface divided by the area of the surface. If P is the pressure exerted on the surface area A and F is the applied force, we have $P = F/A$. In terms of our basic units of measurement, pressure is thus expressed as force per unit area or dynes per square centimeter. Since force is a vector quantity, pressure is also a vector quantity having the same direction as the force, provided we treat the area as a scalar quantity and consider only the component of the force acting perpendicular to the area. In our present treatment of a gas, we assume the pressure of the gas always to be perpendicular to any surface with which the gas is in contact.

A simple way of expressing or representing the pressure in a gas is to note first that a pressure exerted at any point in a gas is transmitted equally in all directions and against all surfaces with which the gas is in contact. If a constant force F is exerted on the piston of the cylinder, the pressure on the piston resulting from this force is transmitted equally in all directions throughout the gas; this is just the pressure exerted by the gas on all the walls of the cylinder. The actual pressure in the gas in the cylinder can now be expressed in terms of the force F and the area A of the piston. Since the force on the piston is F and is distributed over an area A, the pressure on the piston is F/A; this is the pressure at any point in the gas. This means that if a surface were placed at any point in the gas, the gas would exert a force on that surface (in a direction perpendicular to it) equal to F/A times the area of the surface. There is a simple way to keep a constant pressure in the gas in our cylinder. All we need do is place some weights W on the piston. The force on the piston is then always equal to W and the pressure P in the gas is always W/A.

We need only the three measurable quantities P, V, and t to describe the behavior of all gases to a very good approximation. This may seem quite strange and remarkable at first since we have said nothing at all about the kind of gas we are dealing with. One might think that some other quantity should be introduced to take account of the chemical differences among gases. But this additional complication is not necessary because all gases and all mixtures of gases behave the same way physically (except under certain abnormal conditions which we discuss later) and this physical behavior can be described by the same three quantities, P, V, and t, for all gases.

We saw above that all gases behave in the same way and that their behavior can be described in terms of three measurable quantities P, V, and t. However, this holds only if the temperature of the gas is high enough so that no condensation is occurring. In our present discussion we therefore assume this to be so and treat all gases as identical, insofar as physical behavior is concerned.

To distinguish between the detailed properties and exact behavior of real gases and the ideal situation assumed in the foregoing paragraph, we call a gas that behaves in this ideal way an "ideal" or "perfect" gas, keeping in mind that an ideal gas is only an approximation (though a very good one) to a real gas. Further on we give a precise definition of a perfect gas, but we need only note now that the physical behavior of an ideal or perfect gas can be described entirely in terms of the quantities P, V, and t.

These quantities are not independent of each other; they are so related that we can calculate any one of the three quantities if the other two are known. This relationship is expressed symbolically as $P = \mathrm{f}(V,t)$. This equation means that the pressure in a perfect gas is some function f of the volume and temperature of the gas. This statement is also expressed by the equation $\mathrm{f}(P, V, t) = 0$.

Either one of the above equations is called the equation of state of a perfect gas; it holds for all perfect gases, regardless of their chemical nature. To find this equation we proceed in two steps: First, we consider only those processes during which we keep the temperature of the gas constant and see how the pressure changes when we alter only the volume of the gas or vice versa. Such processes are called isothermal processes since the temperature is kept constant.

Robert Boyle, a contemporary of Newton, and the first to study isothermal processes in gases, discovered how the pressure and volume of a perfect gas are related to each other when the gas changes isothermally. He expressed his discovery as an equation now known as Boyle's law of gases. To find this law, we consider a perfect (ideal) gas in a cylinder sitting on a large furnace (which we call a heat reservoir henceforth) at the constant temperature t (centigrade scale). The base of the cylinder is to consist of such material that the temperature of the gas in the cylinder is always the same as that of the heat reservoir, no matter what we do to the gas. This requirement means that the material at the base of the cylinder is to be a perfect heat conductor so that heat flows instantly through this base from the gas to the furnace or vice versa, as may be required, to keep the gas temperature constant. We can now compress the gas in the cylinder or allow it to expand without changing its temperature by altering the position of the piston.

We consider the gas first with the piston of the cylinder in position 1 so that the pressure and volume are P_1 and V_1. If we now place additional weights on the piston, it sinks until the pressure throughout the cylinder has increased to a new value P_2, corresponding to the increased weights, and the volume of the gas becomes V_2 which is smaller than V_1. How does the ratio of the new volume to the old volume (V_2/V_1) compare with the ratio P_2/P_1 of the new pressure to the old pressure? Boyle discovered, by making careful measurements, that these two ratios are the inverse of each other, so that $V_2/V_1 = 1/(P_2/P_1)$, or $V_2/V_1 = P_1/P_2$ for constant temperature t. We may state this somewhat differently by noting that the previous equation is equivalent to $P_1V_1 = P_2V_2$ for constant t.

Since this is true no matter what the initial and final values of the pressure and volume are, we state Boyle's law (which is the content of the foregoing equation) as follows: In any perfect gas whose temperature is kept constant, the product of the pressure and the volume remains constant no matter how the pressure and volume are changed separately. This law may be expressed in simple algebraic fashion as PV = constant. From this relationship we see that if the volume of a perfect gas is decreased by a given factor (e.g., by 2), its pressure must increase by the same factor. If the volume goes from V to V/n the pressure must go from P to nP so that the product PV becomes $(nP)(V/n)$ and thus remains the same.

A very convenient graphical way of representing the changes that occur in the volume and pressure of a gas as time goes on is by means of a so-called P–V diagram. In this graph the pressure P of the gas is plotted along the vertical axis and the volume V is plotted along the horizontal axis, so that a single point, A, gives both the volume and pressure of the gas at some particular moment. Thus, some point B may represent a given quantity of gas at a particular moment when its volume is 2 units and its pressure is 3 units, but this point gives us no information about the temperature of the gas at that moment. If the conditions of the gas are changing, the point on the P–V diagram changes from moment to moment. If we connect all the points on the graph that represent the changing gas over a given period of time, we obtain a curve in the P–V diagram that describes some kind of process that is going on in the gas. If the pressure and volume of the gas are 3 and 5 units respectively at some initial moment, and at some later moment they are 5 and 5 units, respectively, the point B representing the state of the gas at the initial moment and the point F representing it at the final moment (the end of the process) are on the curve, or path, BCDF where C and D are intermediate states representing all the states of the gas as it goes from its initial to its final state. (It is convenient to refer to any particular set of values of P and V as representing a "state" of the gas.) If we are interested only in bringing the gas from its initial state B to its final state F without concern about the intermediate states, we can go from B to F along an infinitude of paths such as BAC′D′F or BAC″D″F. But if we want the gas to pass through a uniquely specified series of intermediate states, we choose only one path in the graph—the one that connects all the P–V points at the specified intermediate states of the gas.

If we now choose the initial point B and the final state F of the gas at random, the products of the pressure and volume for the two states are certainly not the same, which simply means that these two states are not at the same temperature. It follows that the P–V path connecting B and F does not represent an isothermal process in the gas. In fact, the temperature of the gas when it is in a state represented by a point (e.g., D) on the path from B to F is, in general, different from the temperature at B or F. The temperature of the gas at F (its final temperature) is higher than its temperature at B (its initial temperature) if the *PV* product of F is larger than that for B.

If we pick a state W whose temperature, that is, its *PV* value, is the same as that of state A (3, for example), where does it lie on the *P*–*V* diagram? We cannot answer this question if we know only that the *PV* product of W is 3, just as it is for A, because there are many such points, in fact, an infinite continuum. All the points on the *P*–*V* diagram that have the same *PV* value as A lie on a curve called an equilateral hyperbola. If we consider all the points for which the *PV* product is 6, for example, we obtain another equilateral hyperbola and so on. All these curves taken together are called the isothermals of a perfect gas. Each curve gives the relationship between the pressure and volume of a gas for a definite temperature. Whenever the temperature of the gas is changed we go over to another one of these curves.

During an isothermal process the product *PV* remains constant so that Boyle's law describes the behavior of a gas that is undergoing isothermal processes, but it does not tell us what happens to a gas when the temperature changes, and this we must know if we are to understand fully the properties of a perfect gas. We can investigate how a gas behaves when its temperature changes by dividing the problem into two parts and studying each one separately. If we just heat a gas and let it change as it may with no restrictions, both its pressure and volume change, but we now arrange things so that only one of these parameters changes while the other one is constant. We first consider the cylinder containing a definite unchanging quantity of gas with a set of fixed weights on its freely movable piston. By heating the gas (which can be done with a Bunsen burner) we can raise its temperature without changing its pressure since the weights on the piston remain the same. With the temperature of the gas given by a thermometer inside the cylinder, this arrangement permits us to study the way the volume of the gas only is affected by the temperature. This phenomenon was first studied by the French chemist Charles (c. 1800) who found that the volume of a gas changes by 1/273 of its volume at 0°C when the temperature of the gas changes by 1°C; if the temperature of the gas is increased by 1°C, the volume increases by $V_0/273$, where V_0 is the volume of the gas at 0°C, and if the temperature of the gas is reduced by 1°C, its volume decreases by exactly the same amount. We express this important result by the algebraic equation $V_t = V_0[1 + (t/273)]$ where V_t stands for the volume of the gas at the centigrade temperature t and V_0 is its volume at 0°C. The remarkable thing about this relationship is that it is true for all gases regardless of their chemical nature; all gases that behave in this way are called perfect gases. Taking advantage of this behavior of a gas, we can construct a gas thermometer which is very useful because its behavior does not depend on the kind of gas we use. It therefore has universal properties which gives us the absolute temperature directly.

The algebraic equation which shows how the volume of an ideal gas changes with temperature when the pressure is held constant still does not give the complete picture of how a gas changes in general. To obtain this answer, we must now allow all the parameters P, t, V to change freely (without altering the amount of gas we are dealing with) and then see what general algebraic relationship exists among them. Boyle's law and Charles's law are just the first two steps in the derivation of this general algebraic relationship which is called the "equation of state of an ideal gas." With the aid of these two laws we can now derive a third relationship which, together with the other two, lead to the equation of state of an ideal gas. This third relationship deals with the manner in which the pressure of a gas varies with its temperature when the volume of the gas is held constant.

We again consider the gas in the cylinder but this time we clamp the piston so that it is not free to move. The immobility of the piston guarantees the constancy of the volume. If we now heat the gas, only the pressure and temperature change. If P_0 and V_0 are the pressure and volume of this gas at 0°C and the temperature is raised to the value t, the volume is still V_0 but the pressure increases to a new value P_t such that $P_t = P_0(1 + at)$, where a is the coefficient of pressure increase. Using the laws of Boyle and Charles, one can show that $a = 1/273$; in other words, the coefficient of volume expansion of a gas equals its coefficient of pressure increase.

Now that we have discovered how the volume and pressure of an ideal gas change (separately) with respect to temperature, we can derive the equation of state. We saw above that the pressure P of a gas at any temperature t is related to its pressure at 0°C by the equation $P_t = P_0[1 + (t/273)]$ provided the volume of the gas is kept at the constant value V_0, the value of the volume at 0°C. But we also have $V_1 = V_0[1 + (t/273)]$ where V_t is the volume of the gas at the temperature t. From these equations we find that $PV = (P_0V_0/273)(t + 273)$. This equation is the equation of state of a perfect gas since it shows us how the pressure, volume, and temperature of such a gas are related to each other when they are permitted to change freely. We simplify this equation of state somewhat by introducing the absolute temperature. Since $(t + 273)$ is T we have $PV = (P_0V_0/273)T$. Since $P_0V_0/273$ is a constant (it is always the same quantity no matter how P, V, and T change, as long as we keep the same amount of gas in the cylinder), the equation of state can also be written as PV/T = constant or $P_1V_1/T_1 = P_2V_2/T_2$ where the subscripts 1 and 2 refer to two different sets of conditions. This equation tells us that no matter how the pressure, volume, and temperature of a perfect gas change separately, the algebraic combination PV/T always remains the same for a gas as long as the quantity of gas is the same. The equation of state of a perfect gas is very useful in chemistry and physics because it permits us to calculate any one of the three quantities P, V, T of a gas if we know the other two. In many chemical and physical properties in industry the temperature and volume of a gas change, and one then has to know how the pressure changes. The answer is provided by the equation of state.

We saw that the quantity PV/T for an ideal gas is a constant, provided the amount of gas remains the same. We therefore cannot give the constant in the equation of state a definite value unless we specify the amount of gas we have. We may take any amount of gas we please, but it is convenient, from a chemical point of view, to work with a mole of gas, which is that quantity of gas whose mass in grams numerically equals the molecular weight of the gas.

The molecular weight of a gas (or of any substance) is the mass of a single molecule of the gas on a scale on which the oxygen molecule (O_2) is assigned the number 32. More roughly speaking, we may say that the molecular weight of a substance is the number of times the mass of one molecule of the substance is greater than the mass of a single atom of hydrogen. The molecular weight of H_2O is 18, of CO_2 is 44, and so on. In dealing with atoms instead of molecules, we speak of the atomic weights—a scale of weights on which the oxygen atom is given the number 16. We obtain the molecular weight of a molecule by adding up the atomic weights of all the atoms in the molecule. The masses of moles of different gases (different molecular weights) are different. The equation of state of a single mole of any gas is given by the equation $PV/T = \mathrm{R}$ or $PV = \mathrm{R}T$. R is the "gas constant" for a single mole of gas. This number is the same for one mole of any gas, regardless of its chemical nature or molecular weight. Before we give the numerical value of this constant, we note that R is not just a numerical constant but has definite physical dimensions. However, it cannot be expressed in terms of our basic physical quantities, length, time, and force (or mass). The reason for this, as can be seen from the equation of state, is that the temperature T enters into the formula for R, and T itself cannot be expressed in terms of length, time, and mass. T itself must be introduced as a basic dimension and R must therefore be expressed in terms of length, time, mass, and temperature. To obtain the dimensional expression for R we note first that PV is (force/area)(volume). But volume/area = length, so that PV is force times length. Thus, PV has the dimensions of work or energy, and must be expressed as ergs. Since R equals PV/T, its dimensional formula must be ergs/degrees or (gm cm^2/sec^2 degrees).

The numerical value of R can be obtained by introducing the following empirical data into the formula $PV/T = \mathrm{R}$ at 0°C (273°K). At 0°C one mole of any gas, kept at a pressure of 1 atmosphere (atmospheric pressure), occupies a volume of 22.4 liters or 22,400 cm^3. Now a pressure of 1 atmosphere equals the weight of a column of mercury 76 cm high and with a cross section of 1 cm^2. Hence, 1 atmosphere = 76 ρg where ρ, which equals 13.55 gm/cm^3, is the density of mercury, and g is the acceleration of gravity. By inserting numbers we find that 1 atmosphere of pressure is equal to 1.01×10^6 dynes/cm^2 or, roughly, one million dynes per square centimeter. Using this value for P and placing $V = 22{,}400$ and $T = 273$, we obtain a value for R of 8.31×10^7 ergs/°C.

For either more or less than one mole of gas in our container, the equation of state changes. Since we must have the same general relationship among the gas parameters P, T, and V as we did for one mole, the only change that can occur is in the value of the constant. Instead of

R as the constant we must use a number that is smaller or larger than R, depending on whether we have less or more than one mole.

To see how the constant changes with the quantity of gas, we consider two identical containers of gas, standing next to each other, each containing one mole of gas. The pressure, volume, and temperature are the same for both and we have $P = \mathrm{R}T/V$ for each container. This equation for each container can also be written in the form $V = \mathrm{R}T/P$.

We now bring both containers together, placing them in contact to form a single container with twice the volume of each of the individual ones. We remove the dividing wall to obtain a single container in which the pressure and temperature are the same as before but with twice the amount of gas (2 moles) and twice the volume. The equation of state applies to this sample of gas just as it did to the individual samples. Thus, the quotient T/P multiplied by the appropriate gas constant gives the total volume $2V$ of the large container so that $2V =$ (constant) (T/P). But from the previous equation for the single mole of gas in each small container we have $T/P = V/\mathrm{R}$. Hence, $2V =$ constant (V/R) so that the constant $= 2\mathrm{R}$. This relation shows us that if we double the amount of gas, we must use the same equation of state but with the gas constant multiplied by 2. This example illustrates the following general rule: If a container of volume V encloses n moles of a perfect gas at the pressure P and at the absolute temperature T, the equation of state of the gas is $PV = n\mathrm{R}T$ where n is a pure (dimensionless) number. Although this equation applies to a vessel containing a single kind of gas, it also applies to a mixture of gases. Each component of the mixture behaves as though the other components were not present. Each component thus exerts its own pressure on the walls of the container as though it alone occupied the container. The total pressure is then the sum of all these partial pressures. Suppose that we have n_1 moles of gas 1, n_2 moles of gas 2, and so so, and we have k different gases. If P is the total pressure of the mixture, we then have $PV = n_1\mathrm{R}T + n_2\mathrm{R}T + \cdots + n_k\mathrm{R}T$. If we divide each of these quantities by V, then $P = P_1 + P_2 + \cdots + P_{\mathrm{k}}$. This equation is Dalton's law of partial pressures.

The above form of the equation of state of a perfect gas is not always convenient to work with because it contains the volume V. In astrophysical investigations, for example, we deal with the conditions of the stellar gas at various points inside a star and not with the behavior of this gas throughout a given volume. It is therefore important for this purpose to express the equation of state of a gas at a point. We eliminate the volume from the equation of state and introduce instead the density of the gas. Suppose that m is the total mass of the gas, and that its mean molecular weight is μ. We must use the mean molecular weight because, in general, we deal with a mixture of gases in a star, each with its own molecular weight. Owing to the validity of Dalton's law of partial pressures, we treat a mixture of gases just as we treat a single gas, provided we use the average molecular weight of the mixture.

Since m is the total mass of our mixture of gases and μ is its mean molecular weight, we have for the total number n of moles in the mixture $n = m/\mu$ where μ is to be expressed in grams since m is in grams and n is a pure number. Using this equation for n in the equation of state, we have $PV = (m/\mu)\mathrm{R}T$. If we now divide both sides of this equation by V, we obtain $P = (m/V)\,(\mathrm{R}T/\mu)$. Since m/V (the mass per unit volume) is just the density ρ of the gas, we have $P = (\mathrm{R}/\mu)\rho T$.

The form in which we have just written the equation of state is very useful in studying the internal structure of stars like the sun because such stars are very hot, gaseous spheres which radiate energy into space at a fairly constant rate. At each point in such a sphere the gaseous mixture exerts a pressure in all directions, which produces an outward force that tends to cause the star to expand. If no force counteracted the outward forces arising from this gaseous pressure, the star would indeed expand; in fact, it would explode violently, and stars could not exist. But since stars do exist, we know that an inward force just balances the outward gas pressure at each point inside the star. This inward force is just the force of gravity exerted by the total mass of the stellar material on each bit of the stellar gas. This gravitational interaction indicates that the only way a star can be formed from a cloud of gaseous material is if the mass of the cloud is sufficiently large so that the gravitational forces are strong enough to compress the cloud into a gaseous sphere and keep it in equilibrium against the gas pressure.

If the mean molecular weight of the mixture of a hot, gaseous sphere surrounding some point in it is μ, its density at a given point is

ρ, and the temperature there is T, the pressure P of the gaseous mixture at that point is given by $P = (R/\mu)\rho T$. We can draw an interesting conclusion from this equation concerning the relationship of the temperature inside the star to the mean molecular weight of the stellar material. If the molecular weight of the gases that constitute a star were suddenly increased, as would occur if, for example, the hydrogen were replaced by helium, while the geometry (or dimensions) were kept the same, so that the density ρ also remained the same, the pressure P at each point as given by the equation of state would fall because μ is in the denominator of the expression for P. The star would begin to collapse; but this collapse would immediately cause the temperature T to rise at each point, and hence at the center, which would increase the rate at which the energy is generated. The pressure would thus rise sufficiently to prevent further collapse. It is easy, then, to see that a hotter, and therefore more luminous, star results from increasing the mean molecular weight of stellar material.

Before leaving the equation of state of a perfect gas, we consider one more interesting application of it to stars. From what we have already stated, we see that unless the mass of the gaseous cloud from which the star condenses is quite large, the self-gravitational field acting on the cloud is not large enough to keep the gases from dispersing into space. But if we start with a sufficiently large mass the gravitational force can cause the cloud to contract into a spherical structure. This, however, does not mean that the structure remains in such a form; indeed, this does not generally happen unless something else occurs simultaneously. The principle of the conservation of mechanical energy tells us that the total mechanical energy (kinetic plus potential) of a system of particles remains unchanged no matter what the configuration of the particles may be. If the particles do come together at some moment, they must, in time, separate again for they may be pictured as moving in orbits in which energy is conserved. All possible configurations in which the total energy of the system remains the same recur repeatedly, meaning that the collection of particles disperse again.

Keeping this fact in mind, we see that if the gaseous cloud is to contract into a sphere (star) and remain stable, the cloud must lose energy as it contracts. Once the cloud of particles (molecules of gas) has lost energy in the process of contracting, the particles no longer have enough kinetic energy to move away from each other, and they remain together as a sphere under their mutual gravitational attraction.

We are not yet in a position to see just how the gaseous cloud loses energy as it contracts gravitationally, but we can get some understanding of the process if we note that as the cloud contracts, the gravitational potential energies of the molecules diminish. As a result, the kinetic energy of their random motions must increase so that the cloud becomes hotter and radiates energy into space owing to collisions among the molecules. The cloud thus loses energy and the sphere of gas is stable, remaining in its compact form after contraction has occurred.

Using the equation of state of a gas, we can now give an approximate answer to the following question: What is the average temperature throughout a star of mass M, radius R, and mean molecular weight μ if it is in gravitational equilibrium? It is fair to suppose that the average temperature is equal to the temperature about halfway down to the center of the star. It is also reasonable to assume that half the mass of the star lies in the outer envelope above this point. The force of gravity pulling this envelope inward (the weight of this outer envelope) is given roughly by Newton's law of gravity as $[G(M/2)(M/2)]/(3R/4)^2$ where we have assumed (as an approximation) that the entire mass of the outer envelope is concentrated in a thin shell three-fourths of the way out from the center.

The outward pressure exerted by the stellar gas at this halfway point is $(\mathrm{R}/\mu)\rho T$, where T is the average temperature and ρ is the stellar density halfway down. If we multiply this quantity by the area $[4\pi(R/2)^2]$ of the inner envelope, we obtain the outward push of the gases that must support the weight of the outer envelope $(4\pi R^2 \mathrm{R}\rho T/4\mu)$. Since this quantity must equal the weight of the outer envelope, we must have $\pi R^2(\mathrm{R}\rho T/\mu) = GM^2/R^2$. If we now assume, as an approximation, that $(4/3)\pi R^3\rho$ is the total mass M of the star (ρ is taken as the average density) and substitute this quantity for M in the previous equation, we obtain, as an approximate value of the mean temperature (just multiplying both sides by 4/3), $T \approx (16/27)\mu(GM/R)(1/\mathrm{R})$. This value is a bit too large as more accurate calculations prove, but this analysis is presented here to give the reader some understanding of the thermal conditions inside a star; this equation gives 10 million degrees for typical stars like the sun.

We have seen that the equation of state for a single mole of any perfect gas is $PV = RT$, where R is the gas constant for a mole of gas. We also saw that R has the same value for all gases, which can be deduced from the experimental fact that one mole of any gas occupies the same volume as one mole of any other gas if P and T are the same for both gases. The value of PV/T for one mole of gas is thus the same as for one mole of any other gas. The constancy of R for one mole of all gases is thus an experimental fact.

Taking a mole of gas as an ensemble of molecules moving about randomly, we can express the gas law in terms of these molecules whose number is the same for all gases. This assertion can be proved; if N_0 is the number of molecules in one mole of gas whose molecular weight is μ, and N'_0 is that number in a gas whose molecular weight is μ', we have from the definition of a mole $N_0 m = \mu$ and $N'_0 m' = \mu'$ where m is the mass in grams of a single molecule of the first gas and m' is the same for the second gas. If we divide the first of these equations by the second, we obtain $(N_0/N'_0)(m/m') = \mu/\mu'$. But by the definition of molecular weight we must have $m/m' = \mu/\mu'$. Hence, $(N_0/N'_0)(\mu/\mu') = \mu/\mu'$ or $N_0/N'_0 = 1$. This equality of N_0 and N'_0 means that N_0, the number of molecules in a mole of gas, is the same for all gases; its value, which is called Avogadro's number, is 6.03×10^{23}. This number is the reciprocal of the mass, in grams, of the hydrogen atom.

If we start with one mole of gas and remove from it a single molecule, the gas law for the remaining quantity of gas is different from that for the single mole because we have reduced the quantity of gas by one molecule (or by $1/N_0$) of a mole. The gas law for this amount of gas is therefore $PV = (1 - 1/N_0)RT = (N_0 - 1)(R/N_0)(T)$. Placing $R/N_0 = k$, which we may consider as a basic constant of nature (like the gravitational constant G or the speed of light c), we write the gas law for $(1 - 1/N_0)$ moles in the form $PV = (N_0 - 1)(kT)$, which is a very instructive form because it relates the product PV to the absolute temperature T and to the number of molecules, $N_0 - 1$, in the container. If instead of removing one molecule from our mole of gas, we remove n molecules, we are left with $(1 - n/N_0)$ of a mole and the gas law becomes $(N_0 - n)kT$. Again we see that the product PV is expressed in terms of the number of molecules since $(N_0 - n)$ is just the total number of molecules in the container. From this we obtain the following general law for a gas consisting of N diverse molecules (regardless of their chemical nature) occupying a volume V at an absolute temperature T: $PV = NkT$.

The above equation is the most general expression for the gas law (and also the simplest) since it gives us PV directly in terms of the number N of freely moving particles in the gas and the absolute temperature T. Note that every particle that is free to move by itself (whether it is a molecule, an atom, and electron, a nucleus, or a tiny grain of dust) must be counted in calculating N.

We can now give a deeper meaning to the constant k which is known as the Boltzmann constant and whose value is 1.380×10^{-16} gm cm^2/sec^2 deg. It is the gas constant for a single molecule. To see this we note that we increase the product PV by the amount kT whenever we introduce an additional molecule into our gas without altering T. We might say that for a given absolute temperature T and volume V every molecule carries with it the same capacity to contribute to the total pressure and this capacity is measured by the universal constant k. The total pressure is the sum of the equal pressure contributions of each particle in the gas. From the previous equation we see that if P, V, and T are the same for two gases in two different containers, the N must be the same in both containers. We can now express the pressure of a gas in any container without specific reference to the volume of the container by introducing the particle density of the gas, which we define as the number of particles (or molecules) in a unit volume of the gas. If the particle density is n so that $n = N/V$ (by definition), then $P = (N/V)(kT)$ or $P = nkT$. The advantage of this form of the gas law, which is the simplest, is that we can calculate the pressure (for a given T) just from a knowledge of the number of molecules per unit volume, without reference to the total volume of the gas.

Going back now to the form of the gas law, $P = \mathrm{R}\rho T/\mu$, which is used in astrophysics, we introduce k into it instead of R and write $P = (N_0/\mu)(k\rho T)$ or $P = k\rho T/\mu H$, where H is the reciprocal of N_0 and is just the mass in grams of a proton. Its numerical value is 1.66×10^{-24}.

We saw that we can express the pressure in a gas in terms of the absolute temperature T of the gas and the number of molecules N but this expression for the pressure does not tell us how the pressure is related to the dynamical behavior of the molecules of the gas. We can now find this relationship by applying Newton's law of motion to the molecules of the gas. Since these laws tell us that the pressure in the gas arises from the random motions of the molecules colliding against any surface placed in the gas, we can calculate this pressure by determining the number of collisions that occur per second per square centimeter and then multiplying this number by the momentum transferred to the wall by each such collision. This procedure is just Newton's second law of motion applied to an ensemble of particles moving randomly.

We consider a container with molecules in it moving around randomly. To simplify this analysis, we assume that all the molecules have the same mass m (the final result is also valid for a mixture of molecules with different masses) and that the average speed of a molecule is v. Since the pressure of the gas stems from the bombardment of the walls of the container by the molecules, we consider first the nature of the collision of a molecule with a wall.

On the average, the kinetic energy of a molecule after it bounces off a wall is the same as before it strikes the wall. If this were not so, the molecules in the container would either lose or gain kinetic energy continuously. In the first case, the temperature of the walls of the container would steadily increase until all the molecules in the container had lost all their kinetic energy and were at rest on the floor; in the second case, the walls of the container would get colder and colder. Since neither of these things happens, we may assume that the energy of a molecule remains unchanged during a collision with a wall.

To find the pressure against a wall A, we calculate the number of collisions that occur in a direction at right angles to the wall A. If a molecule hits the wall obliquely, the only component of its momentum that contributes to the pressure against the wall is the component perpendicular to the wall. Since the molecules in the vessel are moving about randomly, collisions against the wall A are occurring from all directions. One might therefore suppose that we have to consider every possible kind of collision and calculate what part of each such collision is at right angles to A before we can calculate the pressure; but this calculation is not necessary because we can use a shortcut. Since the molecules are moving randomly, we may assume that at any moment one-third of the total number in the container are moving at right angles to A (either directly toward it or directly away from it) and the other two-thirds are moving parallel to A. If n is the number of molecules per unit volume, then $n/3$ molecules in each unit volume are moving at right angles to A. Consider now only those molecules in each unit volume moving directly toward A. This number is $n/6$ since half of the number $n/3$ are moving to the right and half are moving to the left. Thus, the number of molecules in each unit of volume that concern us for purposes of calculating the pressure is $n/6$. The pressure may now be calculated as follows: We construct inside the container a cylinder of unit cross-sectional area whose unit area base B is in contact with the wall A, and let v (the average molecular speed) be the length of this cylinder so that the volume of this cylinder is v. The total number of molecules in this cylinder moving directly toward A is $n/6$ (the number per unit volume) multiplied by the volume of the cylinder which is just v as already noted. The number of molecules in this cylinder moving directly toward A is $nv/6$. But this quantity is also exactly the number of molecules per second that strike a unit area of A at right angles to A because each of the $nv/6$ molecules in A is moving directly toward A with a speed v, and each lies within the distance v of A. All of them reach A within 1 second and hit the unit area B.

But each of these molecules has the momentum mv before it hits the wall and the momentum $-mv$ after it rebounds. Each molecule thus suffers a total change of momentum $mv-(-mv)$ or $2mv$, and this amount is the momentum each molecule transfers to the unit area of A during a collision. Since $n/6$ such collisions occur per second, the total momentum transferred to a unit area of the wall per second is $(2mv)(nv/6)$ or $(n/3)(mv^2)$. This equation gives the pressure P of the gas against the wall. We thus have $P = (n/3)(mv^2)$ which can be written in the form $P = \frac{2}{3}n(mv^2/2)$; we see that the pressure in a perfect gas depends on the average kinetic energy $(mv^2/2)$ of a molecule in the gas. Alternatively, we may write that the pressure is equal to two-thirds of the number of molecules per unit volume times the average kinetic energy of a single molecule. Since $n(mv^2/2)$ is the total kinetic energy of the molecules in a unit volume (the kinetic energy density), which we label e, we have $P = \frac{2}{3}e$. If V is the total volume of the container, we also have $PV = \frac{2}{3}E$, where $E = eV$ is the total kinetic energy of the molecules in the container. Thus, the pressure of a perfect gas is determined by the kinetic energy of its molecules.

If we compare the expression we just obtained for the pressure of a perfect gas $P = (n/3)(mv^2)$ and the expression in the previous section $P = nkt$, we obtain the very important result $kT = mv^2/3$ or $\frac{3}{2}kT =$

$\frac{1}{2}mv^2$. This result now gives us a much deeper physical insight into the meaning of the absolute temperature T. We see that T is two-thirds of the average kinetic energy of a molecule divided by the universal Boltzmann constant k.

We can now deduce another important consequence of the above expression that relates the absolute temperature of a gas and the average kinetic energy of its molecules. Consider the average velocity **v** as a vector and let v_x, v_y, v_z be its components in some rectangular Cartesian coordinate system X, Y, Z. We then have from the theorem of Pythagoras $v^2 = v_x^2 + v_y^2 + v_z^2$ which leads to $\frac{3}{2}kT = \frac{1}{2}mv_x^2 + \frac{1}{2}mv_y^2 + \frac{1}{2}mv_z^2$. However, since the molecular motions are random, v_x, v_y, v_z are equal on the average and we may write $\frac{1}{2}mv_x^2 = \frac{1}{2}mv_y^2 = \frac{1}{2}mv_z^2 = \frac{1}{2}kT$.

These relationships express a very famous theorem known as the theorem of the equipartition of energy which tells us that in an ensemble of freely moving molecules at the absolute temperature T each of the three mutually perpendicular components of motion of each molecule has on the average the same amount of kinetic energy, $\frac{1}{2}kT$. This theorem has the following consequence: If work is done on a gas at a temperature T or if it is supplied with energy in any way, not only is this energy distributed equally among all the molecules, but each component of motion of each molecule receives, on the average, the same amount of energy, namely $\frac{1}{2}kT$.

Although we have deduced this theorem for the kinetic energy of randomly moving particles of a monatomic gas, it has a much wider application than that. Indeed, it is true in a very general way for all modes of energy and not just for translational kinetic energy. To explain this statement we introduce the concept of the degrees of freedom of motion of a particle. First, consider the translation of a particle and resolve this translational motion into its three mutually perpendicular components in some coordinate system X, Y, Z. Since these three components of the motion are independent of each other, we say that a particle in translational motion has three degrees of freedom. Suppose now that in addition to its translational motion the particle can also vibrate and rotate (this would be so if the molecules of our gas consisted of two or three atoms). It then has vibrational and rotational degrees of freedom as well as translational degrees of freedom. We now define a degree of freedom as any independent mode of motion that a particle can have. This definition leads us to the following generalized theorem of the equipartition of energy: If a given amount of energy is supplied to a system of particles at a temperature T, each of which has n degrees of freedom, then each of these n degrees of freedom receives, on the average, the same amount of energy, $\frac{1}{2}kT$.

From the way we have deduced this theorem, it is clear that it follows from Newton's second law of motion since the distribution of energy among the molecules in a gas results from the molecular collisions which produce energy interchanges among the molecules in accordance with Newton's law of motion. If we should discover at some time that the equipartition theorem breaks down, we must conclude that Newton's second law of motion breaks down also for molecules.

In this chapter we introduced the equation of state of a perfect gas (the gas law) from two different points of view: the macroscopic and the microscopic (or molecular) points of view. From the macroscopic point of view we introduced the parameters P, V, and T which we can define without reference to molecules. From the microscopic point of view we introduced the average kinetic energy of a molecule in place of T. The microscopic point of view leads to the introduction of a new universal constant of nature, the Boltzmann constant k, which must take its place with c (the speed of light in a vacuum) and G (the gravitational constant) in nature's "hall of fame" of constants.

CHAPTER 13

The Laws of Thermodynamics

Creation sleeps. 'Tis as the general pulse.
Of life stood still, and Nature made a pause;
An awful pause! prophetic of her end.

—EDWARD YOUNG, *Night Thoughts*

In the previous chapter we began our study of the structure of matter by considering bulk matter in its most unstructured state, namely in its perfect gas state. We saw there that the general properties and gross behavior of this state of matter can be understood in terms of three measurable macroscopic quantities: the pressure, the temperature, and the volume of the gas, and we derived the algebraic relationship among these three quantities called the equation of state of a perfect gas. We noted in our discussion of a gas that any particular set of numerical values of the three parameters P, V, T defines what we call a state of the gas, but owing to the equation of state, which connects the three gas parameters P, V, T algebraically, only two of them need to be specified to define any state of the gas. A state can thus be defined by specifying (P, V) or (P, T) or (T, V).

Suppose now that we have two different states of a given quantity (a given mass) of gas. At least one of the three quantities P, V, T must change when the gas goes from one of these states to the other. But the algebraic combination PV/T does not change. We may therefore say that the state of a gas changes whenever P, or V, or T changes; if any one of these quantities does change, the other two must change in such a way as to keep the algebraic expression PV/T constant.

We again consider a quantity of gas in a cylinder with a movable piston. Since P, V, and T have definite values, the gas is in a definite state which we might specify as S_0. We can now change this state in a number of different ways. We can place an additional weight on the piston or remove one of the weights thereby altering the pressure P. We can also raise or lower the piston, thereby changing the volume. Finally, we can heat the gas (using a Bunsen flame) or churn it about mechanically (do work on it) and thus raise its temperature. Each of these actions leads to a different change of state, and if we perform a number of such changes successively, the gas ends up in some final state which, in general, differs from its initial state (a different set of P, V, T values). Such a series of successive states can be represented by a path in a P–V graph as we have already explained. Many different paths lead from an initial state A to the same final state B.

Consider now a closed path or a cycle such as AC′D′BDCA which brings the system back to its initial state. Along such a path P, V, and T all change but on returning to point A, these three quantities take on their initial values. This is also true of any algebraic combination of P, V, and T such as PV or PV^2 or PV/T, and so on. Such an algebraic combination of these three basic quantities is called a state function of the gas because this combination always takes on the same value whenever the gas is in the same state. As an example of a state function that is the same for all states of a gas we have the combination PV/T which equals the gas constant R, but this state function cannot be used to label a state since it does not differentiate between any two states. On the other hand, the state function PV does change, in general, but it does not change along a path at each point of which T is constant (along an isothermal path).

Since we can construct (algebraically) any number of combinations of P, V, and T, an infinitude of state functions exists, but only a few of them are physically significant or useful in the study of gases. In this section we deal with one of these combinations which we must introduce and understand before we can formulate the first law of thermodynamics. Later in this chapter we introduce still another state function, which leads us to (and is useful in formulating) the second law of thermodynamics.

To see the nature of the state function that we introduce to express the first law of thermodynamics, we return for a moment to the principle of conservation of energy. Two forms of mechanical energy (kinetic and potential) entered into our statement of the principle of conservation of mechanical energy, which says that if no work is done on a system of bodies, or if it does no work against an external force, then its total mechanical energy remains the same. The important thing here is that the system must be isolated so that it has no chance of doing work on some other system and thus losing some of its energy. Now a system can lose energy in two apparently different ways by doing work. In one of these ways, the lost energy shows up as mechanical energy again but associated with some other system. If a perfectly elastic sphere (e.g., a highly tempered steel ball bearing) in motion were to hit an identical stationary sphere head-on, the first sphere would lose all its energy (it would come to rest) and the second sphere would move off with the same speed and in the same direction as the first sphere. The second sphere would thus take up all the kinetic energy that the first sphere had initially. Here we have an example of nonconservation of energy for the first sphere, but we can see just where the energy of the first sphere has gone; it has been transferred to the second sphere so that there is no loss of energy at all if we take both spheres into account. In this simple case of two perfectly elastic mechanical bodies we can see just how the total mechanical energy of the two bodies is conserved; whatever energy one body loses, the other gains.

Consider now another case of the loss of energy of a body. We set the body (it can be a sphere or have any other shape) moving on a surface by pushing it (we do work on it and give it kinetic energy), and observe that as soon as we let it go, it begins to slow down and finally comes to rest. We say that it has been brought to rest by the force of friction generated by the contact between the body and the surface. The body does work against this force of friction and thus gives up its energy. The difference between this and the previous situation involving the two elastic spheres is that we cannot see what happens to the kinetic energy of the moving body in this case. We observe the moving body losing kinetic energy, but we see no other body gaining either kinetic or potential energy. Owing to this "mysterious" disappearance of kinetic energy, scientists, in general, before the middle of the 19th century, believed that energy is not conserved in a situation where frictional forces are involved. But Rumford, Leibniz, Lavoisier, and others had noted that the temperature rises at the point of contact between two rubbing bodies and they believed that the loss of kinetic energy of these bodies can be explained by the heat that is generated by the rubbing action. Leibniz and Lavoisier (far ahead of their times) suggested that the kinetic energy that is lost by a moving body working against friction, such as in the example given above, is transferred to the atoms of the rubbing bodies, and Rumford surmised that the mechanical energy lost in friction is converted into heat. But at that time nobody knew the nature of heat nor understood how heat and energy are related. Heat was thought to be a fluid substance (caloric) that can neither be created nor destroyed but flows from one place to another and from one body to another. Only when it was understood that heat is a form of energy (one of the forms that energy assumes when it is in transit from one physical system to another) was the science of thermodynamics born with the statement of the first law of thermodynamics, which is an extension of the principle of the conservation of energy to take into account heat and, in fact, all forms of energy.

Before stating the first law of thermodynamics precisely we consider an example of another mechanical system which is losing mechanical energy, but in a situation in which we can more readily see where this lost mechanical energy goes. At the same time, we see that this lost mechanical energy is related in some manner to the parameters P, V, T that describe the state of a gas. The mechanical system to which we refer is a pendulum swinging freely under atmospheric conditions. We saw that if such a pendulum were swinging in a vacuum and if there were no friction at the point where the string of the pendulum is attached, the pendulum would swing forever owing to the conservation of mechanical energy. But this does not happen if the pendulum is swinging in air even if there is no friction where the string is attached. Owing to the friction between the bob of the pendulum and the surrounding air the pendulum loses energy and comes to rest.

To see what happens to this energy, we surround the pendulum completely by rigid walls that permit no circulation of air and prevent heat from leaving or entering the surrounded region. If we now record the readings of a very sensitive thermometer placed in the enclosed region, we observe that it rises steadily as the pendulum slows down. This shows us that the pendulum is giving up something to the surrounding air which causes the temperature of the air to rise. Since we know that the pendulum is losing energy, we deduce that energy flows from the pendulum into the surrounding air thus causing the temperature of the air to rise. Accepting the increasing temperature of the air as proof that the mechanical energy of the pendulum is not lost when it leaves the pendulum but is transferred in some form or another to the air, we extend the principle of the conservation of energy to include nonmechanical systems, such as gases, as well as mechanical systems. This principle now states that the total energy of any physical system is always conserved provided we include in the total energy not only the mechanical energy of the moving parts but also any energy in the surrounding medium or energy contained in any form (other than mechanical) in the objects that constitute the system.

What does this principle mean for our pendulum and the air surrounding it? The pendulum comes to rest so that the entire system of pendulum and surrounding air no longer has any mechanical energy, but the temperature of the air is higher than it was initially, showing that the air has gained energy if an increase in temperature means a gain in energy. We shall see that this is so and also that the same amount of mechanical energy transferred to a gas always produces the same change in temperature for a given amount of gas under the same conditions. As we shall see, this statement is essentially the content of the first law of thermodynamics, which we state more precisely later.

We now consider somewhat more carefully the qualitative difference between the energy in the gas after the pendulum has come to rest and the mechanical energy that was present in the pendulum. There is an obvious difference in the sense that we can "see" the mechanical energy in the pendulum in its motion (kinetic energy) and its height above the ground (potential energy). But we cannot "see" the energy in the gas. After the pendulum has come to rest the air surrounding the pendulum looks the same as it did before, even though it has received energy from the pendulum. Owing to this hidden quality of the energy of the air, we call it internal energy and label it with the letter U. By internal energy we mean any form of energy, other than bulk mechanical energy, that is contained in the gas. We see from what we have said that the internal energy of a gas is related in some way to the temperature of the gas, and we surmise that the internal energy is the state function of the gas that we spoke of in the previous section. We consider this point more fully when we state and discuss the first law of thermodynamics.

We come now to another important qualitative difference between the initial mechanical energy of the pendulum and the final internal energy of the air surrounding the pendulum. The mechanical energy of the pendulum is concentrated in the bob of the pendulum and is therefore highly organized. We can use it all (if there were no friction at all) to do work for us so that it is completely useful and available. Thus, the moving bob of the pendulum can drive a nail into a wall of the container if so directed. The internal energy of the gas (the air), however, is not a concentrated, highly organized form of energy but is distributed throughout the entire volume of air so that not all of it is available for doing work, in contradistinction to the mechanical energy of the pendulum.

To show this more clearly we insert a small piston with weight w at the top of the cylinder from which we suspend our pendulum, and suppose that the height of the piston above the top of the cylinder is h_0 before the pendulum is set oscillating. This height is such that the total volume V of the air in the container is related to the pressure P (determined by the weight on the piston) and the temperature T of the gas (air) by the gas law $PV/T = \mathrm{R}$ (if we assume that there is just one mole of air in the container). We now set the pendulum swinging and note that the piston rises as the pendulum slows down, while at the same time the temperature of the air increases. The gas does work by lifting the weight on the piston as the pendulum slows down. This work comes from the mechanical energy lost by the pendulum. But not all of this mechanical energy goes back into mechanical energy again; that is, into the work of lifting the piston, for when the pendulum comes to rest, the temperature of the air is higher than it was originally and the work done on the weight resting on the piston (as measured by the difference in the initial height h_0 of the piston and its final height after the pendulum has come to rest) is less than the total mechanical energy lost by the pendulum. We can understand why this is so if we keep in mind that the temperature of the air finally is higher than it was initially, showing that although some of the mechanical energy lost by the pendulum is reconverted into mechanical energy again (lifting the weight on the piston) some of it remains hidden in the gas as internal energy. We come back to this point and its importance when we consider the second law of thermodynamics and the relationship of this law to the convertibility of one form of energy into another.

If we return to the kinetic picture of a gas we can explain the qualitative difference between the internal energy of the gas and that of the pendulum from a microscopic or a molecular point of view. As the pendulum oscillates through the air, the molecules of the air collide with the bob randomly and, on the average, rob the pendulum of its energy. Some molecules (those that are moving in the same direction as the bob and collide with it from behind) give the bob some of their own energy, but those that are moving against the motion of the bob and collide with it from in front take energy from the pendulum. Since, on the average, more forward than rear collisions occur per second, the bob loses energy to the molecules steadily and finally comes to rest. We see that all the mechanical energy of the bob goes over into the kinetic energy of the individual molecules of the air. But since these molecules are moving about randomly, the energy lost by the pendulum is now disorganized since it is distributed over these many randomly moving molecules.

We saw above that there is a state function of a gas (its internal energy U) which changes when the gas interacts with a mechanical system. If the internal energy of the gas is taken into account, then the total energy of the gas and the mechanical system combined (internal energy and mechanical energy) is conserved. This statement is essentially the content of the first law of thermodynamics, which we now state more precisely. We return to the gas in the cylinder capped by a movable piston on which a weight w is resting so that the pressure P in the gas is just w/A where A is the area of the piston. That this gas has internal energy is easily seen when we remove some of the weight. The piston then rises carrying the remaining weights upward and the temperature of the gas (as shown by the thermometer) falls. The gas does work on the weight (it increases its potential energy) and gives up some of its internal energy as indicated by the drop in the temperature of the gas.

To arrive at a precise formulation of the first law and at the same time introduce the concept of heat, we place a Bunsen flame under the cylinder and observe that two things occur. The piston rises (showing that work is being done) and the temperature rises. Some kind of energy has passed from the Bunsen flame into the gas ultimately showing up partly as mechanical work (the raising of the piston) and partly as internal gas energy. We call the energy that passes from the flame into the gas heat. Thus, heat is a form of energy and it is often considered to be the same form of energy as internal energy. But here we distinguish between heat and internal energy by reserving the term "heat" for internal energy that is in transit (or flowing) from one system to another.

Let a small amount of heat ΔQ pass from the flame into the gas and let ΔU be the resulting small increase in the internal energy of the gas. Further, let ΔW be the small amount of work done by the gas on the piston. The first law of thermodynamics then states that ΔQ equals the sum of ΔU and ΔW, or $\Delta Q = \Delta U + \Delta W$.

The above equation is a precise algebraic statement of the first law of thermodynamics and it tells us something remarkable that took humanity a very long time to understand and to use, namely, that work can be obtained from heat. For that reason we may refer to this equation as a law of optimism because it holds out great promise for the release of humanity from the drudgery of backbreaking physical labor (a promise which has already been realized to some but not to its fullest extent).

Before scientists recognized heat as a form of energy and discovered the first law of thermodynamics, physical work was performed by water (the waterwheel), by wind (the windmill and sails), by beasts of burden, or by man himself. This condition greatly limited the development of technology, encouraged slavery, and imposed drudgery on mankind in general. The first law of thermodynamics, with its statement that work can be obtained from heat, presented the vista of great technological developments leading to a world free of physical drudgery. Since heat can be transformed into work and there are many sources of heat, it appeared, in the early years of the study of heat, that the possibilities of extracting work from heat were practically endless. The most practical and efficient method that has been developed for obtaining work from heat is the heat or thermal engine (the steam engine, the internal combustion engine, the diesel engine, and so on), and from the study of these engines as long ago as 1820, even before heat was recognized as a form of energy and the first law was discovered, Sadi Carnot, the great French heat engineer, showed that there is a limitation to how much heat can be transformed into work by an engine. We consider this point more fully when we introduce the second law of thermodynamics and discuss the concept of the efficiency of a heat engine.

The first law itself does not appear to place any restriction on the transformation of heat into work, but since heat, as we have defined it, is energy in a state of flow or in transit, we see that certain restrictions on the transformation of heat into work exist. We can get work from heat only if we can get heat to flow through the engine that is to do the work, and this flow, as we shall see, can be achieved only if a temperature difference between the source of the heat and the engine exists. It is no more possible to get work from the hot air all around us on a very hot summer's day than it is to get work from the cold air on a cold winter's day since in neither case is there a flow of heat.

The first law tells us that we can transform heat completely into work only if we can prevent the internal energy of the engine and its environment from increasing when heat is supplied, which in general is impossible. We also note from the first law that we can get work from the internal energy of a gas under appropriate conditions. Even if $\Delta Q = 0$ (no flow of heat), we have from the first law $\Delta W = -\Delta U$, which means that the system can do work by giving up some of its internal energy as indicated by the minus sign on the right-hand side of the above equation. But here, too, this can occur only under very special circumstances. If the gas in our cylinder with no Bunsen flame applied to it is under high pressure (higher than the pressure of the surrounding atmosphere) and we remove some of the weight on the piston, the gas expands, thus raising the piston and doing work. The gas then loses internal energy as shown by the drop in its temperature.

We expressed the first law of thermodynamics in terms of heat flowing into a system, the increase in the internal energy of the system, and the work done by the system, without specifying the nature of the system. We now apply these ideas to a perfect gas, which is a system for which we can write down the expression for the internal energy U explicitly in terms of the temperature and also the expression for the work done by the gas. We consider first the internal energy and arrange things in such a way that the gas in the cylinder can do no work when heat flows into it. We clamp the piston so that it cannot move and the gas cannot expand and do work. The heat that flows into the gas from the Bunsen flame now goes entirely into increasing the internal energy of the gas so that we have $\Delta Q = \Delta U$ if an amount of heat ΔQ flows into the gas and ΔU is the increase in its internal energy. This process increases the temperature of the gas, and it is now of interest to see how the increase in temperature of the gas is related to the increase in the internal energy.

If ΔT is the amount by which the temperature of the gas increases when an amount of heat ΔQ is supplied to the gas when the piston is clamped so that there is no change in the volume of the gas, then we see from the previous equation that the quantity $\Delta Q/\Delta T$ (calculated for a fixed volume) is the rate of increase of the internal energy of the gas per degree rise in its temperature. But this quantity also has a different interpretation. We know that we can, in general, change the temperature of any substance (solid, liquid, or gas) by heating it, provided the substance is not passing from the solid to the liquid phase or from the liquid to the gaseous phase. If we heat 1 gram of substance and increase its temperature by 1°K (Kelvin, or centigrade), the quantity of heat supplied is called the specific heat of the substance being heated. Thus, the specific heat of water at atmospheric pressure and at 14°C is 1 calorie (4.185×10^7 ergs) which, by definition, is the unit of heat energy. Measurements show that the specific heats of different substances vary greatly, with water having a higher specific heat than that of most other substances.

Since the volume of solids and liquids does not change much when such substances are heated, we do not have to differentiate between the specific heat at constant volume and the specific heat when the volume is allowed to change for such substances. For gases, however, we must draw such a distinction because the volume of a gas can change considerably if allowed to do so, when it is heated. We therefore introduce two different specific heats of a given gas: the specific heat at constant volume c_v and the specific heat at constant pressure c_p. We now place 1 gram of a given perfect gas in our cylinder and send an amount of heat ΔQ into it, keeping its volume constant while doing so. If the temperature of this gas rises by the amount ΔT, then, by definition, $c_v = \Delta Q/\Delta T$. But from the first law we see that this means $\Delta U/\Delta T = c_v$ or $\Delta U = c_v \Delta T$. We see that the specific heat at constant volume of a perfect gas is just the rate at which the internal energy of a gas increases per degree rise in absolute temperature. The specific heat, of course, differs from one gas to another.

If the specific heat at constant volume of a perfect gas is assumed to be the same for all temperatures, in other words, independent of the temperature of the gas (an assumption which is valid for Newtonian mechanics if Newton's law of motion $F = ma$ holds), then $U = c_v T +$

constant, where the constant is introduced because the energy of a system is determined only up to an arbitrary constant (a number that is the same for all states of the gas). Since only differences in energy are considered in physical problems, this constant drops out in all calculations. We see that the internal energy of a gas is proportional to its absolute temperature. If we place the constant in the above equation equal to zero, which is always possible by choosing an appropriate frame of reference, we have $U = c_v T$. If we introduce the specific heat c_v into the algebraic expression of the first law, we have (for 1 gram of gas) $\Delta Q = c_v \Delta T + \Delta W$.

To write the expression ΔW for a perfect gas, we do not keep the volume constant when heat is supplied to the gas but allow it to expand by a small amount so that the piston rises by an amount Δh. When this happens the weight w is lifted through this distance by the gas which then does work $w\Delta h$. Hence, $\Delta W = w\Delta h$. But $w/A = P$ where P is the pressure in the gas and A is the area of the piston. On multiplying and dividing the right-hand side by A, which changes nothing, we get $\Delta W = PA(\Delta h)$. But $A\Delta h$ (area times height) is just the change in the volume of the gas resulting from the rising of the piston. If we call this change in volume ΔV, we finally have $W = P\Delta V$ for the small amount of work done by a gas when its volume expands oy the amount ΔV at pressure P. Replacing ΔW by $P\Delta V$ in the expression for ΔQ, we now obtain (the first law expressed entirely in terms of Q, P, T, and V for 1 gram of a perfect gas) $\Delta Q = c_v \Delta T + P\Delta V$.

We now consider the specific heat at constant pressure, c_p (the heat required to change the temperature of 1 gram of gas by 1° when the pressure is kept constant), and see how it is related to c_v. We note that c_v is the heat required to change the temperature by 1° when the gas does no work at all (the volume of the gas is constant). None of the heat that goes into the gas is used up in doing work. All of it is available for raising the temperature of the gas. The quantity c_p, on the other hand, is the heat required to raise the temperature of 1 gram of the gas while the gas is expanding against an outside force (the weight on the piston) and is therefore doing work. Not all of the heat is available for raising the temperature of the gas: some of it is needed to do the work of raising the piston so that more heat is needed to increase the temperature of the gas by the amount ΔT than would be needed if the volume of the gas were constant and it did no work.

We may present the difference between c_v and c_p somewhat more descriptively if we first supply enough heat to the gas (which is at temperature T and pressure P) to raise its temperature by an amount ΔT, while we keep the volume of the gas constant (we do not permit the piston to rise). The heat required is then $c_v \Delta T$ by the definition of c_v. The pressure in the gas is now higher than it was initially because of the increase in temperature. If we use the equation of state for a single gram of gas, which is $PV = \mathrm{R}T/\mu$, we see that the pressure increases by the amount $\Delta P = \mathrm{R}\Delta T/\mu$ during this process. We now allow the gas to expand, without allowing any heat to flow into or out of the cylinder, until the pressure of the gas is back to its initial value P. During this expansion the gas pushes the piston upward and does work so that its temperature drops below the value $T + \Delta T$. We must therefore supply some additional heat after this expansion to bring the temperature of the gas back to the value $T + \Delta T$. If we add this heat to $c_v \Delta T$, we obtain $c_p \Delta T$ so that c_p is indeed larger than c_v.

A simple way to deduce the relationship between c_p and c_v for a perfect gas is to use the first law and note that $\Delta Q = c_p \Delta T$ if P is constant (the gas expands as the heat ΔQ is added). But from the first law we also have (for 1 gram of gas and for constant pressure) $\Delta Q = c_v \Delta T + P\Delta V$. Hence, (using $\Delta Q = c_p \Delta T$) $c_p \Delta T = c_v \Delta T + P\Delta V$. But from the gas law (for 1 gram of gas) $P\Delta V = \mathrm{R}\Delta \mathrm{T}/\mu$ and $\mathrm{R}\Delta \mathrm{T}/\mu = (c_p - c_v)\Delta T$. We can now rewrite the gas law in terms of c_p and c_v so that $PV = (c_p - c_v)T$. This is a very instructive and useful way of writing the gas law because we shall use it to show that Newton's law of motion $F = ma$ breaks down when we apply this law to molecules.

We have now discovered two ways of writing the equation of state of a perfect gas (for 1 gram of gas) but instead of working with c_p and c_v directly, we work with the ratio of specific heats γ which we define as $\gamma = c_p/c_v$. Since $c_p > c_v$, γ is always larger than 1.

The gas law can now be expressed as $PV = (\gamma - 1)U$. This very general relationship in which the temperature no longer appears explicitly but in which the ratio of the specific heats γ does, holds for any quantity of gas even though we derived it for 1 gram. As we shall see, the appearance of γ in this equation of state of a perfect gas permits us to check Newton's law of motion $F = ma$ for the molecules of a gas, and we shall see that this law of motion breaks down for such particles.

We now write the gas law in a form which does not contain V explicitly which necessitates that we introduce the energy density u in a gas. This quantity is just the energy per unit volume or U/V. From the previous equation we therefore have $P = (\gamma - 1)(U/V)$ or $P = (\gamma - 1)u$. We now analyze this relationship between the pressure and the energy density of a perfect gas from the microscopic point of view (from the kinetic point of view) of a gas using the relationship previously described. We saw that if a perfect gas consists of N freely moving molecules that exert no forces on each other but suffer random collisions, we then have $PV = NkT$. Comparing this equation with $PV = (\gamma - 1)U$, we have $NkT = (\gamma - 1)U$ or $U = NkT/(\gamma - 1)$. This equation relates the total internal energy in a gas to the total number of molecules in the gas, to the absolute temperature of the gas, and to the ratio of specific heats of the gas. If v is the average speed of a molecule and m is its mass, the average kinetic energy of a molecule is $\frac{1}{2}mv^2$ and the total internal energy U is $(N/2)mv^2$ (provided that we assume that the only energy a molecule can have is the kinetic energy associated with its three degrees of random translational velocity). From this reasoning we see that $T = (\gamma - 1)/2k(mv^2)$. This relationship tells us that the absolute temperature of a gas depends on the ratio of the specific heats of a gas, on Boltzmann's constant k, and on the average kinetic energy of a molecule of the gas. If we compare this expression with the expression $kT = mv^2/3$, obtained using Newton's law of motion (the force exerted by a molecule bouncing off a wall is the time rate of change of momentum of the molecule), or if we compare the expression $PV = (\gamma - 1)U$, derived above, with the expression $PV = \frac{2}{3}E$ obtained previously (where E is the total kinetic energy of the molecules of the gas), we see first that U and E are not the same, in general. In fact, $(\gamma - 1)U = \frac{2}{3}E$ or $U = \frac{2}{3}[E/(\gamma - 1)]$.

It is important to make the above distinction between U and E because U, in general, can consist of a number of different parts, one of which is E, the kinetic energy. In addition to the kinetic energy of translational motion, a molecule may vibrate, rotate, or have some other kinds of internal motion. But although these vibrations, rotations, and internal molecular motions all contribute to U, they do not affect the temperature of the gas. Only the kinetic energy E contributes to the temperature, and that is why U and E differ by a factor that contains γ.

We saw above that there is a distinction between the total internal energy U of a gas and the total kinetic energy E of all its molecules, and we discovered that these two quantities are related to each other by the factor $(\gamma - 1)$ where γ equals the ratio of the specific heats of the gas. If we solve the last equation for $\gamma - 1$, we have $\gamma = \frac{2}{3}(E/U) + 1$. This result is extremely important because it relates a measurable macroscopic characteristic of a gas (its ratio of specific heats) to two quantities E and U that can be deduced from the basic laws of Newtonian mechanics. This method permits us to check Newtonian mechanics by studying the properties of gases. We introduce again the concept of the degrees of freedom of a particle or a system of particles. We saw previously that a degree of freedom is a mode of motion that is available to a particle or system of particles. The various modes of motion that we deal with in treating a gas as an ensemble of molecules are translational, rotational, vibrational, and the internal motions of molecules.

Suppose now that the molecules of our gas are pictured as compact spheres with no internal structure, so that they have only translational degrees of freedom. These spheres are the individual atoms as well as the molecules of our gas, so that we have what we call a monatomic gas (each molecule of the gas contains only one atom). In this case all the internal energy of the gas is in the form of the translational kinetic energy of the individual monatomic molecules and we thus have $U = E$. From the previous equation for γ we obtain $\gamma = \frac{2}{3}(E/U) + 1 = 5/3 = 1.667$. According to classical Newtonian theory the ratio of the specific heats of all monatomic gases should be 5/3. For a monatomic gas the relation $PV = (\gamma - 1)U$ gives $PV = \frac{2}{3}U$ which is the result we obtained using the kinetic theory.

We can check this result experimentally by measuring the ratio of the specific heats of the noble or inert gases such as helium, krypton, and argon which are monatomic. These measurements give the following values for γ: He = 1.66; Kr = 1.69; Ar = 1.668, which are in good agreement with the theory. So classical mechanics holds for monatomic gases.

We now consider a gas consisting of diatomic molecules such as O_2, H_2, or N_2 (each molecule of the gas consists of two atoms held together by electrical forces). We picture such diatomic molecules as tiny dumbbell structures. But such a structure, in addition to moving about, rotates and vibrates. Only part of its internal energy is kinetic energy or translational motion; it also has rotational and vibrational energy. How is the total energy U of a diatomic molecule, on the average, divided into a translational kinetic energy part E and into a rotational and vibrational part? We use the theorem of equipartition of energy that we introduced previously and, which, as we saw, is a consequence of Newton's law of motion. The equipartition theorem tells us that if energy is supplied to a gas (we do work on it by compressing it or else we heat it), this energy is distributed equally (on the average) among all molecules and each independent mode of motion (degree of freedom) of each molecule acquires, on the average, an amount of energy equal to $\frac{1}{2}kT$ where T is the absolute temperature of the gas. Since there are just three independent translational modes of motion (degrees of freedom) the total average kinetic energy E of a molecule is just $\frac{3}{2}kT$ as we have already noted. Hence, the total internal energy U equals $\frac{3}{2}kT$ plus $\frac{1}{2}kT$ multiplied by as many rotational and vibrational degrees of freedom as there exist. In short, $U = E +$ (rotational and vibrational energy) $= \frac{3}{2}kT + (n/2)kT$, where n is the number of rotational and vibrational degrees of freedom. The diatomic molecule has two rotational degrees of freedom since it can rotate about a vertical axis and a horizontal axis (both perpendicular to the line connecting the two atoms of the molecule). We neglect the rotation about the line connecting the two atoms of the molecule because these two atoms may be pictured as points and there is no energy associated with the rotation of a point. There are also two vibrational degrees of freedom of the diatomic molecule. We may picture the two atoms of the molecule as though they were attached by a spring which can vibrate lengthwise. We thus have two energy modes associated with such spring-like vibrations: the kinetic energy of the vibration and the potential energy. Taking the rotational and vibrational energies into account we see that there are four degrees of freedom in addition to the three transitional degrees of freedom so that $n = 4$ and $U = \frac{7}{2}kT$. Using this equation and noting that $E = \frac{3}{2}kT$, we obtain from the equation $\gamma = \frac{2}{3}(E/U) + 1$ the value $\gamma = 2/7 + 1$ or $\gamma = 9/7 = 1.286$ for diatomic molecules.

This value does not agree with the experimental data as obtained from the measurements of γ for various diatomic molecules. We have the following measured values: $H_2 = 1.404$; $O_2 = 1.399$; $Br_2 = 1.32$; $Hl = 1.40$; $I_2 = 1.30$. The experimental value closest to the theoretical value 1.286 is that for iodine, but the other values are way off. Something is wrong with the theory and there is no way to patch things up classically to bring the theory into agreement with the facts. To try to bring the theoretical and experimental values of γ into closer agreement we might, for example, argue that the atoms in a diatomic molecule are so strongly bound and the molecule is so rigid that it can barely vibrate. This argument does not help, however, because Newtonian theory says that vibrations must still occur and the amount of energy associated with such vibrations (however restricted these vibrations may be) must still be kT per molecule. Suppose, however, that we assume that diatomic molecules are infinitely rigid (which is a physical impossibility). Vibrations are then eliminated and there are then only 5 degrees of freedom (3 translational and 2 rotational) so that $U = \frac{5}{2}kT$ and $E = \frac{3}{2}kT$. We then obtain from the theory $\gamma = \frac{2}{5} + 1 = 7/5 = 1.40$. This value agrees pretty well with the measured values of γ for H_2, O_2, and HI, but it is way off for Br_2 and I_2. The classical theory is definitely wrong.

Things get even worse for complex molecules with many internal degrees of freedom (all kinds of rotations and vibrations). As the number of degrees of freedom increases, γ should, according to the classical theory, get closer and closer to 1. In fact, if the total number of degrees of freedom of a molecule is n (which includes translational, rotational, and vibrational degrees), then classical theory predicts that $\gamma = 2/n + 1$, which approaches 1 as n gets very large. The molecule ethane, C_2H_6, for example, consists of eight atoms with many degrees of freedom (about 24 in all). Its value of γ should be about 1 theoretically. Measurements show, however, that the ratio of specific heats for ethane is 1.22, 22% larger than 1.

Another discrepancy arises between the values of γ predicted by classical theory and the measured values when one considers the dependence of γ on temperature. According to classical theory, as we have noted, γ is the same for all temperatures. But actual measurements show that γ rises for every gas as the temperature of the gas decreases. As the temperature of the gas is increased, the measured value of γ diminishes and approaches the predicted theoretical value at high temperatures. Thus, the measured value of γ for diatomic hydrogen (H_2) decreases from 1.6 at -185°C to the theoretical value 1.3 at 2000°C in complete contradiction to Newtonian theory. This discrepancy is the first indication that Newton's laws of motion cannot be applied (except approximately) to the motions of atoms and molecules. This difficulty between the kinetic theory of matter (and of gases in particular) and Newtonian mechanics was recognized as long ago as 1859, when the great British theoretical physicist James Clerk Maxwell, who discovered the electromagnetic theory of light, stated that the classical way in which energy is distributed among various modes of translational and rotational motion "could not possibly satisfy the known relation between the two specific heats." In a later lecture Maxwell referred to this difficulty again and said "I have now put before you what I consider to be the greatest difficulty yet encountered by the molecular theory."

This difficulty with the specific heats of gases was but one of a number of apparently insoluble problems that faced classical physics near the end of the 19th century. The discrepancy between the constancy of the measured value of the speed of light for differently moving observers and its nonconstancy as deduced from the Newtonian concepts of absolute space and absolute time led Einstein to the theory of relativity; this discrepancy between the observed and the theoretical values of the ratio of specific heats of gases was finally eliminated by the introduction of the quantum theory, which is based on the concept that action is quantized (comes in indivisible units or lumps). But the quantum theory did not arise from any attempt to explain or eliminate the difficulty associated with the classical theory of specific heats. In fact, it stemmed from another discrepancy between the classical theory of radiation (light) and its observed behavior. It is a curious historical fact that the two great theories of modern physics, the theory of relativity and the quantum theory, had their origin (at about the same time) in the inability of classical physics to explain the behavior of light. Each of these theories introduced a new constant into the laws of nature; the theory of relativity introduced the speed of light c and the quantum theory introduced the quantum of action h (Planck's constant). We shall see just what role the quantum of action plays in physics when we analyze in detail the quality of the radiation emitted through a small aperture of a hot furnace. But before we can do that analysis, we must study the restrictions that nature places on the transformation of heat into work which leads us to the second law of thermodynamics.

Before leaving the subject of the specific heat of gases, we consider an important type of process that can go on in a gas in which the ratio of specific heats γ plays a crucial role. We have already studied the behavior of a gas in a cylinder which is in contact with a heat reservoir at a constant temperature T in such a way that the temperature of the gas remains constant under all conditions. The gas is then said to undergo isothermal processes and Boyle's law, $PV =$ constant, is obeyed. If one plots P against V for isothermal processes for different values of the constant (actually the temperature), one obtains a series of parallel hyperbolas. We now take the same cylinder of gas and insulate the walls of the cylinder very carefully so that no heat can flow into or out of the gas; if we now compress the gas by pushing down on the piston or if we allow the gas to expand under these conditions, the temperature changes, as we deduce from the first law. In this case, then, the processes the gas undergoes are not isothermal. We call all processes, during which no heat can enter or leave a gas, adiabatic processes and we seek the relationship between P and V for such processes. Adiabatic processes are important in any phenomena that happen very rapidly (like explosions) because changes occur so quickly in the gas that there is no time for heat to enter or leave the gas while it is expanding or contracting. In a diesel engine, for example, the gas in the cylinder is compressed by a piston so rapidly that heat cannot escape from the cylinder, so that the temperature in the cylinder rises very quickly and the gas explodes, driving the piston back. Hence, diesel engines require no spark plugs. Adiabatic processes occur inside stars and in the earth's atmosphere so that if we are to understand such stellar and atmospheric phenomena, we must understand adiabatic processes.

To apply the first law of thermodynamics to an adiabatic process in a gas we set $\Delta Q = 0$ (no flow of heat) and obtain $0 = \Delta U + \Delta W$ if we consider only very small changes in the gas (shown by Δ). Using the expression previously derived from ΔV and ΔW we obtain $\Delta T/T = -(\gamma - 1)(\Delta V/V)$. This equation shows us how the temperature of a gas changes with volume during an adiabatic process. We see that because of the minus sign the temperature increases when the volume diminishes and vice versa if we consider only very small changes in volume and temperature. This equation is also directly deducible from the general form of the first law $0 = \Delta U + \Delta W$ for an adiabatic process, which leads to $-\Delta W = \Delta U$. This relation shows us that if we do a small amount of work on the gas (we compress it by pushing the piston down slightly) so that ΔW appears with a minus sign, the internal energy of the gas increases (ΔU appears with a plus sign). But if the internal energy of the gas increases, its temperature must rise. We also have $\Delta W = -\Delta U$ showing that if the gas does an amount of work ΔW, its internal energy decreases by an amount ΔU (the minus sign next to ΔU tells us that U decreases) and the temperature drops.

The amount by which the temperature changes when the volume is changed by the small amount ΔV is given by $\Delta T = -(\gamma - 1)(\Delta V/V)T$. Thus, in a monatomic gas [$\gamma = 5/3$ so that $(\gamma - 1)$ is 2/3] at 0°C (273°K) the temperature changes by (2/3)(273/100) (roughly 2°) if the volume changes adiabatically by 1% ($\Delta V/V$ is 1/100). If we lower the piston and reduce the volume adiabatically by 1%, the temperature of the gas rises by about 2° (it goes from 0°C to 2°C). If we allow the gas to push the piston up and thus expand and do work while V increases by 1%, the temperature of the gas drops by about 2°C.

We can see the great importance of this relation for atmospheric phenomena and climatic changes. Consider a high-pressure region in the atmosphere (pressure greater than standard) in which the air is being compressed adiabatically. If air with a great deal of water vapor and water particles in the form of clouds (moist air or saturated air) enters this region, it is compressed, its temperature rises, and evaporation occurs so that there is no rain. On the other hand, if moist air enters a low-pressure area, it expands and cools quickly so that condensation and rain occur. This is why, in general, a rising barometer (a gauge that measures atmospheric pressure) means good weather and a falling barometer means bad weather.

From the relationship between $\Delta T/T$ and $\Delta V/V$ for an adiabatic process, which tells us that the percentage increase in T equals $(\gamma - 1)$ times the percentage decrease in V, we can deduce that for a perfect gas $TV^{(\gamma-1)}$ = constant (for adiabatic processes). If instead of working with T and V as our independent macroscopic gas variables we work with P and V, we obtain from the above equation and the gas law (replace T in the above equation by PV/constant)PV^{γ} = constant. Finally, if we work with P and T, we obtain $P^{(1-\gamma)}T^{\gamma}$ = constant. The constants on the right-hand sides of the three adiabatic relationships are, of course, not the same since they are equal to different algebraic combinations of P, V, and T but that does not matter; only the forms of the equations are important.

It is interesting to compare the graphical representation of adiabatic and isothermal processes. We picture a family of isothermal curves (hyperbolas) in the PV plane (pressure plotted against volume) and note that if the state of a gas is represented by a point on one of those curves, this point stays on the same curve as long as the temperature of the gas remains constant (Boyle's law). We now picture pressure plotted against volume on the same PV plane as the perfect gas undergoes adiabatic changes, keeping the isothermal curves in the graph for comparison purposes. We thus have two sets of intersecting curves, isothermals and adiabatics. Each isothermal curve represents a different value of the temperature of the gas and along the adiabatic curves the temperature changes. We observe that no two adiabatic curves cut each other nor do any two isothermals. But the adiabatic curves cut the isothermals. The adiabatic curves are steeper than the isothermals because the exponent γ attached to V in the adiabatic relationship is larger than 1. For an isothermal we have $P = \text{constant}/V$ whereas for the adiabatic we have $P = \text{constant}/V^{\gamma}$. For a monatomic gas this quantity becomes $P = \text{constant}/V^{5/3} = \text{constant}\ /V^{1.67}$. Thus, the pressure changes more pronouncedly with volume in an adiabatic process than in an isothermal process.

Suppose that we have a gas in a cylinder which goes through a series of changes that are alternately isothermal and adiabatic until finally returning to its initial state. The gas starts initially with its state given by point A (a point of intersection of an isothermal and an adiabatic). It then changes isothermally so that the point defining its state moves along the isothermal curve in which the isothermal cuts another adiabatic. From there the gas changes adiabatically along this adiabatic curve to the point C where it cuts another isothermal and it then changes isothermally to the point D along this isothermal. From D it moves along a second adiabatic curve back to its original state A. The gas has thus performed a complete cycle consisting of two isothermal processes and two adiabatic ones. Such a series of alternate isothermal and adiabatic processes is called a Carnot cycle; this kind of cycle, the physical basis of heat engines, determines the efficiencies of such engines.

The first law of thermodynamics places no restrictions on the transformation of work into heat or heat into work. All it demands is that there be an exact balance between the total initial and the total final energy. And yet a closer comparison of the natural transformation

→ heat indicates that in the natural course of events the transformation of work into heat is favored over the transformation of heat into work; the former appears to be a more natural process than the latter. Going back to the swinging pendulum in the cylinder of gas we note that all the mechanical energy of the pendulum disappears as it does work on the surrounding gas and all of it ultimately (and quite naturally, with no help from an external agency) is transformed into "heat" or internal energy. We also saw that we can arrange things to retrieve some of the energy that the pendulum lost to the surrounding air by allowing this air to expand and raise a piston, thus doing work, but this work is always less than the mechanical energy lost by the pendulum. This fact shows us that in the natural course of events mechanical energy or work is always transformed into heat or internal energy but the reverse process does not happen naturally. The transformation of heat or internal energy into work happens only under very special or contrived circumstances and then only in a limited way so that it is never complete. The transformation of work into heat is not a naturally reversible process. Since the second law of thermodynamics stems from the existence of irreversible processes in nature we discuss the whole question of reversible and irreversible processes before we consider the second law of thermodynamics. Natural processes are generally irreversible but under special circumstances we can control things so as to make a process completely reversible.

Before we discuss irreversible processes in general, we consider the conditions that must be fulfilled for a process to be reversible. To that end we return to the simple mechanical system of the two colliding, perfectly elastic spheres that we introduced previously. Suppose that sphere A is at rest on a smooth surface at a definite distance from a perfectly reflecting wall C and let an identical sphere B roll toward A so that there is a head-on collision of the two spheres (we neglect all frictional losses of energy resulting from the contact of the spheres with each other or with the surface on which they are rolling). After the collision sphere A rolls in the same direction as and with the same speed that B had before the collision and B comes to rest. Thus, A and B reverse their roles. A then strikes the wall C, and is reflected back to B with the same speed it had after the initial collision. After striking B, A comes to rest and B regains its original speed and rolls toward a wall D. After colliding with wall D (which we assume to be perfectly reflecting) sphere B returns with the same speed ready to strike A again. Here we have a complete and perfect reversal of a series of events.

If we had a motion picture film of these events, we could run it forwards or backwards without the viewers knowing the difference; this situation is quite different from one's reaction to a film of ordinary events that is run backwards. We are immediately aware, when we see such a film run backwards, that the order of events is wrong because they do not correspond to the way events normally unfold in time. We see time reversed in such a film and we reject it as unreal precisely because we know that time proceeds in only one direction in accordance with our experience. Before we can consider in detail why the flow of time in only one direction is associated with irreversible processes in nature, we must analyze the difference between reversible and irreversible processes on a macroscopic scale as in a gas.

Before going to a gas, however, we study some other interesting principles of reversible processes which illustrate the conditions that must exist, in general, if a process is to be reversible. Consider a saturated solution of sugar in water with a lump of some undissolved sugar in the water. This system of sugar and water (which is to be contained in a cylinder with a piston compressing the air above the surface of the water) is in equilibrium as long as the ambient temperature and pressure are constant. By equilibrium we mean that the rate at which the undissolved sugar is going into solution (molecules of the undissolved sugar are constantly separating from the mass of solid sugar and becoming part of the dissolved sugar) exactly equals the rate at which molecules of the sugar in solution are leaving the solution and attaching themselves to the undissolved mass so that everything remains the same from a macroscopic point of view. We can alter the pressure above the surface of the sugar solution as needed by lowering or raising the piston. If we compress the piston slightly, we increase the pressure on the saturated air (a mixture of air and water vapor) above the solution and some of the water vapor condenses thus decreasing the concentration of the sugar solution which is then no longer saturated. Hence, the undissolved sugar molecules go into solution faster than those in solution leave and this process continues until saturation is again reached. More sugar molecules are then in solution and fewer in the undissolved mass than there were initially. If, howev-

er, we now raise the piston slightly to its original position, some water evaporates and the solution becomes supersaturated. Hence, precipitation of sugar molecules from the solution onto the undissolved sugar occurs until saturation and equilibrium are restored. Under these conditions the process "sugar goes into solution" is reversible because by changing the ambient condition ever so slightly in one direction or the other we can have molecules of sugar going into solution or leaving the solution.

As another example of a reversible process we consider a mixture of ice and water under standard pressure and at 0°C. If these conditions are fixed, the ice and water remain in equilibrium in the sense that the rate at which molecules of ice are changing into water (the rate of melting) is exactly equal to the rate at which molecules of water are becoming ice (the rate of freezing). Thus, everything remains the same. If we now reduce the ambient temperature ever so slightly below 0°C, the rate of freezing exceeds that of melting and some of the water gradually freezes. But we can reverse things by raising the temperature slightly above 0°C. Some of the ice then gradually melts. So by changing the temperature up or down slightly we can initiate a process in one direction or the other. We can achieve the same thing by raising or lowering the ambient air pressure. A slight decrease in pressure (keeping the temperature at 0°C) causes some of the water to freeze. Thus, the process "freezing $\Leftrightarrow$ melting" at 0°C is reversible since we can make it go in one direction or in the reverse direction by changing conditions slightly.

As a very simple example of a mechanical reversible process we consider a gas in a cylinder (with a piston) at a definite pressure so that mechanical equilibrium exists. If we now compress the piston slightly and very slowly, so that equilibrium exists at each stage of the slight compression, we do work and the internal energy of the gas increases. If we now release the piston the gas does work on the piston which returns slowly to its initial height and the internal energy of the gas returns to its initial value so that everything is back to its initial state and the entire process is reversed.

If we carefully consider the three examples of reversible processes we have just described, we see that the feature that is common to all of them is "equilibrium." As a general rule, then, we may say that processes during which the system is in equilibrium are reversible.

But equilibrium can exist only during a process that occurs so slowly that conditions at each stage differ only infinitesimally from those of the preceding state. In general, infinitesimally slow processes are reversible whereas rapid processes are irreversible. From these considerations we see that processes that occur naturally all around us (we may refer to these as spontaneous processes) are irreversible. Indeed, no real process can ever be reversible, but by making processes go very slowly in systems that are always close to equilibrium we can approximate reversibility.

Now that we have discussed the conditions that must exist in macroscopic systems that are undergoing reversible processes, we can see what links irreversible processes to the unidirectional flow of time. We can best do this by going back to the microscopic (molecular or kinetic) features of a process. Again we consider the collisions of spheres (which we may picture as representing the molecules of a gas), but this time we deal not with the collision of two spheres but rather with a single sphere colliding with a large number of others. A good example is the initial impact of the cue ball on a billiard table with the other 15 balls stacked together in a triangular arrangement. Immediately after the collision all 16 balls are moving about more or less randomly and we see at once that an exact reversal of all the events following the collision will not occur. It is not that such a reversal cannot occur in principle (there is no physical law that forbids it; the Newtonian laws of motion are identical for processes proceeding in a given direction or in the opposite direction) but rather that the probability for an exact reversal is so minute as to be insignificant. One could, in principle, bring about an exact reversal at any moment after the initial collision by simultaneously reversing the velocities of all the balls on the table so that they all instantaneously move with their same speed but in directions opposite to their original directions.

If this is not done and the billiard balls are allowed to move freely after the initial collision (we neglect friction and any energy lost during collisions) they collide with the walls of the billiard table, bounce off and move about in a completely random pattern. The chance that these random motions will, in time, bring all the billiard balls back to their initial state (the state that existed just before the cue ball collided with the other 15 balls) is so small that we dismiss it completely from our considerations and simply state that the billiard balls will not revert to their original configuration. From these considerations we see that irreversibility stems from the involvement of a large number of particles in any event. Since all matter consists of tiny molecules and atoms, and even a small amount of matter contains vast numbers of these particles, we can see why ordinary events in nature are irreversible and why the reversal of such ordinary events conflicts with our sense of the direction of the flow of time.

We saw above that we must distinguish between two kinds of processes that can, in principle, occur in nature: spontaneous or natural processes which are, in general, irreversible, and reversible processes, which can occur only under highly controlled conditions in which the interacting systems involved are always in a state of equilibrium and the processes going on are proceeding extremely slowly (at an infinitesimal rate). If a motion picture of a "spontaneous" process is taken and then run backwards we see a sequence of events that do not occur in nature. On the other hand, if we take a motion picture of the sequence of events that occur in a reversible process and run it backwards, we see nothing happening on the screen that is strange or contrary to our experiences. The recognition of the difference between reversible and irreversible events in nature led to the second law of thermodynamics. But the second law does more than merely state that two kinds of processes can occur; it states, in fact, that irreversible processes are the rule in nature and that these are the kinds of processes that tend to go on in a changing isolated system.

Since the second law states that irreversible processes occur naturally in a system, we can formulate this law precisely in terms of a typical irreversible process. Moreover, since the second law is a statement of the existence of irreversible processes, it makes no difference as far as the content of the law is concerned which irreversible processes we use to express this law. Since a gas (which is initially confined to part of a larger volume) immediately and spontaneously expands to fill the entire volume if it is freed from its confinement, we can say that the second law of thermodynamics is equivalent to the following statement: A gas expands spontaneously into a vacuum; a spontaneous contraction of a gas into a smaller volume never occurs. We could also say that the second law is equivalent to the statement that "two different gases in contact diffuse into each other spontaneously and become a uniform mixture; the separation of a uniform mixture of gases into two separate gases does not occur spontaneously."

We can state a number of other irreversible processes each of which can be used to formulate the second law, but instead of that we concentrate here on the two irreversible processes that were actually used by two of the great 19th century physicists whose names are associated with the second law. Both of these irreversible processes are concerned with the flow of heat and are therefore more closely related to the thermodynamics that we have been discussing than is the diffusion of gases. One of these two irreversible processes places a restriction on the spontaneous flow of heat from one body to another. This process was used by the German physicist Rudolf Clausius to formulate the second law. The other process, which places a restriction on the transformation of heat into work, was used by the British physicist Lord Kelvin to formulate the second law.

Clausius's formulation of the second law of thermodynamics is based on the following observation (an empirical truth): If two bodies at different temperatures are placed in thermal contact, heat always flows spontaneously from the body at the higher temperature to the one at the lower temperature. With this fact in mind, Clausius stated the second law of thermodynamics as follows: A transformation whose only final result is the net transference of a finite amount of heat from a body at a given temperature to a body at a higher temperature is impossible. Notice that this statement does not say that heat cannot go spontaneously from a cool body to a hotter one. It says that at the end of any spontaneous process involving two bodies, there is no net gain of heat by the hotter body (or net loss of heat by the colder body). The earth radiates heat to the sun just as the sun radiates heat to the earth but in a given interval of time the earth gets much more heat from the sun than the sun does from the earth. Differently stated, if a net spontaneous flow of heat from body A to body B occurs, then the temperature of A must be higher than that of B. Although the net spontaneous flow of heat is always from the hotter body to the colder body, we can reverse the process by applying work. We can compress the gas very quickly (adiabatically) and then conduct into a warmer gas the heat that is generated. We thus reverse the flow of heat forcing it to go from the colder to the warmer gas.

Since Clausius's statement of the second law is not a quantitative statement, it does not lend itself easily to a mathematical formulation of the second law which is what we now seek; even though it expresses the second law in terms of an important irreversible process, it does not give us a quantitative measure of irreversibility. To express the second law in a useful quantitative form we must introduce a measurable physical quantity that is related in some way to irreversibility; Kelvin's statement of the second law shows us how such a quantity can be introduced. The reason that Kelvin's formulation of the second law, as stated below, is so useful is that it implies a measurable quantity in nature which increases whenever an irreversible process occurs; Kelvin's statement of the second law shows us the nature of this quantity: The transformation of heat completely into work with everything else remaining the same in the universe is impossible.

Before discussing how the above statement leads to the measurable quantity that we spoke about and thus to a mathematical formulation of the second law, we emphasize the importance of the last part of Kelvin's statement, for without it the first part of the statement is incorrect. We consider a gas in a cylinder placed on a heat reservoir (a furnace) that is at the same temperature as the gas. If the gas is allowed to expand slowly (the weight on the piston of the cylinder is removed bit by bit) thus lifting the piston, heat flows slowly from the reservoir into the gas. But the temperature of the gas always remains the same as that of the furnace so that the internal energy of the gas also remains the same. Hence, by the first law of thermodynamics, all the heat from the furnace is transformed into work. This result, however, does not contradict Kelvin's statement of the second law because after the work has been done on the piston, the gas is not the same as it was to begin with. The volume of the gas is larger than it was originally so that "everything else" does not remain the same, as required by Kelvin's statement.

We see that Kelvin's statement of the second law really tells us that there is a quantity in nature that always increases whenever irreversible processes occur. Since, according to Kelvin's statement, heat can never be completely transformed into work, a certain amount of the energy that can do work always becomes unavailable to us. But there is no restriction on the transformation of work or mechanical energy into heat. Hence, the amount of unavailable energy in the universe always increases, or put differently, the available energy in the universe constantly decreases.

Before we introduce the specific numerical quantity that can be used to express the second law algebraically, we note that the concept that the amount of available energy (energy that can be converted into work) in an isolated system must always decrease (at best, it can remain constant, but then only if the processes going on in the system are reversible, which in practice is a physical impossibility) enables us to determine whether or not further changes in an isolated system are possible. Certain restrictions on the changes in a system are imposed by the conservation laws that we have already learned. The total momentum, angular momentum, and energy of an isolated system cannot change. But we now have a further restriction because changes can go on in an isolated system only if these changes result in a decrease in the available energy. Energy will be available, however, only if there are temperature differences so that a net quantity of heat can flow from one point to another. As time passes, changes in the system continue until all temperature differences are ironed out and all the energy is degraded into an unavailable form.

Before leaving this general discussion of the second law of thermodynamics and going on to the concept of entropy which is the state function that we use to express the second law algebraically, we note that the second law, with its emphasis on irreversibility, establishes a hierarchy of energy forms. Certain forms of energy can be transformed completely into other forms, but the reverse is not true. Thus, certain forms of energy are to be preferred to other forms as far as the usefulness of energy goes. When a more useful form of energy is transformed into a less useful form, the energy is degraded. From this point of view the second law states that in any isolated system all irreversible processes result in a degradation of energy. Taking the universe as a whole we see that the energy of the universe is constantly being degraded.

If two forms of energy can be transformed into each other, we consider them of the same grade, but otherwise we consider that grade higher which can be completely transformed into the other grade. Thus, mechanical energy (both kinetic and potential) is a higher grade of energy than heat, whereas mechanical energy and electrical energy are of the same grade, both being the highest grade of energy. Heat that is at a higher temperature is a higher grade of energy than the same quantity of heat (or internal energy) at a lower temperature. In connection with this fact it should be noted that whenever a flow of heat is used to do some work, the complete degradation that normally occurs is delayed because some of the heat is transformed into mechanical energy which is a higher grade of energy than heat. But this interruption of the degradation of the total energy is only temporary because in the end all the mechanical energy itself that is obtained from the heat is finally degraded unless it is transformed into the potential energy of a permanent structure like the Empire State Building or into the mechanical energy of an artificial satellite circling the moon or earth.

Since the degradation of energy is always accompanied by heat, we must avoid the generation of heat as much as we can in obtaining useful forms of energy. In many instances as in the steam engine or in generating electricity, we can obtain useful energy only by first generating heat (burning coal or using nuclear fission) to send into a boiler, but we must then try to convert the maximum possible heat from the boiler into mechanical energy. This process can be done by thermal engines, but there is a natural limitation on the efficiency of such engines.

When burning occurs, the available chemical energy released by the burning atoms (oxygen and carbon) is degraded into heat, and carbon dioxide is released. This degradation is reversed by green plants which use carbon dioxide and water to form chlorophyll and oxygen, but plants can do this only by using light which itself is a high grade of energy. The light is degraded by passing from a high-temperature source to a low-temperature sink where it is absorbed. In any case, whether an engine reverses to some extent the degradation of heat or plants reverse the degradation of chemical energy, the net amount of energy that is degraded always exceeds the amount that is transformed to a higher grade.

We saw above that the second law of thermodynamics can be expressed in terms of the continual increase of the unavailable energy in an isolated system. We now express this increase of the unavailable energy in terms of a state function of the system. To see what this expression means we return to our perfect gas in the cylinder and note that a particular state of the gas is given by specifying any two of the three quantities P, V, T so that two of them define a state of the system, the third then being given for this state by the equation of state of the gas. As we have already noted, a state function is any measurable parameter associated with the gas that has the same numerical value when the gas returns to the same state. Each of the three quantities P,

V, and T is a state function, but each by itself does not define a state. We may, indeed, have two or more different states all with the same value of P, V, or T or the same values of any state function in general. The temperature which is a state function can have the same value for a whole series of isothermal states (states for which PV = constant). The same is true of the internal energy U which is a state function. But each different state is associated with only one value of a given state function.

We now use the quantities V and T to define our states and consider the transformation of our gas from one state A (given by a set of values V_1, T_1) to another state B (given by V_2, T_2). The gas can go from A to B in an infinitude of ways, some of which are reversible and some irreversible. During these various transformations from A to B the gas can undergo many different kinds of changes that, in general, involve compressions, expansions, and the loss or gain of heat. Can we now find a state function which is larger for state B than it is for state A if the available energy in state B is less than in state A (the unavailable energy is to increase when the system goes to state B) or vice versa?

Let the initial state A of the gas be one in which it is confined in a given volume under a high pressure (let us say twice the atmospheric pressure) with its temperature equal to that of the surroundings. Let the final state B of the gas be one in which the pressure is the atmospheric pressure and the volume of the gas is twice its original volume with its temperature unchanged (an isothermal expansion governed by Boyle's law). The passage of the gas from state A to state B can occur in many ways, but we consider just two ways, one of which is completely irreversible and the other of which is reversible at all stages. The irreversible transformation from A to B shows us that energy that is available in state A is degraded and becomes unavailable in state B so that the total amount of unavailable energy in state B is greater than it was in state A. The reversible transformation of the gas from state A to state B thus shows us how to find a state function of the gas (the entropy) that increases when the gas goes from A to B.

In the irreversible transformation we simply allow the gas to expand freely (we remove the confining walls of the gas or punch holes in them) into a container with twice the original volume of the gas. The temperature and internal energy of the gas are exactly the same as they were initially (the free expansion of a perfect gas leaves the temperature unaltered) but the gas has lost its ability to do work. In state A the gas is under a high pressure and therefore can do work by expanding against a piston with weights on it. Energy is therefore available in state A to do work. If we allow the gas to expand freely we do not use this available energy and it is no longer available in state B. The total available energy has thus been decreased by the amount of work the gas could have done if it had expanded against a piston.

Now let us start from the beginning again with the gas in state A and arrange things so that it goes over to state B reversibly. We must allow the expansion to proceed very slowly, which we can achieve by having the gas, as it expands, push against a vertical piston with weights on it. Since we want the temperature to remain constant (the same as the environment) we must, while the gas is expanding, keep the cylinder of gas in thermal contact with its surroundings in such a way that heat can flow freely from the surroundings into the gas. We do this by making the walls of the cylinder perfect heat conductors (silver or copper). As the gas expands against the piston it does work by drawing upon its internal energy, and its temperature would therefore fall if it were insulated from its surroundings. But this does not occur because as soon as there is any slight drop of the temperature in the gas, enough heat flows into the cylinder from the surroundings to bring the temperature back to its initial value. The heat flowing into the gas is converted into the work that raises the piston. We see then that the total available energy that is present in the gas in state A can be measured by the total heat Q that flows into the gas from its surroundings when the gas is expanding reversibly and doing work. Hence, Q also equals the total energy that becomes unavailable when the gas expands irreversibly (free expansion) from A to B without doing work.

We have obtained a quantity (the heat that flows into a system when the system is undergoing a reversible change at constant temperature) which can be used to measure the increase in the unavailable energy (or decrease in the available energy). But Q itself is not a state function and hence is not the entropy, but it is related to the change in entropy. A state function is a quantity that is always the same when a system is in a given state regardless of how the system reaches that state (by reversible or irreversible processes), but the heat that flows into or out of a system depends not only on the final state but also on the kind of transformation the system experiences. In the case of our expanding gas, no heat flows into or out of the gas during its free irreversible expansion. On the other hand, the flow of heat is a maximum when the expansion occurs reversibly at constant temperature. Hence, though the end state B is the same in both cases, the amount of heat Q that flows into the gas varies from 0 for the completely irreversible process to a maximum for the reversible expansion at constant T. Q itself cannot be a state function.

However, a state function which does just what we want can be constructed from Q and T. We consider a system (like our gas) which is undergoing a reversible process during which all kinds of changes may occur (P, V, and T may change and heat may be leaving or entering the gas). Suppose that we follow the system while it is changing very slightly (infinitesimally); in fact, so slightly that T remains constant during this change. Suppose also that an infinitesimal amount of heat ΔQ flows into (or out of) the system during this small change (which is reversible because it is very small). We then consider the quantity $\Delta Q/T$, which is negative if heat flows out of the system and is positive if heat flows into the system. Clausius showed that this quantity gives the infinitesimal change of a state function S, which Clausius called the entropy of the system.

Associated with every isolated system in nature is a state function S called the entropy which always increases when irreversible processes occur in the system. If reversible processes occur, the entropy remains constant, but the entropy of an isolated system can never decrease. If the system changes reversibly by an infinitesimal amount which results in an exchange of heat ΔQ between the system and its surroundings while the temperature of the system is T, then the entropy of the system changes by the amount $\Delta Q/T$. This change in entropy is negative (a decrease) if the system loses heat, but it is positive (an increase) if the system gains heat.

We defined the change in the entropy of a system for very small (infinitesimal) changes of the system using a formula proposed by Clausius. Suppose now that we have a system which starts in some state A, as represented by a point in the PV diagram, and then passes over to some other state B via a series of intermediate states. As we have already stated, the system can take an infinite number of paths (series of intermediate states) from the initial state A to the final state

B, but the change in the entropy of the system, by the very definition of entropy as a state function, is the same regardless of the path (the series of intermediate states) along which the system goes from A to B. Hence, we can find the total change in entropy (which must apply to all paths) if we know how to calculate the entropy along any special path. But we do know how to do that calculation because Clausius has shown that for a very small change in the state of the system involving the exchange of heat ΔQ the entropy change is $\Delta Q/T$.

We take the system from the initial state A at temperature T_0 to the final state B at temperature T_f in a series of small, slow steps with the temperature of the system changing very slightly at the end of each step. Since each of these steps is small, the system remains in equilibrium during each step which is thus reversible. Hence, all of these steps taken together constitute a reversible path from A to B, and we can calculate the change in entropy from A to B along this entire path by first calculating the step-by-step changes using the Clausius formula and then adding up all these changes. Clausius showed if one follows this procedure, the result is always the same no matter which reversible path is used.

To be specific and to describe the process exactly, we break up the reversible path into a series of small reversible steps 1, 2, 3, etc. Let the initial temperature of the system (the temperature in state A) be T_0, and let T_1 be its temperature after step 1, T_2 its temperature after step 2, etc., with T_f its final temperature in state B. If ΔQ is the heat exchanged by the system with its surroundings during step 1, ΔQ_1 the heat exchanged during step 2, etc., then the total change in entropy when the system goes from state A to state B is $S_B - S_A = \Delta Q_0/T_0 + \Delta Q_1/T_1 + \cdots + \Delta Q_f/T_f$, where ΔQ_i is negative if the system loses heat at step i and positive if it gains heat. This equation is called the Clausius sum along the path from A to B.

We see that the Clausius sum gives the change in entropy when a system goes from an initial to a final state only if this sum is evaluated along a reversible path. In fact, one can show that the Clausius sum evaluated along a path between two states of a system depends upon the path. For irreversible paths the value of the sum can range from 0 up to (but not including) a maximum value which the sum takes on only along a reversible path. We may therefore say that the change in entropy of a system that goes from some initial state to some final state is independent of the way the system goes from the initial to the final state and is given by the largest value that one can obtain for the Clausius sum; this maximum value is obtained along any reversible path connecting the initial and the final states. The value of the Clausius sum along any irreversible path is always less than the change in entropy of the system. If a system is taken along any closed path (reversible or not) so that it returns to its initial state (this is called a cycle), the change in entropy of the system is zero. The Clausius sum is also zero if the cycle is a reversible one, but it is negative if the cycle is irreversible at any stage.

To illustrate some of the points discussed in the previous paragraphs and to show how the Clausius sum gives the change of entropy if it is calculated along any reversible path but is always less than the entropy change when it is evaluated along an irreversible path, we consider both a reversible and an irreversible isothermal expansion of a perfect gas at temperature T. The reversible expansion can be carried out by keeping the cylinder in which the gas is confined in constant contact with a heat reservoir at temperature T while the gas is expanding very slowly and doing work by lifting a piston with weights on it. We achieve this result by removing the piston weights bit by bit. When we remove a bit of the weight, the gas expands slightly and does a small amount of work. Its temperature would therefore fall if it were not for the small amount of heat ΔQ that flows immediately from the heat reservoir into the gas. This small amount of heat thus compensates for the internal energy that the gas loses in pushing up the piston so that the temperature of the gas is restored to its initial value. We see that the heat that flows slowly from the reservoir into the gas is transformed into the work done on the piston.

If Q is the total heat that flows from the reservoir into the gas as the gas expands at constant temperature T, then the Clausius sum is just Q/T. Hence, the total change in entropy of the gas (in this case an increase) is also Q/T since the heat flows into the gas reversibly. But Q is also the work done by the gas as it expands. Hence, $S - S_0 =$ work done$/T$ where S_0 is the entropy of the gas before it begins its expansion and S is the entropy after the expansion. Note that in this case the change in entropy of the heat reservoir is just $-Q/T$ (since it loses heat Q) so that the total change in the entropy of the gas and reservoir together is zero as it should be because the whole process is reversible.

We now allow the same gas to expand freely and hence irreversibly, without doing any work, from its initial volume V_0 to its final volume V. The temperature in this case also remains unchanged because the gas is perfect. The change in the entropy of the gas is, of course, exactly the same as before since the gas starts from the same initial state and ends up in the same final state as in the reversible expansion. In this case, however, the Clausius sum is zero since there is no flow of heat from the reservoir into the gas. Thus, the Clausius sum in this irreversible example does not give the change in entropy of the gas. We can make the Clausius sum take on any value we wish from 0 up to its maximum value, which equals the work done divided by T, by arranging things so that the total expansion consists partly of a free expansion (irreversible) and partly of a controlled expansion against a piston (reversible).

Another example of entropy change shows that the entropy of an isolated system always increases when an irreversible flow of heat occurs (a flow from a higher temperature to a lower temperature). For our system we take two bodies 1 and 2 at different temperatures and in thermal contact. If T_1 is the temperature of body 1 and T_2 the temperature of body 2, with $T_1 > T_2$, then heat flows spontaneously from 1 to 2. If a small amount of heat ΔQ leaves body 1, its entropy diminishes by the amount $\Delta Q/T_1$. If this quantity of heat flows into body 2, its entropy increases by the amount $\Delta Q/T_2$. The Clausius sum in this case is $-\Delta Q/T_1 + \Delta Q/T_2$ and this equation must be the change in entropy of the two bodies considered as a single system so that $\Delta S = \Delta Q(1/T_2 - 1/T_1)$. But $T_1 > T_2$ so $1/T_2 > 1/T_1$ and the quantity in parentheses is positive. Hence, ΔS is positive which means that entropy has increased, and we have established our point.

One more interesting characteristic of the entropy is that in an isolated system the entropy can never get smaller but always tends to reach the largest possible value. The most probable state of a system is the one for which the entropy of the system has its largest value. Once this state is reached, no further spontaneous change can occur in the state of the system.

Previously we described a reversible cycle for a perfect gas consisting of two isothermal and two adiabatic processes performed very slowly. This cycle was first studied by Carnot and is known as a Carnot cycle; the cycle operates between two heat reservoirs, one at temperature T_3 and the other at temperature T_2 where T_3 is larger than T_2.

We start with our gas in its initial state A at temperature T_3 and with its pressure and volume corresponding to the P, V values of the point A on the PV diagram. The gas now expands isothermally along the isothermal T_3 to the point B doing work (raising the movable piston which exerts a pressure on the gas) as it absorbs an amount of heat Q_3 from the reservoir at temperature T_3 (a boiler) with which it is in thermal contact. The entropy of the gas thus changes (increases) by the amount Q_3/T_3. It now continues expanding reversibly and adiabatically (still doing work) from B to C along an adiabatic a_3. During this reversible adiabatic expansion the entropy of the gas remains constant (ΔQ is always zero along an adiabatic since the gas neither gains nor loses heat) and its temperature drops from T_3 to T_2 (it does work at the expense of its internal energy). The gas is now compressed isothermally from B to C (work is done on it) along the isothermal T_2 (it is in thermal contact with the heat reservoir T_2, which is now a heat sink) while an amount of heat Q_2 flows from the gas into the heat reservoir T_2. The entropy of the gas thus changes (decreases) by the amount $-Q_2/T_2$. Finally, the gas is compressed still further reversibly (more work is done on it) and adiabatically along an adiabatic a_2 until it is back to A and its temperature is again T_3. During this adiabatic compression no change in entropy occurs. The total change in entropy of the gas during this cycle is $(Q_3/T_3)-(Q_2/T_2)$. But this total change must equal zero since the gas is back to its original state and all state functions including the entropy must take on their same initial values so that $(Q_3/T_3)-(Q_2/T_2) = 0$ or $Q_3/Q_2 = T_3/T_2$. This relationship has an important application to the efficiency of a heat engine. We again consider the Carnot cycle of a perfect gas and note that it also represents the changes that occur in a perfect heat engine since it describes a system (our perfect gas) that operates reversibly between a source of heat and a heat sink while doing work. Such a system is the most efficient kind of heat engine because it operates reversibly (no friction); its efficiency represents the highest possible efficiency that can be achieved.

By definition the efficiency of a heat engine is the work done by the engine divided by the heat absorbed by the engine from the heat reservoir. From the conservation of energy the work done is just $Q_3 - Q_2$ and the heat absorbed is Q_3. The efficiency of the heat engine $= 1 - (Q_2/Q_3)$ which, the entropy principle tells us, is equivalent to $1 -$

(T_2/T_3). This very important result shows us that no heat engine, however efficiently it is operated (the elimination of all friction), can have a 100% efficiency. The efficiency of a perfect heat engine is smaller than 1 by the fraction T_2/T_3. The smaller we make this fraction, the higher the efficiency, but we can do this only by making T_3 very high or T_2 very low, and both of these possibilities are severely restricted in practice. Thus, T_3 (the temperature of the heat source) cannot be made too high because it is difficult to find materials that can withstand very high temperatures (which are, in general, difficult to achieve in any case). We cannot do very much with T_2 either because that is generally the temperature of the environment (or possibly of the cool water of a stream) and there is no way we can reduce this temperature.

As an example of how the efficiency formula is applied, we consider a steam engine whose power is obtained from a boiler at a temperature of 227°C so that $T_3 = 500°K$. If the temperature of the environment is 27°C so that $T_2 = 300°K$, we see that under ideal conditions (no friction) and for very slow operation of the engine we have efficiency $= 1 - (300/500) = 2/5 = 40\%$. Thus, in the ideal case the efficiency cannot exceed 40%. This limitation means that only 40% of the heat that the boiler supplies to the engine is available to do work; 60% of this heat cannot be used and is simply transferred to the environment whose temperature is thus slightly increased. As a result, all heat engines contribute to thermal pollution.

Suppose now that we run our Carnot cycle in the reverse direction D-C-B-A-D starting from state D (again very slowly so that the process is reversible just as before) so that we now remove a certain amount of heat from the reservoir at a lower temperature T_2 and give up a certain amount to the reservoir at the higher temperature T_3. In this case our Carnot cycle behaves like a heat pump (a refrigerator) which takes heat from a cold source and transfers it to a warm sink. To see what is involved here we note that if Q_2 is the heat removed from the cold source at temperature T_2 and Q_3 is the heat given up to the warm sink at temperature T_3, then the work W that must be done on the Carnot cycle to achieve this result is $Q_3 - Q_2$. The important thing in this case is to compare this result with the heat Q_2 taken from the cold source since this comparison tells us how much work is required to achieve a certain degree of refrigeration. We have $(Q_3 - Q_2)/Q_2 = (T_3 - T_2)/T_2 = W/Q_2$. The smaller the difference in temperature between the body we are refrigerating and the surroundings, the smaller is the work required to remove a given amount of heat. An infinite amount of work is needed to extract a finite amount of heat from a body whose absolute temperature is zero.

Before leaving the subject of thermodynamics we note that a third law of thermodynamics was first proposed early in the 20th century (1906) by the German physical chemist Nernst and it is often referred to as Nernst's heat theorem. Although the theoretical basis for this third law is the quantum theory, Nernst discovered the law empirically by studying chemical reactions at various temperatures. In our definition and discussion of entropy, we noted that only differences of entropy are meaningful in classical physics because an arbitrary additive constant is contained in the definition of the entropy which cannot be specified classically. This constant can be calculated, however, from the Nernst heat theorem which states that the entropy of any system at absolute zero is zero.

CHAPTER 14

The Laws of Radiation

A plausible impossibility is always preferable to an unconvincing possibility.

—ARISTOTLE, *Poetics*

In Chapters 12 and 13 we presented a detailed discussion of the properties of gases and the laws of thermodynamics in preparation for our study of the structure of matter and the forces (other than gravity) that produce the great variety of structures that constitute the matter all around us. Our motive was to begin by considering matter in its most unstructured state (a gas) to see if we could discover anything about the constituent basic particles of which all matter is constructed. Having learned something about the nature of the particles from our studies of gases, we next try to discover how these particles interact with each other to form liquids and solids. We know empirically that if the temperature of a gas is lowered and its density is increased, a point is reached at which the gas begins to condense. This means that the constituent particles of the gas begin to stick together owing to the attractive force between them. We must understand these forces if we are to understand how material structures are formed.

We have seen that the overall properties of perfect gases (e.g., the equation of state) can be understood from a kinetic point of view by introducing the molecular concept, as was done by such 19th century scientists as Avogadro, Maxwell, and Boltzmann. If we assume that a gas consists of randomly moving and perfectly elastic noninteracting spheres (the 19th century concept of a gas) whose motions are governed by Newton's classical laws, we can deduce the equation of state of a perfect gas as well as other gaseous properties; but we also run into a contradiction between the predictions of classical Newtonian theory and the empirical results when we consider the ratio of specific heats of gases. As we noted in Chapter 13, this contradiction arises when we treat a molecule realistically as a structure that has internal degrees of freedom, and then calculate classically the energy of the gas associated with these degrees of freedom. We also noted that this difficulty with the ratio of specific heats stems from the same cause that gave rise to the disagreements that 19th century physicists found between the observed and the classically predicted properties of the radiation emitted by a hot body.

This difficulty has a direct bearing on our analysis and understanding of the structure of molecules and atoms (which we must have if we are to study the structure of matter in general) since an analysis of the way radiation and matter interact can ultimately reveal to us the basic forces that keep the particles in atoms and molecules together. The reason for this is that we cannot measure directly (as we can gravitational forces) the forces between the constituent particles of an atom or a molecule but must infer the nature of these forces from the way molecules and atoms absorb, scatter, or emit radiation. Another way to learn about such forces is to bombard atoms and molecules with other particles and see what happens to these bombarding particles after their collision with the molecules and atoms in matter. Although the "bombarding-with-particles" procedure for studying the forces inside atoms and molecules is today the dominant tool, as well as an extremely powerful one, in the investigation of the structure of atomic and subatomic (nuclear) structures, it appeared on the scientific scene relatively recently, requiring, as it does, an incredibly advanced technology. However, Lord Rutherford and his assistants used this technology in 1911, when they bombarded very thin strips of gold foil with the high-speed alpha particles (the nuclei of helium atoms) that are spontaneously emitted from the radioactive nuclei of uranium atoms. These alpha-particle scattering experiments, as they are called, led to the modern planetary (or nuclear) picture of the atom.

Long before subatomic particles such as alpha particles were introduced as projectiles to probe matter, scientists were using the absorption and the emission of light by matter to analyze the structure of matter and to deduce that the forces involved are electromagnetic. This use of light as a tool to study matter requires only the simplest type of optical instrument, the spectroscope, which, as we shall see, is undoubtedly the single most powerful and useful scientific tool ever devised by man because it opened the door not only to our understanding of the structure of molecules and atoms but also to an understanding of the structure and evolution of stars, the properties of the interstellar medium, and the structure of the universe itself. No other instrument in the history of science has revealed so much to us and for so small a cost and with so small an effort.

Since matter can emit and absorb radiation (light, in particular), we surmise, even without a detailed and exact understanding of the processes involved, that radiation interacts with matter in some way and that this interaction depends on the structure of matter. We can learn something about this structure by studying the emitted radiation just as we can deduce something about the internal structure of a piano by analyzing the sounds it emits when its keys are struck. But for us to do this analysis we must first understand the nature of light itself (or those of its properties that are pertinent to our problem).

The world of Newtonian physics was a world consisting of particles of matter moving about in accordance with well-established (from the classical point of view) and precise laws of motion. These particles have, under appropriate conditions, certain well-defined kinematical and dynamical properties such as velocity, acceleration, momentum, angular momentum, and energy. Although matter in Newtonian physics was pictured (when necessary) as existing without any of these properties, the notion that velocity, acceleration, momentum, angular momentum, and energy could exist independently and outside of matter was unheard of and would probably have been rejected out of hand if it had been proposed. And yet phenomena which appear to be quite independent of matter and which have a velocity of their own do exist, and these phenomena have been known from the moment man could first comprehend what he saw because light is just such a phenomenon.

Thousands of years elapsed before humanity became aware of light as an entity that had to be dealt with just as matter has to be dealt with. The idea that light possesses physical attributes and can react physically on a body came to man only very slowly because there appeared to be nothing palpable about light. Until the work of Kepler on the formation of optical images by the eye of the observer (known as Kepler's dioptrics, this work is considered to be the precursor of modern optics), there was hardly any science of optics and the way the eye "sees" an object was completely misunderstood. People found it difficult to understand how all the perceptive features of an object as large as a mountain or building can "enter the eye" and how such an object can be seen in its totality with no apparent diminution in size. Although people in Europe had been wearing spectacles to correct their vision since the time of Roger Bacon, who is said to have invented eyeglasses in the 13th century (the records show that the Chinese had been using spectacles for hundreds of years before that time), just how lenses correct vision was not understood until Kepler explained this phenomenon in terms of the refraction of light. Kepler did not discover the correct law of refraction (which Snell and Descartes separately did later) but he did use refraction to explain correctly the way optical features are formed by lenses. Newton knew of these discoveries and he also knew that light travels with a finite speed (as first discovered by Roemer) so that he looked upon light as he did upon other physical phenomena—something to be explained or understood in terms of the basic laws that he had formulated. In particular, since light has a finite speed, he was inclined to view it as something that is associated with particles (matter of some sort, albeit of an invisible nature) since velocity without matter was unthinkable. Newton was also aware that two kinds of dynamical phenomena are associated with the motions of material particles. One is the usual motion of groups of particles, stemming from the inertial properties of these particles, which produce the various kinematical and dynamical phenomena that we see all around us. The other, however, is a more subtle kind of motion in which the average positions of the particles of matter that are moving do not change but in which a pulse of some kind is transported from place to place in a more or less regular fashion. This type of motion is called a wave or wave motion, examples of which are the waves on the surface of water when a stone is dropped in the water and the sound waves that are propagated through the air. The undulatory motion of the surface of the water that transports surface water waves can easily be seen, but the oscillatory motion of the air particles that transport sound waves is not directly visible, but one can deduce this motion indirectly. Nevertheless, Newton knew the relationship of the propagation of sound waves to the motion of air particles. Since light can travel through the apparently empty space between stars, its motion, from Newton's point of view, could not be a wave phenomenon since there is nothing in interstellar space that can oscillate as do the water and air particles that are associated with water and sound waves. This absence, in itself, probably predisposed Newton against a wave theory of light and propelled him toward a corpuscular theory. But we shall now see that Newton had what must have appeared to him to be a very powerful argument against the wave theory and in favor of the corpuscular theory.

To understand this argument we consider a wave passing through a small aperture or past a sharp edge. Sound waves and water waves, as we can observe, bend (are diffracted) into what we might call the shadow area of an intervening object. In other words, the waves do not travel in a straight line but bend around corners of objects. If a sound (or water) wave starts from some source O, it is received not only by an observer at any point C which is not "hidden" from O by a wall AB, but also by observers at points D and E who are "behind" the wall. Newton was aware of this bending or diffraction phenomenon of waves, as we all are, from the simple observation that we do not have to be in the line of sight of a person to hear him. The sound waves bend around walls permitting us to hear people in other rooms. But when Newton examined the shadow cast by the sharp edge of an object onto a screen when he illuminated the object with light from a distant point source, he found the edge of the shadow to be very sharp (or at least it appeared to be so to him). He concluded from this observation that light does not bend around corners (shows no diffraction) as sound does, and that it therefore is not a wave. He assumed instead, on the basis of what appeared to him to be straight line motion, that light consists of corpuscles which have inertia and momentum and which therefore obey his laws of motion. This conclusion was the genesis of the Newtonian corpuscular theory of light.

The deductions from the Newtonian corpuscular theory force us to discard it; but, as we shall see later, this difficulty does not mean that all corpuscular theories of light are wrong. We begin by considering the propagation of light from one medium (let us say a vacuum) into another medium (e.g., glass or water). We know from direct observation that, in general, when light passes from the vacuum into a denser medium (or vice versa), its direction of motion changes. That is why when part of a straight stick is in water and part of it is out of water, the stick appears to be bent (with the bend appearing just at the surface of the water). We can understand this phenomenon if we analyze the behavior of a beam of light passing from air (or a vacuum) into glass. If i is the angle of incidence of the beam [the angle the beam of light makes with the normal (the perpendicular to the surface) at the point O where the beam strikes the glass surface], then this angle decreases by a definite amount when the beam enters the glass. The beam bends toward the normal. When the beam has passed into the glass, we no longer refer to this angle as the angle of incidence but rather as the angle of refraction r and we say that when light passes from a less dense to a denser medium, its path is refracted (bent) toward the normal (the angle r is less than the angle i) and if light passes from a dense into a less dense medium, it is refracted away from the normal. The relationship between i and r, which is known as the law of refraction, was discovered by Snell, not long after Kepler had done his pioneering work in optics, and independently by Descartes. Snell discovered that the sine of the angle i is proportional to the sine of the angle r which is written as $(\sin i/\sin r) = n$ (Snell's law), where n is called the index of refraction of the transparent substance into which the light passes from the vacuum. This number is always larger than 1 and increases as the density of the transparent medium increases. It is larger for glass (about 1.5) than it is for water (about 1.33) and is about 2.4 for diamond. We shall see that n equals the speed of light in a vacuum divided by the speed of light in the dense medium (which is not the same for all colors, as it is in a vacuum).

If we accept the Newtonian corpuscular theory of light, Snell's law forces us to conclude that light travels faster in a dense medium than in a vacuum. We apply the principle of the conservation of momentum to Newton's light corpuscles as they travel from the vacuum into the dense medium. Let c be the speed of these corpuscles in the vacuum and let v be their speed in the medium. If m is the mass of such a corpuscle, its momentum in the vacuum is the vector $m\mathbf{c}$ whereas in the medium its vector momentum is $m\mathbf{v}$. Now according to Newton's second law of motion the light corpuscle experiences a change in its momentum as it crosses the surface of the medium because the particles of the medium exert a force on the corpuscle. But only the component of the momentum that is perpendicular to the surface (the normal component of the momentum) is altered, because the force on the Newtonian corpuscles is perpendicular to the surface. The transverse component of the momentum (the component parallel to the surface) is not affected because there is no net transverse force on the Newtonian corpuscle (the corpuscle is pulled equally in all transverse directions because there are equal numbers of molecules of the medium all around it on the surface as it enters; but there are no molecules above it as it enters, so that there is nothing to compensate the force at right angles to the surface exerted on the corpuscle by the molecules of the medium below it). Hence, the transverse component of the momentum before the corpuscle enters the medium ($mc \sin i$) must be the same as the transverse component after the corpuscle enters ($mv \sin r$). As a result, we have $mc \sin i = mv \sin r$ or $(\sin i/\sin r) = v/c$.

But we know from observation that $\sin i$ is larger than $\sin r$ because $i > r$. Hence, $v/c > 1$ or $v > c$. Thus, the Newtonian corpuscle of light would have to travel faster in glass than in a vacuum to give agreement with the observations. But measurements (long after Newton) of the speed of light in various media show that c is always larger than v in contradiction to the deductions from Newtonian corpuscular theory. Newton thought he could explain the greater speed of the corpuscle as it passes across the surface of the medium by postulating forces acting inwardly on the corpuscle as it passes across the surface of the medium, but this belief was not correct. The particles (molecules or atoms) in the medium interact with the light and slow it down. Although we must reject the Newtonian corpuscular theory of light, there is another kind of corpuscular theory first proposed by Planck and fully developed by Einstein (the photon concept) which we must accept. When we introduce the concept of the photon later, we shall see that its particle properties are such that one obtains the correct law of refraction. But before we go on to the quantum or photon theory, we discuss the wave theory of light which was first proposed by Christian Huygens, a contemporary of Newton's and a great physicist in his own right, who had independently of Newton discovered the laws of motion, but whose genius was greatly overshadowed by Newton's brilliance.

A wave may be defined in its most general sense as any periodic and continuous disturbance that is propagated from point to point with a definite speed. By periodic we mean that an observer stationed at any point finds that the test particle he is using to detect the wave (e.g., a piece of cork on water if we are dealing with water waves) repeats the same pattern of motions, however complicated, in response to the wave over and over again. By continuous we mean that there is no spatial discontinuity or break in the wave; if test particles were placed in a continuous array along the path of the wave they would all be moving at the same time but not all in step with each other. Since the motion of a test particle (the cork on the water) represents kinetic energy, it is clear from our definition of wave motion that a wave represents the propagation of kinetic energy and not the propagation of matter. The matter, as represented by the test particle, fluctuates about a mean position without going anywhere.

We differentiate between two kinds of waves: transverse waves and longitudinal waves. In transverse waves the motions of the test particles are at right angles to the direction of propagation of the energy being carried by the wave—that is, at right angles to the direction of propagation of the wave itself. Thus, the cork on the surface of the water oscillates vertically (perpendicular to the surface of the water) while the water wave itself moves horizontally. The surface water wave is a transverse wave. For a longitudinal wave the test particles oscillate back and forth parallel to the direction of the propagation of the wave. A longitudinal wave consists of alternate compressions and expansions (or rarefactions). Sound propagated through air is a good example of a longitudinal wave.

The three physical parameters (measurable quantities) that are associated with a wave are the wavelength, λ, the frequency, ν, and the speed of the wave, c. The wavelength is the distance between two successive crests. It is also the distance from any given test particle to the next one in line (to the right or left of it) which is exactly in step with the given test particle. The motions of two such test particles are in phase. If we start from a given test particle, then every time we shift our position by a distance of one wavelength, we meet another test particle whose motion is in phase with that of the first test particle.

By the frequency of a wave we mean the number of crests (or troughs) that pass a given point per second, or the number of vibrations per second of the test particle. Whereas the physical dimension of wavelength is a length (we express it in centimeters) the physical dimension of frequency is the reciprocal of time. We therefore express frequency as $(\text{second})^{-1}$ or 1/second. The speed of the wave is the distance that any section of the wave (e.g., a crest or a trough) travels in a unit time. The speed is the number of crests that pass a given point in a unit time (the frequency) multiplied by the distance between two successive crests (the wavelength). The speed is the product of the wavelength and the frequency, which gives us the basic relationship for wave motion $\lambda\nu = c$.

In general, the speed of a wave in a given medium may depend on the wavelength, but we limit ourselves, for the time being, to waves whose speeds are the same for all wavelengths. The smaller the wavelength, the higher is the frequency and vice versa.

In our definition of wave motion, we placed no restriction on the shape of the wave (the repetitious pattern of the disturbance) but we now restrict ourselves to the simplest kind of wave pattern—a sinusoidal wave which is represented by a succession of equally spaced crests and troughs. The great French mathematician and physicist Fourier showed in the 19th century that every wave pattern (however complex) can be broken down into a sum (known as a Fourier series) of sinusoidal wave patterns of varying amplitudes. This concept makes it easy to study waves because all we need to do is study the sinusoidal wave pattern, from which all wave properties are deducible.

We now extend our considerations of a wave to include the three dimensions of space. Instead of thinking of a wave moving along a single line, we picture the same kind of wave propagated in all directions so that it has a three-dimensional aspect. The waves that emanate from a small source of sound in a uniform gaseous medium (e.g., in still air at a constant temperature) move out uniformly in all directions. On the other hand, the waves on the surface of a body of water have only a two-dimensional aspect and a wave propagated along a string is just one-dimensional. We noted in discussing the one-dimensional propagation of a wave that all the points such as A or B which are homologously situated on a given wave pattern are of the same phase. For one-dimensional waves this analysis gives us an infinite set of points lying on a line and spaced along the line at intervals of one wavelength. For a three-dimensional wave (a wave being propagated in all directions) we introduce the concept of the wave front by using the concept of points of equal phase as follows: Consider any point at some instant in the region through which the wave is advancing. At this point the part of the wave passing through the point is in a certain phase at the moment being considered such as the maximum phase (a crest). Now consider all other points in the neighborhood of the given point that are at the same phase as that of the given point (crest phase) but are not on the line through the given point along which the wave is advancing. All such points of equal phase lie on a surface called the wave front. A line drawn perpendicular to a wave front at any point gives the direction of advance of the wave at that point.

Two extreme, and important kinds of wave fronts occur—the plane wave front and the spherical wave front (plane and spherical waves). In the plane wave all the points of the front, advance in the same direction (along parallel lines) so that points of equal phase lie on planes. Plane waves are the kinds of waves one receives in a homogeneous medium from sources that are at very great distances like the stars. In a spherical wave all points of equal phase lie on concentric spheres whose common center coincides with the source of the waves. A point source in a homogeneous medium generates spherical waves, as observed at all points at finite distances from the point source. All points at equal distances from the point source lie on a spherical wave front, and the direction of propagation of the spherical wave through any such point coincides with the line from the source to that point (the radius of the spherical wave front).

An important wave phenomenon is interference which occurs when two wave trains moving in the same direction with the same velocity but, in general, with different frequencies and different amplitudes (the amplitude of the wave is the height of the wave crest) combine to form a single resultant wave. This resultant wave is obtained by adding the elevations of the two separate waves geometrically (or arithmetically). Where a wave elevation coincides with a wave depression we subtract the depression from the elevation to obtain the resultant wave. If two waves of equal amplitude and equal wavelength are advancing along the same line but in opposite directions, the interference between the two waves gives standing waves. The resultant standing wave does not advance but consists of equally spaced regions of maximum agitation and equally spaced regions of zero agitation. If water waves are reflected from a wall, the reflected waves and the incident waves combine to form standing waves. We then find that along certain equidistant lines (separated by half a wavelength) the oscillation of the water is a maximum whereas at intermediate parallel lines the oscillations are zero. These lines are called nodal lines. The water waves do not advance; the surface of the water rises and falls periodically.

At about the time that Newton was proposing his corpuscular theory of light, Christian Huygens, a great physicist, mathematician, and scientist in his own right, was presenting a wave theory of light which was finally accepted instead of Newton's corpuscular theory because it accounts for refraction, diffraction, and interference (which Newton's corpuscular theory cannot do). Moreover, in explaining refraction, Huygens's wave theory gives the correct relationship of the speed of light in a vacuum to its speed in a denser medium. Huygens was led to his wave theory on observing that when two beams of light cross, they do not interfere with each other which agrees with what one would expect for waves but not for particles. He reasoned that if the beams consisted of corpuscles, the corpuscles would collide and lateral scattering would occur so that two crossing beams of light would disturb each other, contrary to what happens.

To account for the propagation of waves as a succession of wave fronts, Huygens introduced a simple geometrical procedure for constructing the new wave front that arises from any given wave front at some short time later. This method makes it possible to obtain all the equiphase surfaces (wave fronts) of a wave at all points of a medium. This procedure, which is known as Huygens's construction, is a powerful method for following the propagation of a wave through any inhomogeneous, nonisotropic medium or when the wave passes across the interface separating one medium from another. The various properties of waves such as reflection and refraction and interference and diffraction (which were discovered and demonstrated experimentally by Fresnel in France and Young in England about a century after Huygens) can easily be deduced from Huygens's construction.

Consider a light source S in an isotropic (but not necessarily homogeneous) medium and consider the wave front of the light at some moment. What is the wave front at some short interval of time Δt later? The wave front will have advanced to a new position and its shape in general will be different. To see how to deduce the new wave front we first consider the simple case of a homogeneous medium, which means that the speed of the light is the same at all points of the medium. Since the medium is also isotropic (the speed is the same in all directions) it is clear that all the wave fronts are spherical surfaces since the wave spreads out uniformly in all directions. But suppose now that a spherical wave front passes into a region where the speed may change from point to point (an inhomogeneous medium) and let W_1 be the spherical wave front just before the wave enters the inhomogeneous region to the left. It is clear that the wave front becomes distorted because the wave advances more rapidly at one point than at another. Huygens's construction permits us to find the new wave front W_2 after a short interval of time Δt.

Huygens proposed the very remarkable (and correct) hypothesis that each point on a wave front is a source of secondary spherical wavelets that spread out in all directions. The new wave front is then the surface that results from the interference of all the secondary waves arising from the old wave front.

The actual Huygens's construction that we use to obtain the new wave front after a time Δt is the following: Consider any point such as P_3 on the old wave front and let v_3 be the speed of the light at that point. Using $v_3\Delta t$ as a radius (the distance a wavelet from the point has advanced in the time Δt) and P_3 as a center, construct a spherical surface about P_3. Repeat this procedure for other points such as P_1, P_2, P_4 on the surface of the old wave front (using the correct speed of the wave at each point), thus obtaining a group of wavelets. Note that the wavelets are not all at the same distances from the old wave front because the speeds of the wavelets are different at different points. The new wave front of the advancing wave, after the time Δt has elapsed, is that surface which touches all the constructed wavelets at just one point. Such a surface is called the envelope of the wavelets. In general, the shape of the new wave front differs from that of the old one because the speed of the wave, in general, is not the same at the various points of the old wave front. The advance of the old wave front is irregular, resulting in a new wave front that has a different shape from that of the old wave front.

We can now use Huygens's construction to show that Snell's law of refraction holds for waves passing across a surface from one medium to another, provided the waves travel more slowly in the dense medium than they do in the less dense medium. A plane wave advancing, at speed c (in the vacuum), in the direction which makes the angle i with the normal to the surface and is bent toward the normal on entering the medium. It then moves on through the medium at the smaller speed v in the new direction. We want to deduce the relationship between the angle of incidence i, the angle of refraction r, and the two speeds c and v, using Huygens's wave theory of light and keeping in mind that v is less than c.

In the vacuum the wave front of the advancing plane is some plane AB, which makes the angle i with the surface, and in the medium the new wave front, shortly after the point A has struck the surface, is a plane DC (since the shape of the wave is not distorted), which makes the angle with the surface. To obtain DC (D is the point on the new wave front A reaches), we use Huygens's construction as follows: We noted that DC is the wave front in the medium when the point B on the old wave front has reached the point C on the interface between the vacuum and the medium (when B has moved a distance BC). This happens at a time $\Delta t = BC/c = AD/v$ after A has reached this interface. Hence, A will have advanced a distance $v\Delta t$ or $v(BC)/c$ into the medium. This distance is the distance AD. Using AD as a radius, we construct the small wavelet around the point A and then draw the line from C tangent to this wavelet. This line is the new wave front DC in the medium.

We know that $BC = AC \sin i$ and $AD = AC \sin r$ and that if we divide the first of these equations by the second, we obtain $BC/AD = \sin i/\sin r$. But $BC = c\Delta t$ and $AD = v\Delta t$ so that $c/v = \sin i/\sin r$. Since $v < c$ we must have $\sin r < \sin i$ and $r < i$ in agreement with observation. If we introduce the index of refraction which we denote as n, we have Snell's law $\sin i/\sin r = c/v = n$, with the proper relationship between the speed of light in the vacuum and its speed in a material medium.

We have seen that three physical quantities (parameters) are associated with a light wave in a vacuum: the wavelength λ, the frequency ν, and the speed of the wave c, and these three quantities are connected by the basic equation $\lambda\nu = c$. Since the speed of the wave changes from c to a smaller value v when the wave passes into a material medium, the product of the wavelength of the wave and its frequency in the medium must equal v and hence must be smaller than c. It follows that either λ or ν (or possibly both) must get smaller when the light enters the medium. It can be shown that the frequency remains the same, whereas the wavelength diminishes by the same fraction that the speed of the light does. Thus, $\lambda_{(\text{in medium})} = (v/c)\lambda_{(\text{in vacuum})}$.

Since the wavelength of a beam of light changes when the beam passes from the vacuum into a medium because the speed of the light changes, it is natural to ask whether all wavelengths change by the same amount. The answer is no because the speed of light in a given medium is not the same for all wavelengths. This is an important property of light which distinguishes a transparent medium from the vacuum. In the vacuum the speed of light is the same for all wavelengths but in a medium the speed depends on the wavelength. In general, the shorter the wavelength, the slower the light moves. This phenomenon, called the dispersion of light, is very important because it enables us to break a mixed beam of light into its constituent wavelengths. To see why this is so we go back again to Huygens's wave picture and consider an advancing plane wave in the vacuum consisting of a whole range of wavelengths. The wave front AB (in the vacuum) is the same for all of these wavelengths because all the waves (regardless of wavelength) are advancing at the same speed c. In the medium, there is no common speed; the bluer the light, the more slowly it travels, and the more it is bent. Violet light, on entering the medium, travels perpendicular to its new wave front CV whereas red light travels perpendicular to its new wave front CR, where the two planes CV and CR are tilted with respect to each other by an angle which depends on how much faster the red light travels than the violet light does. The original mixed beam in the vacuum with wave front AB is thus split up into a sequence of beams. In this way, we separate the original beam into its different components.

This procedure was first utilized by Newton, who used a prism to break white light up into its component colors. When white light enters at one face of a glass prism in the shape of a triangle ABC, various colored beams leave the opposite face. When the white light enters the prism along the face AB, the rays of different colors are bent by different amounts. On leaving the prism from the face BC, these rays are bent still more so that the red ray waves move off in one direction whereas the blue ray waves move off in a different direction. These rays can be collected on a screen placed at a definite distance from the prism. Newton was the first to carry out this experiment and to demonstrate conclusively that white light is a mixture of all the colors ranging from red to violet. The continuous array of colors into which a prism spreads white light is called the optical spectrum. Each color corresponds to a definite wavelength.

This simple phenomenon is the basis of the prism spectroscope which has played such an important role in the development of astronomy, physics, chemistry, and biology. We can use the refraction of light by a prism to compare the wavelengths of different colors. However, we cannot use a prism to measure the wavelength. To do that we must use an interference phenomenon such as Young first described.

Although we have discussed here only the optical spectrum (the spectrum of visible light), we extend this idea to the entire domain of electromagnetic radiation which includes gamma rays, X rays, ultraviolet rays, infrared, heat, and radio waves; these topics will be discussed later.

We saw that white light consists of a mixture of waves of different wavelengths which correspond to the various colors of the optical spectrum. To measure the wavelength of a definite wave, we use the interference pattern of waves of light that results when a plane wave passes through two closely spaced slits in a screen placed perpendicular to the direction of propagation of the light. This phenomenon was first correctly analyzed by Thomas Young in the early part of the 19th century.

We consider how the light that leaves the two slits, separated by the distance d, is distributed on this second screen at the distance L from the first screen. If light consisted of Newtonian-type corpuscles, moving according to Newton's laws of motion, some particles (about half) would pass through slit S_1 and some (about an equal number) would pass through slit S_2. Those passing through S_1 would cluster around the point B on the second screen opposite S_1 and those passing through S_2 would cluster around the point C on this screen opposite S_2, but very few of these corpuscles would collect around the central point O midway between B and C. In other words, points B and C would be the brightest points on the collecting screen, whereas the central point O would be dim. This situation does not occur in the real world. The point O is the brightest point on the screen and as one moves away from O up or down, the brightness falls off to zero and then increases again. In fact, one finds, starting from point O, that a series of maxima and minima in the brightness is present on the collecting screen.

Although this result is quite at variance with the predictions from the corpuscular theory, we can easily deduce it from the wave theory using Huygens's construction, as Young did. According to Huygens's theory we must picture a train of concentric wavelets originating from each slit and moving out toward the collecting screen. The effect of these two trains of waves at any point on this screen depends on whether the two trains interfere destructively (annihilate each other) or constructively (reinforce each other) at the given point. Since the point O is equidistant from S_1 and S_2, the two wave trains reach O exactly in phase (crest over crest and trough under trough) and reinforcement occurs. Thus, O is a bright point on the screen. However, the distances of points other than O from S_1 are not the same as their distances from S_2 so that the two wave trains are, in general, out of phase at these points.

To analyze this situation in detail we consider the point A on the collecting screen at the distance y from O. If X_1 is the distance of A from S_1 and X_2 is its distance from S_2, the difference between X_2 and X_1 determines whether the wave trains are in phase or out of phase at A. If $X_2 - X_1 = \lambda/2$ (the two distances differ by half a wavelength), then the two trains interfere destructively, because the crest of one is over the trough of the other, and A is then a dark spot. This is also true if $X_2 - X_1$ equals any odd multiple of half a wavelength ($\frac{3}{2}\lambda$, $\frac{5}{2}\lambda$, etc.). However, if $X_2 - X_1 = \lambda$, or 2λ, or 3λ, etc., A is a bright spot because one of the trains is shifted by one, two, or three, etc., wavelengths with respect to the other and reinforcement occurs. From the geometry of the slits and the two screens one deduces the formula $y = (X_2 - X_1)(L/d)$. At points on the screen for which y equals $(\lambda/2)(L/d)$, $(3\lambda/2)(L/d)$, $(5\lambda/2)(L/d)$, etc., we have dark spots, but for points which equal $\lambda(L/d)$, $2\lambda(L/d)$, $3\lambda(L/d)$, etc., we have bright spots.

The formula for $(X_2 - X_1)$ in terms of y, d, and L permits us to measure the wavelengths of light of a given color. All we need do is allow a parallel beam of this light to pass through the two slits and then find the point where the first minimum (dark spot) occurs. If y is its distance from O (which can be measured) we have $\lambda/2 = yd/L$, or $\lambda = 2yd/L$.

This device has enabled physicists to measure the wavelengths of light of various colors with great precision. In this way we have found that the wavelengths of the visible colors are extremely small. They are, in fact, so small that it is convenient to express them in terms of a special unit of length called the angstrom (Å). This unit is so small that one hundred million of them equal one centimeter (1 cm = 10^8 Å). The wavelength of red light is about 7500 Å whereas that of violet light is about 3800 Å. As we go to wavelengths longer than 7500 Å, we pass into the infrared and finally into the radio region of the electromagnetic spectrum where the wavelengths may be of the order of kilometers. Below 3800 Å we pass into the ultraviolet region and finally past the X-ray region into the gamma ray domain where the wavelengths are of the order of 1 Å or smaller.

In introducing our discussion of electromagnetic radiation at the beginning of this chapter, we stated that we can get some understanding of the structure of matter by analyzing the way in which radiation and matter interact. That matter and radiation do interact (as our experience and observations confirm) means that some kind of interactive force between these two physical entities exists and that this force is the very one that governs the structure of matter itself down to its molecular and atomic dimensions; it is essential that we study the interaction between radiation and matter very carefully.

Even without having any scientific curiosity, we cannot avoid observing a wide range of phenomena that depend on the interaction of radiation and matter. All of these phenomena (taken together) can be understood in terms of four broad categories of interaction: (1) the emission of radiation by matter; (2) the absorption of radiation by matter; (3) the reflection of radiation by matter; and (4) the transmission of radiation through matter. Since emission and absorption of radiation have given us the deepest insight into the structure of matter and the forces that give rise to this structure, we consider these two phenomena in some detail.

We know from experience that various objects all around us emit radiation of different kinds and at different rates. A candle flame sends out a yellowish beam of rather low intensity whereas a glowing electric light filament sends out an intense beam of white light. A blowtorch, an alcohol flame, and a welding arc emit intense bluish light, whereas the electric filament in a heater or a toaster emits a red glow. Finally, a hot radiator or a hot laundry iron emits radiation that we can feel but cannot see. Although the differences in the quality of the emitted radiation in the various examples cited here are very striking, they share a common characteristic which is more important than these differences.

To find this common characteristic we pass the radiation in each case through a prism and obtain a spectrum. We then find that the spectrum, except in the case of the hot iron, consists of a continuous array of colors with no break between them. As we move from the red end of the spectrum to the violet end, each color merges into its neighboring color gradually and imperceptibly so that we cannot say just where one color ends and the other begins. The only differences we find in the spectra of the various examples cited are in the intensities of the various colors that are present. A spectrum of this sort, called a continuous spectrum, is emitted when a solid, a liquid, or a dense gas is brought to incandescence. Nothing about a continuous spectrum can tell us anything about the chemical nature of the hot body that is emitting this spectrum. The continuous spectra emitted by incandescent iron, tin, mercury, carbon, uranium, glass, and so on are quite similar in their overall nature; their differences can be accounted for by the differences in the temperatures of these various hot bodies. Although a continuous spectrum does not give us any information about the chemical nature of the emitting body, two other types of spectra do. These two spectra are the bright-line emission spectrum, obtained from a very tenuous (low density) gas, and the dark-line absorption spectrum, obtained when radiation with a continuous spectrum passes through a cool tenuous gas. In each case, the spectrum is characterized by lines that are more or less sharp and occupy definite positions in the spectrum. The particular lines that are present (their positions in the spectrum) and their intensities for both the bright-line emission spectrum and the dark-line absorption spectrum depend on the chemical nature of the gas. As we change from hydrogen, to helium, to oxygen, and so on, the spectral lines change, but as long as we are dealing with the same gas, we always obtain the same lines.

To obtain a bright-line emission spectrum we evacuate a glass tube as thoroughly as we can, introduce into the tube a very small quantity of the gas we are studying, and put a high voltage (about 10,000 volts) across the electrode ends of the tube. The whole tube then begins to glow with a color that is typical of the gas in the tube. This phenomenon is quite similar to the kind of light we get from the fluorescent tubes that are used so extensively today for lighting purposes. When the light from this glowing tube is sent through a prism, the spectrum, instead of consisting of a continuous array of colors, consists of a discrete set of colored lines separated by dark bands. All the light in this type of spectrum is concentrated in these discrete lines;

there is no radiation between the lines. Since each line occupies a definite position in the spectrum, it has a definite wavelength and a definite frequency which must, in some way, be related to the structure of the atom emitting this radiation.

To obtain the absorption spectrum of the gas we send the radiation from an incandescent solid (a hot electric filament) through the tube described above when there is no electric voltage across the tube (when the gas in the tube is not glowing). If we send the light through a prism after it has passed through the cool gas, we obtain a continuous spectrum upon which a set of dark lines is superimposed. These dark lines occupy exactly the same positions in this spectrum as do the bright lines in the previous spectrum. We see that the dark lines of the absorption spectrum of a given gas are as characteristic of the atoms of this gas as are the bright lines of the emission spectrum of this gas. It is interesting to note that the spectra of most stars are dark-line absorption spectra which tells us that a star consists of a hot gaseous sphere surrounded by a cool gaseous atmosphere.

It may seem strange that the spectrum of a hot solid tells us nothing about the chemistry of the solid, whereas the spectrum of a tenuous gas does, but a little thought and an analysis of the physical situation tells us why this is true. A substance emits radiation because the individual particles in the substance (the atoms or molecules) vibrate in some manner or other. Since each type of atom or molecule has its own distinct modes of vibration (like a particular violin string) we expect the emitted radiation to consist of a distinct set of frequencies (or wavelengths) corresponding to these atomic vibrational modes. Such a correspondence, as manifested by the distinct spectral lines, exists for a tenuous gas because the individual atoms are so far apart that they vibrate independently of each other and thus emit radiation with their own characteristic frequencies. In a solid, however, the atoms are closely packed and influence each other as they vibrate. Owing to this influence, interference between vibrational modes occurs (higher vibrational harmonics are excited) so that there is a smearing out of frequencies into a continuous distribution and we obtain a continuous spectrum.

Although we must turn to the bright-line emission spectrum (or the absorption spectrum) of a gas to unravel the mystery of the structure of the atom, we devote part of this chapter to a detailed analysis of the continuous spectrum. A study of the work done by Max Planck in this area leads us to the second of the two great revolutions (the theory of relativity was the other) that occurred in physics at the beginning of the 20th century. In the remainder of the chapter we consider the continuous spectrum of the radiation emitted by various kinds of surfaces and by a furnace (through a small opening) at an absolute temperature T.

At the end of the 19th century the wave theory (as opposed to the Newtonian corpuscular theory) of light and of electromagnetic radiation in general was universally accepted by scientists and firmly supported by refraction and diffraction phenomena. Moreover, the wave theory received great theoretical support at the hands of James Clerk Maxwell, whose mathematical investigations of the electromagnetic field led him to a wave equation for the propagation of electromagnetic disturbances, which also accurately describes the way light is propagated. This electromagnetic wave theory of light was firmly established in the late 1890s when Heinrich Hertz demonstrated experimentally that he could produce Maxwellian electromagnetic waves (we now call them radio waves) by causing electric currents to surge back and forth through properly designed electric circuits, and that these waves behave just like light.

To physicists at that period this situation seemed to be a very happy state of affairs because the universe seemed to be properly divided into two interacting but quite different entities—matter on the one hand and radiation (energy) on the other. Matter manifests itself in the form of a finite number of discrete localizable particles (e.g., atoms, molecules) which cluster together, whereas radiation manifests itself as a continuous, unbroken wave that fills all of space and cannot be localized. Radiation originates from matter, spreading out in all directions at the speed c the moment it is born. Ultimately, it may meet other matter and be absorbed so that its existence is transitory whereas the existence of matter is permanent. This esthetically satisfying picture of the material universe appeared capable of explaining all observable phenomena, and yet at the very moment that it was achieving some of its greatest successes, the wave theory of radiation was being undermined by investigations conducted in two different areas. One of these dealt with the analysis of the continuous spectrum emitted by hot bodies and the other dealt with the emission of charged particles (electrons) from metallic surfaces when light strikes these surfaces. Although the continuous spectra emitted by various hot solids at the same temperature are essentially the same, certain slight, but important differences exist that depend not on the differences in the chemical compositions of these hot bodies but on the nature of their surfaces. As we shall see, one special kind of surface emits radiation that can be described in terms of general laws more readily than radiation emitted by other kinds of surfaces can; we consider the radiation of such surfaces in detail.

We first study the behavior of radiation in general when it strikes an opaque surface. The incident beam, designated as I, interacts with this surface, and some of this radiation, designated A, is absorbed; the rest of the radiant energy in the incident beam is reflected and is designated R. When we say that a beam of light is reflected from a surface, we mean that the wavelength (and hence the frequency) of the beam R that comes off the surface is exactly the same as that of the incident beam I. In other words, the reflection process (the reflecting surface) has no effect on the wave characteristics of the beam being reflected. By the same token, a beam that is reflected from a surface has no effect on the reflecting surface.

The situation is quite different, however, for absorption since electromagnetic energy is absorbed and that means that the surface gains energy in the process. Owing to this gain the temperature of the surface rises when it absorbs radiation and the more effectively (or efficiently) it absorbs radiation, the higher its temperature rises when a given amount of radiation strikes it in a given time. The rate of absorption of radiation by a surface is therefore related to the rate of emission of radiation by this surface because the more rapidly the surface absorbs radiation, the hotter it gets (the higher its temperature), and the more rapidly it emits or radiates. A surface that absorbs very efficiently also radiates very efficiently for a given temperature.

To avoid confusion here we must differentiate clearly between reflection from a surface and the reemission of radiant energy after the surface has absorbed radiation. When a surface reflects the radiation that strikes it, the reflected radiation is identical in character (frequency or wavelength) to the incident radiation. But when a surface absorbs radiation and then reradiates it, the reradiated energy in no way resembles the incident radiation. The energy that the surface radiates is not concentrated in a single frequency (color) but is spread out over a continuous range of frequencies with most of the reradiated energy concentrated in frequencies considerably lower than that of the incident radiation. Absorption and reradiation therefore constitute a process by which radiation of any given frequency is spread out among all frequencies. Reflection does not spread out the radiation at all. The way the absorbed energy is reradiated by a surface does not depend in any way on the nature of the incident radiation; it depends only on the nature and the temperature of the absorbing surface.

We saw that when radiation is incident on a surface, part of the radiation is reflected and part is absorbed. Obviously, the absorbed and the reflected radiation complement each other; the more that is reflected, the less that is absorbed and vice versa. We know from experience that not all surfaces reflect or absorb equally well. All surfaces fall between two extreme types, at one end of which is the perfect reflecting surface (a highly polished silver surface), and at the other end of which is the perfect absorber (a "perfect black body"). A perfect reflector is a surface that absorbs no radiation at all, but reflects all frequencies equally well. Such a surface experiences no increase in temperature when radiation strikes it, and it emits no radiation. A perfect absorber is a surface that absorbs all frequencies of incident radiation equally well and reflects none of the incident radiation. As this surface absorbs radiation of any frequency, its temperature rises and it radiates (in all frequencies) at an ever-increasing rate. Its temperature continues to rise until the rate at which it is emitting radiation (in all frequencies) exactly equals the rate at which it is absorbing the incident radiation. A good reflector is a poor absorber (and a poor emitter) of radiation whereas a good absorber is a poor reflector but a good emitter of radiation.

Consider now 1 cm^2 of any given surface at a given temperature. This surface is emitting radiation of any given frequency ν at a definite rate; we call this rate of emission in the given frequency E_ν. The subscript ν attached to E indicates that the rate of emission (for the given temperature) varies from frequency to frequency. This variation is in complete accordance with our experience because we know that a warm body radiates infrared waves much more effectively than it does ultraviolet waves or radio waves. On the other hand, if this same body were very hot (like the surface of the sun) it would radiate more intensely in the yellow and green frequencies than in the red. Let A_ν be the rate at which this same square centimeter of surface absorbs radiation of frequency ν. Since absorption is just the inverse process to emission, we expect A_ν and E_ν to be related to each other in some way; this relation is expressed by a general law discovered by G. R. Kirchhoff near the end of the 19th century. He found that the ratio E_ν/A_ν for radiation of frequency ν for a surface at temperature T is the same for all surfaces, regardless of their nature; it depends on the frequency ν of the incident radiation and on the absolute temperature T of the surface. If we call this ratio B_ν, we may write $E_\nu/A_\nu = B_\nu$ or $E_\nu = B_\nu A_\nu$. The rate of emission by a surface is proportional to its rate of absorption; the faster it absorbs a given frequency, the faster it emits that frequency at any given temperature. The proportionality factor, i.e., the quantity B_ν, which is the same for all surfaces, determines how rapid the emission is as compared to the absorption rate. From this reasoning it follows that B_ν, which depends on both the frequency ν of the incident radiation and the temperature of the absorbing and emitting surface, is a very important physical quantity. If we know how this quantity, which is a universal function, depends on frequency and temperature, we can determine the rate of emission of a surface for any frequency from its rate of absorption which is easy to measure. This fact prompted physicists of the 19th century to try to determine the mathematical form of $B_\nu(T)$ from the known basic laws of physics. The letter T in parentheses is introduced to indicate that $B_\nu(T)$ depends on the absolute temperature of the radiating surface.

Although it was easy enough for physicists to determine $B_\nu(T)$ experimentally, all attempts to deduce its mathematical dependence on the frequency ν and temperature T from the laws of classical physics failed; not until Planck introduced the quantum concept of radiation was this basic problem solved in a most revolutionary and unexpected way. To see just what Planck did we must examine $B_\nu(T)$ more carefully and analyze its physical meaning.

We have seen that a given surface at a definite absolute temperature T emits or radiates and absorbs energy over a continuous range of frequencies, and the rate at which it emits energy (the number of ergs per second per square centimeter) in a given frequency is proportional to the rate at which it absorbs this same frequency at the same temperature. We now consider how we are to measure the proportionality factor $B_\nu(T)$ and how we are to deduce its mathematical form, i.e., its algebraic dependence on ν and T.

We can answer these two questions by considering the emission of radiation by surfaces of the two extreme types we described previously. The perfectly reflecting surface absorbs no radiation so that for it $A_\nu = 0$ which means that $E_\nu = 0$ also. A perfectly reflecting surface emits no radiation at all regardless of its temperature. Obviously, there are no perfect reflectors in nature because all real bodies radiate, even though some may radiate very slowly. Since both E_ν and A_ν equal zero for a perfect reflector, such surfaces cannot be used to determine $B_\nu(T)$, but surfaces at the other extreme (perfect absorbers) can be used. Since a perfect absorber completely absorbs radiation of all frequencies that strike it, we place $A_\nu = 1$ for all such surfaces (A_ν ranges from 0 to 1, where 0 means no absorption and 1 means complete absorption). We then see that $E_\nu = B_\nu(T)$ for such surfaces and we have thus obtained a physical interpretation or meaning for $B_\nu(T)$; it is the rate at which 1 cm^2 of a perfect absorber at the absolute temperature T emits radiation of frequency ν.

At this point it is convenient to describe a perfect absorber somewhat differently. If a surface completely absorbs all the radiation (regardless of frequency) that strikes it, it must appear perfectly black because it reflects no light into our eyes. Hence, a perfect absorber is also a "perfect black body," and we shall refer to it in that way from now on. The physical quantity $B_\nu(T)$ is therefore the rate at which a perfect black body emits radiation of frequency ν when its temperature is T. We should be able to measure $B_\nu(T)$ by heating a perfect black body to the temperature T and then analyzing the radiation that it emits by sending this radiation through a prism of the right kind. We thus obtain a continuous spectrum, and we can measure the intensity in each part of the spectrum (for each wavelength) by means of a sensitive photometer. The values given by this photometer for the various values of the frequency ν (or the wavelength) as we move the photometer from one part of the spectrum to the other are precisely the quantities $B_\nu(T)$ that we seek.

Although this procedure appears to be fairly simple in principle, we immediately run up against the very practical problem of finding a perfect black body. A good approximation to such a body is any flat black surface like carbon black or charcoal, but it is by no means a perfect black body since an object like a piece of charcoal, which looks black, may not absorb all the nonvisible radiation such as the ultraviolet or the infrared. In fact, we know that this absorption gap exists which means that incandescent charcoal or carbon black does not emit true black-body radiation. This fact did not daunt the experimental physicists who were working in this field at the time because they soon discovered a clever artifice for producing black-body radiation without using a physical body.

To understand how this artifice was introduced and exactly what it is, we recall that according to Kirchhoff's law of radiation the more effectively a body absorbs radiation, the more effectively it emits radiation. If we want to construct a device that radiates like a perfect black body (a perfect radiator) we must construct a device that absorbs perfectly and we obtain such a device when we punch a small hole (window) in the wall of a hollow container. This small window behaves just like a black body, and the smaller the size of the window, the blacker the container. This fact agrees completely with our experience; we know from direct observation that broken windows in old buildings look quite black as do the doorways to church interiors. We can see why a small window in the wall of a cavity behaves like a black body if we study a beam of radiation that impinges on such a window. Regardless of the wavelength (or frequency) of this radiation, it is trapped in the cavity once it enters the window so that practically none of it (in the form in which it entered) ever gets out again. The beam suffers multiple reflections and absorptions along the inner walls of the cavity until all of it is absorbed. Hence, the small window behaves in every respect as though it itself had absorbed this radiation. In other words, a small window in the wall of a cavity behaves like a perfect black body insofar as absorption goes. But this fact means, according to Kirchhoff's law, that it must radiate like a perfect black body.

The above analysis of the behavior of a small aperture shows us how to obtain black-body radiation. We start with a completely enclosed cavity whose walls are opaque and which we can heat to any desired absolute temperature T (essentially a furnace). When we heat the walls to some temperature T, the interior surfaces of the walls radiate energy into the cavity until the cavity is filled with this radiation and the temperature in the cavity itself is also T. When the temperature is the same throughout the cavity and equal to the temperature of the walls, namely T, the radiation in the cavity is in thermodynamic equilibrium with the walls in the following sense: the rate at which each square centimeter of the interior walls is emitting radiation into the cavity exactly equals the rate at which it is absorbing radiation from the cavity. The cavity (or furnace) is filled with a fixed amount of radiation that does not vary as long as T remains constant. If T is altered, the amount of radiation in the cavity increases (if T increases) or decreases (if T decreases). The radiation in the cavity is black-body or thermal radiation of temperature T.

To study black-body or thermal radiation of temperature T, all we need to do is punch a small hole ($1\ cm^2$) in the wall of our furnace, and analyze the radiation emitted per second from this hole. The amount of this radiation emitted in the frequency ν is the quantity $B_\nu(T)$ that we seek.

We have seen that the radiation coming from a small opening in a furnace at the absolute temperature T is black-body radiation. If we send this radiation through a prism, we can spread it out into its various component frequencies and then measure the intensity of each of these frequencies and thus obtain $B_\nu(T)$. Such experiments were done very carefully near the end of the 19th century and the results were exhibited in a graphical form in which the quantity $B_\nu(T)$ (the intensity of a given frequency) is plotted against the frequency ν. The empirical curve that is thus obtained for a given temperature of the furnace gives us the shape of the black-body radiation curve defined by the function $B_\nu(T)$. The plotted curves give us the observed or empirical geometrical form of the function $B_\nu(T)$, but we cannot deduce the exact mathematical expression for $B_\nu(T)$ from these curves because this expression is not simple enough for us to establish a precise correspondence between some particular formula and the observed curves. The curves are not accurate enough for this purpose, but the knowledge that the correct formula when plotted must give us a curve that has the general shape of the empirical curves was very helpful in enabling physicists to deduce the correct radiation law.

Before considering the various attempts (and their failures) to deduce the expression $B_\nu(T)$ from the basic classical laws of physics (mechanics, electricity, thermodynamics) we see that we can learn some important characteristics of black-body radiation from the empirical curves. Each curve has a maximum value for some particular wavelength which means that there is always a wavelength (or frequency) in the emitted radiation in which more energy is concentrated than in any other wavelength. We label this wavelength λ_m where the subscript m stands for maximum. The various component wavelengths in the emitted radiation are distributed, in varying amounts, around this maximum value. We have here a situation that is similar to (but by no means the same as) the distribution of any variable characteristics among members of a group of objects possessing this characteristic in varying degrees. When grades are distributed (strictly according to merit) among a random group of students in a particular class, we find most of the grades concentrated around a certain average grade (let us say the grade C) with very few low grades (such as F) and very few high grades (such as A). Such a distribution is called a normal distribution, which we do not have in black-body radiation; the energy distribution given by the black-body radiation curves are not symmetrical about the wavelength λ_m of maximum intensity. The intensities in the very short wavelengths fall off much more rapidly than those in the long wavelengths. It is much more difficult to obtain visible radiation from a furnace that it is to obtain infrared and heat rays. The temperature of a furnace must exceed 900°K (absolute) before it emits detectable (measurable quantities of) visible radiation.

The empirical black-body curves tell us two more things which are related to the way the shapes of the curves depend on the absolute temperature T of the black-body radiation (i.e., on the absolute temperature of the cavity or the furnace). First, as the temperature of the furnace is increased, the curve describing the rate of radiation is shifted upward on the graph (the intensities of all frequencies are increased) so that the curve has a larger area under it. Second, the wavelength of maximum intensity is shifted to the left (toward smaller wavelengths) toward the bluer colors. Because of this shift of the wavelength λ_m, this phenomenon is known as the displacement law of black-body radiation.

To see the significance of the first empirical geometrical feature of the curves, we note that the area under the radiation curve is just the total radiation (energy) emitted by the furnace per unit time in all frequencies (all wavelengths). Considering a small range $\Delta\lambda$ of wavelengths, we see that the area of the radiation slab defined by $\Delta\lambda$ is just $B_\nu(T)\Delta\lambda$. But this quantity is just the radiation emitted in the wavelength range $\Delta\lambda$. Hence, the total area under the curve, which is just the sum of all such slabs, is the total radiation emitted per second in all wavelengths.

Although one cannot deduce the exact mathematical expression for $B_\nu(T)$ from the empirical radiation curves obtained from the observed data, two definite deductions can be made from these curves. First, a careful examination of the curves shows that the relationship between λ_m and the temperature T is a reciprocal one. From the observed data alone one can show that λ_m varies inversely with T; or ν_m, the frequency of the radiation of maximum intensity, is proportional to T. Thus, $\lambda_m = \text{constant}/T$ where the constant has the numerical value 0.290 and its units (or dimensions) are degrees centimeters. Since this relationship was deduced from theoretical considerations by W. Wien near the end of the 19th century, it is known as the Wien displacement law. It tells us that the wavelength in which the maximum amount of black-body radiation is concentrated decreases as the absolute temperature of the furnace increases. That is why the hotter a furnace is, the bluer the radiation emitted by it looks. If at a furnace temperature of 4000°K the radiation of maximum intensity emitted by the furnace has a wavelength of 7500 Å (red light), then the radiation of maximum intensity has a wavelength of about 3700 Å (violet light) when the temperature of the furnace is 8000°K.

Second, one can show, as the 19th century physicist J. Stefan did empirically, that the area under any one of the radiation curves is proportional to the fourth power of the absolute temperature T associated with that curve (i.e., the temperature of the furnace). This is known as Stefan's law, which is written as $B = \text{constant } T^4$, where B is the total energy emitted per sec per cm^2 by the furnace in all wavelengths, and the constant has the numerical value 5.67×10^{-5} and its units are ergs/cm^2 deg^4 sec. Since this law was later derived theoretically by Boltzmann, it is also known as the Stefan–Boltzmann law. This law tells us that if the temperature of a furnace is doubled, the rate at which it emits energy in all frequencies goes up by a factor of 16, and if the temperature is tripled, the rate of emission increases by a factor of 81, and so on.

We can best understand the physical significance of Wien's and Stefan's laws by observing the radiation emanating per second from a 1 cm^2 window in a furnace as we increase the temperature of the furnace. When the temperature of the furnace is below 900°K, the amount of radiation emitted per second in the visible range of the spectrum is too small to be seen—the furnace is nonluminous, although some infrared radiation is being emitted. Above 900°K visible radiation begins to appear, with the reds coming in first and then the yellows, the greens, the blues, and the violets, with the intensities of all of these colors increasing rapidly as the temperature of the furnace increases. At about 1500°K enough of the visible colors is present to give the eye the impression of dazzling whiteness even though the wavelength of the radiation of maximum intensity is longer than the infrared wavelength. At a temperature of about 4000°K the maximum intensity (as already noted) is in the visible part of the spectrum (red), and at still higher temperatures the blue, the violet, the ultraviolet, and finally, the X rays dominate.

This rough description of the emission of radiation by a small window in a furnace already shows us how useful the Wien displacement law and the Stefan law are to astronomers because we can use these laws to determine first the temperatures of stars and then their luminosities if we know their sizes. By analyzing the radiation emitted by the sun and finding the wavelength λ_m for which the intensity of this radiation is a maximum (it is about 5000 Å), we deduce (using Wien's law) that the sun's surface layers are at a temperature of about 5700°K. The temperature of a star like Rigel, on the other hand, is about 12,000°K because it radiates with maximum intensity at a wavelength of about 2000 Å. Since the temperature of a star like Rigel is somewhat more than twice the temperature of the sun, we see that each square centimeter of Rigel's surface is emitting about 25 times as much energy as each square centimeter of the sun's surface. If we knew the size of Rigel compared to the sun's size, we could find its luminosity.

The exact mathematical expression for $B_\nu(T)$ is too complicated for us to deduce from the empirical curves obtained from actual measurements of the radiation emitted by a black body. These curves are accurate enough to permit us to deduce the Wien displacement law and the Stefan law for the integrated radiation, but that is the best one can do empirically. To find the expression for $B_\nu(T)$ (the intensity for any frequency) one must use the theories that are available, and so physicists spent many hours trying to deduce $B_\nu(T)$ by pure reasoning from the classical Newtonian and Maxwellian theories of matter, radiation, and thermodynamics, but without success. We shall see that this failure of the classical laws to lead to a correct expression for the intensity of the black-body radiation is related to the discrepancy between the classical theory of the specific heats of gases and the observed ratio of the specific heats of gases that we have already discussed.

To see what is involved in trying to deduce $B_\nu(T)$ from classical physics, we work not with $B_\nu(T)$ directly, but rather with the energy density of radiation $u_\nu(T)$ which is related to $B_\nu(T)$. To get at the idea of the energy density of radiation, we close the small window in our furnace and just consider the radiation trapped in the cavity itself. This radiation, which is at absolute temperature T and is in equilibrium with the walls of the furnace (or the cavity), fills the cavity completely. If we were stationed at any point inside the cavity and had appropriate measuring devices, we would find radiation of all frequencies (in varying intensities) moving in all directions toward us and away from us. As we can deduce from the second law of thermodynamics, all directions are equally favored with no particular direction endowed with a greater flow of radiation than any other. The second law also tells us that the intensity of radiation at any point in the cavity must be the same as at any other point since the temperature must be the same at all points. The radiation swirling about in this cavity behaves very much like a gas and we may refer to it as a radiation gas. If the point we are stationed at is at the center of a cube of unit volume (1 cm^3), the total radiation present in this volume at any moment is called the total energy density u of the radiation in the cavity. The amount of radiant energy of frequency ν that is present in the unit volume at any moment is called the energy density of frequency ν and is written as u_ν. If we add up all the u_ν's in the unit volume, we obtain u, the total or integrated energy density.

If we can determine $u_\nu(T)$, we can also find $B_\nu(T)$ since these two quantities are proportional to each other. To see this we punch a window of unit area (1 cm^2) in our furnace again and consider the radiant energy of frequency ν streaming out of this window in 1 second. This quantity is just $B_\nu(T)$. But it is also the amount of radiation contained in a cylinder of unit cross section (1 cm^2) and of length c, where c is the speed of light. But this amount of radiation is cu_ν. Hence, B_ν is essentially equal to cu_ν so that we can concentrate on determining $u_\nu(T)$ instead of $B_\nu(T)$.

Although we cannot deduce the mathematical form of $u_\nu(T)$ from the classical laws of physics, we can deduce the expression for $u(T)$ (the expression for the total energy density, which is the sum total of all the separate u_ν's in our unit volume). We begin by noting that if black-body radiation at temperature T may be treated like a gas at temperature T, then, like a gas, it must exert pressure against the walls of the container. To determine this pressure we picture the walls of the container as being perfectly reflecting so that the radiation simply is reflected from wall to wall without any change in its character or in its intensity. As long as the temperature in the container is T, the total energy density u of the radiation remains constant.

Consider now a unit area of the interior reflecting wall with radiation from the interior of the cavity striking it every second. The amount of radiant energy striking this unit area of the wall in 1 second in a direction at right angles to the wall is just $cu/(3)(2)$. The reason for the factor c is that all the energy in a cylinder of length c will reach the wall in 1 second. The reason for the 3 in the denominator is that there are 3 independent directions in space and only one-third of the radiation on the average in this cylinder is moving exactly at right angles to the wall, either toward the wall or away from it; and only the radiation striking the wall at right angles exerts a pressure on the wall. The reason for the factor 2 in the denominator is that only half the radiation in the cylinder that we have just described is moving away from the wall and half is moving toward it. The unit area of the wall is struck directly every second by an amount of radiant energy $cu/6$.

Since according to the theory of relativity, energy is equivalent to mass, this energy exerts a pressure on the wall, which we can calculate. We note that since energy is equivalent to mass, energy carries momentum with it, and the amount of momentum carried by energy equals this energy, $cu/6$, divided by c, or $u/6$. Reflecting this energy the wall acquires an additional amount of momentum $u/6$. The total change per second in the momentum of the unit area of the wall when it is struck by and immediately reflects radiant energy is therefore just $2u/6$ or $u/3$. But the rate of change of momentum (change per unit time) of an object equals the force on it (Newton's second law of motion). The unit area experiences the force $u/3$. But pressure is force per unit area; hence, the pressure exerted by the black-body radiation on the wall is just $\frac{1}{3}u$. This is a very interesting and remarkable result for if we go back to the thermodynamics of perfect gases, we see that this is similar to the relationship deduced where we saw that for a perfect gas the gas pressure equals $(\gamma - 1)u$ where γ is the ratio of specific heats of the gas and u is the energy density of the gas. If we compare $(\gamma - 1)$ with $\frac{1}{3}$ and equate them, we see that black-body radiation behaves as though it were a gas for which the ratio of specific heats is given by $\gamma - 1 = \frac{1}{3}$ or $\gamma = 4/3$. This result leads us to the speculation that we might be able to treat radiation as though it consisted of corpuscles (similar to the molecules of a gas) for which the ratio of specific heat is 4/3.

If this were true, we would have to assign degrees of freedom to these radiation corpuscles just as we do to the molecules of a gas. We know that if n is the number of degrees of freedom of a molecule in a gas for which the ratio of specific heats is γ, then $\gamma = (2/n) + 1$ or $n = 2/(\gamma - 1)$. If we place $\gamma = 4/3$, we thus obtain $n = 2/(4/3 - 1) = 6$. From this result we see that if our radiation gas consists of some kind of corpuscles, they cannot be simple pellets with no internal structure because such simple pellets (like tiny mass points) have only the three translational degrees of freedom associated with translational motion and not six degrees of freedom. That our radiation corpuscles (if they exist) have six degrees of freedom means they must have some kind of internal motions (like the vibrations of a spring). In fact, we may assume that these corpuscles have vibrational degrees of freedom like a harmonic oscillator in addition to their translational degrees if they exist at all.

We can deduce the dependence of u on T in two different ways, one of which is based on treating black-body radiation as a "perfect gas" whose ratio of specific heats equals 4/3, and the other of which is based on the first law of thermodynamics. In considering the black-body radiation in our cavity as a perfect gas of temperature T exerting a pressure P, we arrived at the relationship $P = \frac{1}{3}u$. This equation led to our assigning the value 4/3 to γ, the ratio of specific heats. But for a perfect gas for which $\gamma = 4/3$ we have the adiabatic relation $PV^{4/3} =$ constant, where V is the volume of the gas. Noting that for a perfect gas $PV =$ (constant)T, we have $V =$ (constant)T/P. Introducing this expression for V into the previous equation for black-body radiation we obtain $PT^{4/3}/P^{4/3} = T^{4/3}/P^{1/3} =$ constant, or $T^{4/3} =$ (constant)$P^{1/3}$. Hence, $P =$ (constant)T^4. But $P = \frac{1}{3}u$ so $u =$ (constant)$T^4 = aT^4$. This relationship shows us that the energy density of black-body radiation varies as the fourth power of the absolute temperature of the furnace (or cavity) containing this radiation. Since u is directly proportional to the total radiant energy B emitted per sec per cm^2, we have here the Stefan–Boltzmann law, which tells us that B is proportional to the fourth power of T. The numerical constant a in the formula for u is known as the radiation constant. Its numerical value is 7.564×10^{-15}

erg/cm^3 deg^4. Since the energy emitted from the furnace is $B = c\mu/4\pi$, we have $B = \sigma T^4$, where $\sigma = ac/4\pi = 5.669 \times 10^{-5}$ erg/cm^2 sec deg.

Previously we deduced the relationship between the total black-body radiation energy density u and the absolute temperature T of this radiation by treating the radiation as a perfect gas for which the ratio of specific heats equals 4/3. This is as far as classical physics can go; it cannot lead us to the correct mathematical form for $u_\nu(T)$ so that some new and revolutionary idea had to be introduced. Planck found the answer in 1900 when he proposed the quantum theory of radiation, which finally led to the photon concept as introduced by Einstein in 1905. To see why Planck found it necessary to depart from the Maxwellian wave theory of radiation and to introduce a corpuscular concept, we consider why the wave theory leads to the wrong law (mathematical expression) for the energy density function $u_\nu(T)$.

Before Planck discovered the correct expression for $u_\nu(T)$, the 19th century British physicist Lord Rayleigh, using classical physics, deduced an expression for $u_\nu(T)$ that is clearly incorrect over the entire range of wavelengths, but matches the empirical black-body radiation curves for large values of λ (the red end of the waves). Rayleigh's deduction is based on the classical law of the equipartition of energy which says that if a gas is at the absolute temperature T, then the average energy associated with each degree of freedom of each molecule in the gas is $\frac{1}{2}kT$, where k is the Boltzmann constant. Although this theorem was deduced for particles moving about randomly, it can be shown to be true in general in classical physics. One can show that if there is any kind of system in thermodynamical equilibrium in a container at temperature T, then each degree of freedom in this system has, on the average, an amount of energy equal to $\frac{1}{2}kT$.

Rayleigh applied this idea to the radiation gas (black-body radiation) in a container. He suggested that the waves in the container can be treated as oscillators (like a collection of vibrating springs) and that since one can assign degrees of freedom to such oscillators, one can also assign degrees of freedom to the radiation waves in the container. Using this idea Rayleigh calculated the number of waves of a given frequency or wavelength that can fit into a unit volume of the container and multiplied this amount by $(n/2)kT$ where n is the number of degrees of freedom associated with a single wave.

To see what this reasoning leads to we give here a rough, over-simplified, but essentially correct description of Rayleigh's procedure. Consider a cube of unit volume in our container and radiation of wavelength λ. The number of standing waves that can lie along each edge of the cube is just $1/(\lambda/2)$ so that the number of such waves to this unit volume is proportional to $(1/\lambda)^3$ or (since $\lambda = c/\nu$) to $8\,\nu^3/c^3$. The reason for dividing the unit length by $\lambda/2$ is that each half wavelength oscillates in a standing wave. Since each of these standing waves can be considered an oscillator, the equipartition theorem tells us that an amount of energy kT is associated with each wave. Thus, the density of the radiation energy of frequency ν is proportional to $(\nu^3/c^3)(kT)$. From this equation one can deduce that the energy emitted by a black body of temperature T in the frequency range $\Delta\nu$ should be proportional to $(\nu^2/c^2)(kT\Delta\nu)$ if classical physics is right. This equation is the distribution function $B_\nu(T)$ that we obtain from classical physics. We see at once that this equation is incorrect for large values of ν although it agrees with observation for small frequencies (long wavelengths); this formula tells us that, regardless of temperature, all furnaces must emit greater and greater quantities of energy as we go to larger and larger frequencies (shorter wavelengths). In fact, the Rayleigh formula gives infinite values for $B_\nu(T)$ as ν goes to infinity. This result (called the "ultraviolet catastrophe") is in distinct disagreement with the empirical data (the observed curves) which show that $B_\nu(T)$ goes to zero as ν increases. Since all attempts to obtain a correct expression for $B_\nu(T)$ (using the laws of classical physics) were fruitless (in fact, it was shown by very general arguments that, however one proceeds, Rayleigh's law is the inescapable consequence of classical physics), it was necessary to turn away from the physics of Newton and Maxwell which is precisely what Planck did when he introduced his quantum hypothesis.

Before considering Planck's great discovery and innovation, we must mention an empirical radiation law proposed by Wien, the discoverer of the displacement law. Wien, being aware of the incorrectness of the Rayleigh law for large values of the frequency, proposed another law, on an empirical basis, which does fit the radiation curves for large values of ν. He had no theoretical justification for this law but noted that it matches the observed values of $B_\nu(T)$ very well as ν increases, and it approaches 0 as ν becomes infinite, which is just what we want. At first it was thought that Wien's radiation law is the correct law because it seems to give the correct values for $B_\nu(T)$ for small values of ν also, but careful measurements show that this is not so. The Wien law gives values for $B_\nu(T)$ that are consistently smaller than the observed values at the long wavelength end of the spectrum (small frequencies). One sees that whereas Rayleigh's law and Wien's law separately match only part of the empirical curve, they reproduce the curve quite well when taken together.

Since classical physics cannot yield a correct radiation law, it was necessary to introduce an entirely new idea even if this idea was in direct contradiction to the classical laws (Maxwell's law) of radiation. Max Planck did precisely that after much hesitation and soul-searching, when he introduced what he called the "quantum theory." Planck did not come to this theory all at once in a single intuitive step, but only after he had carefully studied the mathematical expressions for Wien's and Rayleigh's laws and realized that he could combine both of these expressions algebraically into a single formula that reproduces the entire empirical curve (for all values of the frequency). This effort was an exercise in pure mathematics, without any physical justification except that the final formula thus obtained gives the correct result. On analyzing this correct algebraic formula, Planck saw that he could not possibly deduce it from the laws of classical physics (from Newton's laws of motion and Maxwell's electromagnetic wave theory of light), but that he could deduce it if he assumed the radiant energy to be emitted by the interior walls of the cavity discontinuously in the form of concentrated bundles of energy (energy quanta) rather than continuously in the form of waves, as demanded by Maxwell's theory. This proposal was the origin of the quantum theory, without which we could not possibly understand the structure of atoms, molecules, or nuclei. The quantum theory also clears up the difficulties we encountered in our analysis of the ratio of the specific heats of gases.

We saw that as long as one assumes that a black body emits energy continuously in the form of waves, and that the classical law of the equipartition of energy holds, then Rayleigh's incorrect radiation law is the inevitable consequence. One must assume that energy is emitted discontinuously in the form of lumps (quanta) as Planck did, to obtain the correct radiation law. The difficulty with the Rayleigh law stems from the classical equipartition theorem when it is applied to a continuous distribution of waves. Since all vibrational degrees of freedom are assigned the same amount of energy kT according to this equipartition theorem, and all vibrational degrees of freedom are present, it is clear that according to the classical equipartition theorem more and more energy is present in the form of high-frequency radiation. This result occurs because the smaller the wavelength (large frequencies), the greater is the number of vibrations that can exist in any volume of the cavity. The Planck quantum hypothesis, however, places a severe restriction on the emission of quanta of radiation of high frequency; with such a restriction the radiation curve that one obtains from the quantum theory matches the empirical curve exactly. Planck did not see at once what kind of restriction nature imposes on the emission and absorption of radiation by a black body but he deduced this restriction finally from his analysis of the correct algebraic expression for $B_\nu(T)$, which he had obtained by the clever mathematical trick of combining Rayleigh's law and Wien's radiation law into a single expression involving ν and T. An analysis of this correct mathematical expression, which depends on ν and T in such a way (is a function of ν and T) that it approaches Wien's law for large values of ν, shows that only if one assumes that the amount of energy in a quantum of radiation emitted by a black body is proportional to the frequency ν of this radiation can one deduce the formula that Planck obtained.

Planck pictured the black-body radiation in the cavity as being in constant equilibrium with the atoms in the walls of the container which constantly absorb and re-emit the radiation in the cavity. At the close of the 19th century, when Planck was doing this work, practically nothing was known about the structure of atoms, so Planck could not describe the interaction of the black-body radiation with the walls of the cavity in terms of the actual atomic processes that were occurring. But this problem was no real drawback because Kirchhoff's law of radiation shows that the basic thermodynamic and spectral properties of black-body radiation are entirely independent of the nature of the walls enclosing the cavity. Planck therefore assumed that these walls consisted of harmonic oscillators and that these oscillators absorbed and emitted the black-body radiation in the cavity. This assumption was a very fruitful idea because the classical properties of the harmonic oscillators were well known to Planck and he could see at once what had to be done to these classical features of harmonic oscillators to permit them to emit and absorb radiation discontinuously (in little packets) rather than continuously (in continuous waves) as demanded by the classical theories. Some 17 years later Einstein deduced Planck's correct radiation law from the properties of atoms, but Planck had to content himself with harmonic oscillators.

Planck began by considering the action associated with one complete vibration of a single harmonic oscillator of frequency ν. From the definition of the action of a particle and from the discussion of the action associated with one vibration of a harmonic oscillator, we deduced that this action is E/ν, where E is the energy of the harmonic oscillator and ν is its vibrational frequency. Classical physics says that the energy of a harmonic oscillator of given frequency can vary continuously so that the action can take on all values. In other words, if we adhere to classical physics we must accept the conclusion that the harmonic oscillator's action and energy can vary continuously and the harmonic oscillator can emit and absorb radiation having a continuous range of energies. This conclusion, as we have seen, leads inevitably to Rayleigh's law, which is incorrect. To obtain emission and absorption of energy in lumps (quanta), Planck introduced the revolutionary assumption that the action of the harmonic oscillator can have only a discrete (but infinite) set of values each of which is an integral multiple of a unit of action (a quantum of action). This assumption gives a discrete set of values for the energy E of an oscillator (integral multiples of a unit or quantum of energy) and hence for the energy that the oscillator can absorb or emit. This result leads us to Planck's concept of the unit of action which from now on we call h. Planck proposed the idea, which has since been verified, that the action of a harmonic oscillator occurs in integral multiples of a basic unit or quantum of action h (the famous Planck constant) whose value is 6.7×10^{-27} cm^{2}gm/sec. Since the energy of a harmonic oscillator of frequency ν is the action times the frequency, Planck's assumption is equivalent to the statement that the energy of a harmonic oscillator of frequency ν can have only the values 0, $h\nu$, $2h\nu$, $3h\nu$, . . . , $nh\nu$, and the action can change only by the amount h. This means that such an oscillator emits and absorbs radiant energy in multiples of $h\nu$. Instead of radiant energy being emitted as a continuous wave (which leads to Rayleigh's law), it is emitted in lumps, and the amount of energy in each lump or packet is proportional to the frequency.

To see physically how Planck's assumption leads to the correct distribution law for black-body radiation [the correct expression for

$B_\nu(T)$ or $u_\nu(T)$] we recall that the flaw in Rayleigh's law is associated with the emission of high-frequency radiation; this law favors large values of ν so that the Rayleigh curve goes to infinity as ν increases. But the actual empirical curve reaches a maximum value and then drops down rapidly to very small values as ν increases. This difficulty is now automatically taken care of in Planck's theory since the emission of very-high-frequency radiation means that very-high-energy quanta are emitted, but high-energy quanta cost too much, in terms of the available energy of the black body, to be emitted, except very rarely. In fact, the very high frequencies (frequencies above a certain high value) cannot be emitted at all because the emission of one such quantum would exhaust all the energy of the black body. With increasing temperature of the black body, higher frequencies can and are emitted because according to the first law of thermodynamics, the energy content of a body increases with increasing temperature. Planck's hypothesis causes the high-frequency end of the distribution curve to bend over and approach zero because, according to Planck's assumption, the emission of high frequency quanta of energy becomes more and more prohibitive from an energetic point of view.

The correct algebraic expression for the energy density $u_\nu(T)$ of black-body radiation can be deduced using Planck's quantum hypothesis but we must find the correct law for the equipartition of energy. The incorrectness of Rayleigh's law stems from the classical law that assigns an average amount of energy kT to each oscillator in an ensemble at temperature T regardless of its frequency of oscillation. But the quantum hypothesis clearly denies this outcome for it tells us that nature curtails high-frequency oscillations. It is just as though the degrees of freedom associated with the high-frequency oscillations were in a sense forbidden to participate in the equipartition by the quantum hypothesis. If we recall, we had to assume something similar when we came up against the discrepancy between the observed and the calculated ratio γ of the specific heats of gases. To obtain agreement between the observed and calculated values of γ we had to assume that the internal vibrational and rotational degrees of freedom of the molecules of a gas at low temperatures play no role (were frozen) in the equipartition of energy. Thus, the germ of the idea of a quantum theory was already present in the 19th century kinetic theory of gases. This idea also indicated (correctly) that the quantum theory would ultimately extend its influence far beyond the domain of black-body radiation and would leave its imprint on all of physics, chemistry, astronomy, and biology. With Planck's discovery the constant h was to appear in all the laws of nature.

Since the classical equipartition theorem (the average amount of energy kT per oscillator) leads one astray, Planck had to find a new equipartition theorem consonant with the quantum hypothesis. Since the energy of a harmonic oscillator of frequency ν can have the energies 0, $h\nu$, $2h\nu$, $3h\nu$, . . . , $nh\nu$, we obtain the average energy of any given oscillator at the temperature T by multiplying each of the energies 0, $h\nu$, $2h\nu$, $3h\nu$, etc. by the number of oscillators respectively that have each of these energies and then summing the whole lot. We then divide this sum by the total number of oscillators and thus obtain the average energy per oscillator. If there were altogether 3221 oscillators and if 1000 of these had no energy at all, 1000 had the energy $h\nu$, 200 had the energy $2h\nu$, 20 had the energy $3h\nu$, and 1 had the energy $4h\nu$, the average energy per oscillator would be (1464/3221) $h\nu$. If this example represented an actual distribution, the average energy per oscillator would be about $\frac{1}{3}h\nu$. But this example is not real since the actual number of oscillators with an amount of energy $2h\nu$ or $3h\nu$ or $nh\nu$ for a given temperature T is determined by the temperature. It was shown by Boltzmann that if the absolute temperature of an ensemble of oscillators is T, the number of oscillators with the energy E is proportional to the expression $e^{-E/kT}$, where e is the exponential and equals the infinite sum $1 + 1 + \frac{1}{2} + \cdots$ and k is the Boltzmann constant. Since $E = nh\nu$ according to the quantum hypothesis, we see that the number of oscillators with frequency ν and the energies $h\nu$, $2h\nu$, . . . , $nh\nu$ are proportional respectively to $e^{-h\nu/kT}$, $e^{-2h\nu/kT}$, . . . , $e^{-nh\nu/kT}$, and these are the numbers that we must use in calculating the average energy per oscillator of frequency ν (the oscillators are at the temperature T); one then obtains $h\nu/(e^{h\nu/kT} - 1)$ for the average. This expression is the quantum equipartition law which must be used instead of the classical law kT. One can show mathematically that for small values of the frequency ν, the quantum equipartition law is very nearly equal to the classical law kT. This result is what we expect because we know that the classical law leads to the Rayleigh radiation formula which is correct for small values of ν. For large values of ν, the quantum law leads to the Wien formula, which is also to be expected. To obtain Planck's radiation formula we now multiply the quantum expression for the average energy of an oscillator by the number of oscillators per unit volume. This procedure gives us the following expression for the energy density $u_\nu(T)$ of black-body radiation of frequency ν and at the temperature T: $u_\nu(T) = (8\pi h\nu^3/c^3)(e^{h\nu/kT} - 1)$. This expression is the famous Planck formula which ushered in the modern quantum theory of radiation and matter. It stands with Newton's law of gravity and Einstein's formula $E = mc^2$ as one of the great milestones in man's search for knowledge and for an understanding of the universe.

When Planck introduced his quantum theory in 1900 it was generally neglected by most physicists who looked upon it more as a mathematical curiosity that happens to agree with the observational data than a correct physical theory. It was repugnant to Planck's contemporaries because as a corpuscular theory it went counter to Maxwell's wave theory which successfully explains such diverse optical phenomena as diffraction, polarization, interference, and refraction. Planck himself was unhappy about the corpuscular aspects of his own theory because he could not see how quantum absorption and emission of radiation can possibly be reconciled with Maxwell's wave theory which he was unwilling to discard. As late as 1911 at the first of a series of important international conferences on physics (the famous Solvay Conference), which was devoted to the quantum theory, Planck opposed the idea that radiation consists of corpuscles. He conceded only that oscillators emit radiation in the form of quanta but he insisted that once emitted, the radiation loses its quantum-like character and reverts to a wave form. He argued that "Maxwell's equations retain their validity in surrounding space, but only at sufficient distance from the oscillator. . . . They must be modified inside the oscillator and in its immediate vicinity."

Einstein, an invited participant at that conference, resisted Planck's semiclassical position and insisted that radiation is not only emitted and absorbed in the form of quanta or bundles but exists in that form as it is propagated through space. In 1905, the same year in which he had submitted his famous paper on the special theory of relativity, Einstein had submitted another paper in which he shows theoretically that if black-body radiation is treated as a perfect gas at a given temperature, then it behaves as though it consisted of molecules of radiation, which Einstein called light quanta (photons) and he deduced that the energy of a photon of frequency ν is $h\nu$. He stated this result as follows: "Monochromatic radiation of small energy density (within the validity range of the Wien radiation formula) behaves in thermodynamic theoretical relationships as though it consisted of independent energy quanta (photons) of magnitude $h\nu$." In the same paper Einstein used this photon concept to explain the photoelectric effect. In 1921 Einstein received the Nobel Prize in Physics for this quantum explanation of the photoelectric effect.

One type of convincing analysis which Einstein used in establishing the validity of the concept of the photon deals with the fluctuation of the energy density of black-body radiation when it is in a state of statistical equilibrium. The phenomenon here is similar to the fluctuations that are known to occur in a material gas according to the kinetic theory of gases. Although any gas in a container shows no macroscopic irregularities, such irregularities do show up when we examine the gas microscopically. The reason for these irregularities is that the number of molecules in a very small volume of the gas does not remain constant but fluctuates from moment to moment. To see why this must be so, we consider an element of volume that is so small that the average number of molecules in it is three. If this volume is taken as our unit volume, the average density of the gas is three molecules per unit volume. But owing to the complete randomness of the molecular motions in the gas, the actual number of molecules in this volume element changes from moment to moment. At some moment five molecules may be in this small volume whereas at some other moment no molecules may be there. Such density fluctuations in a gas, like our atmosphere, affect the light passing through it, resulting in a scattering of the short wavelengths (the blue colors). This scattering phenomenon accounts for the blue color of the sky and the red color of the sun when it is setting.

Einstein reasoned that fluctuations occur in the energy density u_ν of black-body radiations just as they occur in a gas. His analysis shows that these fluctuations consist of two parts: one part is completely in accord with the wave picture of radiation, but the other part cannot be explained classically in terms of a wave picture. It can be understood only in terms of a corpuscular picture, according to which each corpuscle of frequency ν must be assigned an amount of energy $h\nu$. Einstein went on later to show that not only is a photon a packet of energy $h\nu$, but it also possesses an amount of momentum $h\nu/c$. The full significance of this last point was not fully understood until the famous Compton effect was discovered (the reddening of photons when they collide with electrons).

When Planck introduced the quantum energy formula $e = h\nu$ to derive his radiation law, he could adduce no theoretical or experimental reason for it. He only knew that it was justified by the final radiation law that stemmed from it. Einstein went further than Planck with his analysis of energy fluctuations, showing that the photons that constitute black-body radiation behave like particles having energy $h\nu$. We can deduce the formula $e = h\nu$ from the Doppler effect for radiation, if we assume photons exist. The Doppler effect tells us that if radiation is emitted from a source which is either receding from or approaching us (it does not matter whether we picture the observer as fixed and the source as moving or vice versa; only the relative motion counts), the frequency of the radiation is different from what it is when the source and observer are fixed with respect to each other. If the observer and source are approaching each other with a speed v, the frequency of the emitted radiation increases by the amount $\nu(v/c)$ where ν is the frequency as measured by an observer fixed relative to the source and c is the speed of light. The reason for this increase in the frequency for the approaching observer is that the observer travels the distance v toward the source in a unit time. He therefore receives v/λ additional waves. But $\lambda = c/\nu$ so that the number of additional waves he receives per second is $(v/c)\nu$ or $(\nu)(v/c)$, which is the increase in frequency. This Doppler effect is also present when light is reflected from a moving mirror. If before the light is reflected, the mirror is moving with speed v opposite to the direction of propagation of the light, the frequency of the reflected light increases from ν to $\nu + 2\nu v/c$. The factor 2 comes in because the Doppler effect acts when the light impinges on the mirror and when the reflected light leaves the mirror.

We now consider black-body radiation in a cylinder of length L and of unit cross-sectional area (the volume of the cylinder is thus L) with perfectly reflecting walls. If U is the total energy of the radiation and u is the energy density, then $U = Lu$ and if there are N photons in this radiation and e is the energy of a photon then $U = eN$. If we compress the radiation in the cylinder by means of a movable piston at one end of the cylinder, we do work on the radiation and thus increase its energy U. If we move the piston an amount Δx parallel to L, we do the work $P\Delta x$ where P is the pressure of the radiation which is just $\frac{1}{3}u$ so that the amount of work we do is $(u/3)\Delta x$. This quantity is the total increase in the energy of the radiation. Hence, the energy of each photon increases by $(u/3N)\Delta x$. We therefore have $\Delta e = (u/3N)\Delta x$. If we multiply and divide the right-hand side by L and note that $Lu = U$ and $U/N = e$ we obtain $\Delta e = (Lu/3NL)(\Delta x) = (U/3N)(\Delta x/L) = (e/3)(\Delta x/L)$ so that $\Delta e/e = \Delta x/3L$. This expression gives the percentage change in the energy of a photon when we do work on the radiation by compressing it.

We now consider the change in the frequency of the photon arising from the Doppler effect when the photon is reflected from the moving piston. The percentage change in the frequency of a photon (the quantity $\Delta\nu/\nu$) caused by a single reflection is just $2v/c$ if the piston is moving at the speed v and the photon is moving parallel to L as deduced previously. Such a photon traverses the length of the cylinder and returns to the surface of the moving piston in a time $2L/c$ so that it suffers $c/2L$ reflections in a unit time. Thus, the percentage change in the frequency of the photon per unit time, while the piston is moving, is $\Delta\nu/\nu = (2v/c)(c/2L) = v/L$. Since the piston takes a time $\Delta x/v$ to travel the distance Δx, the percentage change in the frequency of the photon is just $\Delta\nu/\nu = (V/L)(\Delta x/V) = \Delta x/L$. But any given photon is traveling parallel to L only one-third of the time so that the percentage change in the frequency is just $\Delta\nu/\nu = \Delta x/3L$. This quantity is exactly the same as the percentage change in the energy of the photon. Hence, the energy of a photon must be proportional to its frequency, which is Planck's relationship.

We saw above that we cannot understand the energy fluctuations that occur in black-body radiation unless, as proposed by Einstein, we accept the photon concept and with it the idea that radiation in a container is not distributed continuously throughout the volume of the container, but is concentrated as photons in small regions. It is entirely possible that some volume elements of the container contain no photons whereas other volume elements contain some photons. To bolster his concept of the photon, Einstein showed that it can explain the photoelectric effect which is quite incomprehensible if one accepts the classical wave theory of radiation. This effect was discovered accidently by Heinrich Hertz (the discoverer of electromagnetic waves) who observed that when an electrically charged metal sphere is bathed in ultraviolet radiation, the sphere loses its charge very quickly. He reasoned that this process occurs because the electric field of the ultraviolet radiation accelerates the free electric charges (electrons) on the surface of metal sphere, thus ejecting them from the sphere and causing the sphere to lose its charge.

Following this discovery, Philipp Lenard, at the beginning of the 20th century, investigated the energy with which the ejected electrons leave the metal sphere and found that he could not obtain agreement between the observed energies and those deduced from classical physics (Maxwell's electromagnetic theory of light). Maxwell's wave theory states that the energy in a beam of light is proportional to its intensity. This statement means classically that if a very bright beam of light is directed against a charged metal surface, the ejected electrons should leave the metal with greater speeds (higher kinetic energies) than they would if a faint beam of identical light were used. But this result does not occur. The speed of the ejected electrons does not depend on the intensity (the energy content) of the incident light but only on its color. The experiments of Lenard show that the speed with which the electrons leave the metal surface increases with the increase in the frequency of the light striking the surface. Blue and ultraviolet light eject electrons at greater speeds than does red light. In fact, if the light is too red, no electrons come out of the metal surface regardless of the intensity of the red light. Blue light knocks out electrons at a certain speed regardless of the faintness of the blue light. Increasing the intensity of this blue light has no effect on the speed of the ejected electrons; it simply increases the number of electrons emitted.

Another discrepancy between classical theory and the observations is found in the distribution of the emitted electrons. According to the wave theory of radiation, the incident light is distributed continuously over the irradiated surface. This distribution means that the electrons should be emitted uniformly from all points of the metal surface. This result does not occur; the electrons are ejected from the metal surface only at discrete points, as though the energy of the radiation were concentrated in discrete regions. This result is incomprehensible on the basis of the wave theory but becomes understandable if photons exist. The photons fall in showers and strike the surface at random points, and each photon ejects a single electron. Einstein simply assumed that when a photon of energy $h\nu$ collides with an electron in the metal surface, it transfers all its energy at once to the electron so that the electron acquires the energy $h\nu$. The kinetic energy of the electron (except for a constant that depends on the nature of the surface) is thus $h\nu$ so that $\frac{1}{2}mv^2 = h\nu$ where m is the mass of the electron and v is its speed. If we solve for v, we see that $v = \sqrt{2h\nu/m}$ showing that the speed of the ejected electron increases with the square root of the frequency of the incident light which gives us Einstein's photoelectric formula. After Einstein had proposed this equation for the photoelectric effect, Robert Millikan checked it experimentally and found complete agreement between observation and Einstein's theory.

Although the quantum hypothesis led Planck to the correct algebraic expression for the intensity of the radiation emitted by a black body in each frequency interval, Planck was loathe to accept the full consequences of this theory, arguing that it applied only to the emission and absorption of radiation but did not alter the wave character of light in between. With Einstein's photon hypothesis it became clear, however, that one may not apply the quantum theory to certain processes and not to others. Einstein's correct explanation of the photoelectric effect shows the great power of the quantum theory and the need for it in the domains of physics involving electric charges as well as radiation. But Einstein went further than this in a subsequent paper where he showed that even in phenomena that are not directly related to radiation one can obtain agreement between theory and observation only if the quantum concept is introduced. In this paper he dealt with the theory of the specific heats of solids, and demonstrated that the quantum theory clears up a serious discrepancy that exists between the classical theory of such specific heats and the empirical law discovered experimentally by Dulong and Petit. The empirical law shows that the specific heat of a mole of any solid is 3R (where R is the gas constant) as long as the temperature of the solid is not too low; but that as the temperature is reduced below a certain value, the specific heat decreases continually, approaching zero as the absolute temperature approaches zero.

To see where this result disagrees with the predictions from classical theory, we recall that when heat is added to a system bringing it to a temperature T each degree of freedom of each particle in the system take up (on the average) the same amount of energy $\frac{1}{2}kT$. In a monatomic gas, each molecule has only the three translational degrees of freedom (since there are no forces between molecules) so that the energy associated with each molecule is $\frac{3}{2}kT$, but in a solid each molecule has potential energy as well as kinetic energy. The potential energy stems from the forces between the molecules that cause the molecules to stick together and thus to form a solid. When heat is added to the solid each molecule vibrates about a mean position as though there were elastic forces constantly drawing the molecules back to their centers of oscillations (mean positions). The energy of each molecule thus consists of a kinetic part and a potential part so that each molecule has six degrees of freedom instead of three as in a perfect gas. The average energy associated with each molecule in a solid is (according to classical theory) $\frac{6}{2}kT$ or $3kT$. If N_0 is Avogadro's number (the number of molecules in a mole), then the total internal energy of one mole at temperature T is $3N_0kT$ or $3\text{R}T$ since $N_0k = \text{R}$. The heat required to change the temperature of one mole of any substance by one degree (the specific heat) is just 3R, regardless of the initial temperature of the substance. In other words, classical theory predicts that the molar specific heat of any solid is 3R at all temperatures. This fact agrees perfectly with the empirical law of Dulong and Petit at ordinary and high temperatures but it disagrees with this law at low temperatures. We again have a definite contradiction between classical theory and experiment, and we recall that another such contradiction exists in the theory of the ratio of the specific heats of gases. We now show how the quantum theory, as first explained by Einstein, eliminates these discrepancies between theory and observation.

Einstein pointed out that if a solid is treated as a collection of particles oscillating about their mean positions, then such particles are equivalent to harmonic oscillators and we must, according to the quantum theory, assign to each of these oscillators an average amount of energy equal to $(h\nu)/(e^{h\nu/kT} - 1)$ instead of $3kT$ as given by classical theory. With the quantum expression for the average energy of each molecule one can show that the total energy of a mole is $(N_0h\nu)/(e^{h\nu/kT} - 1)$ and the heat required to raise the temperature of one mole by one degree (the molar heat) depends on T, going to zero as T goes to zero so that the specific heat does indeed go to zero as T goes to zero. The reason for this outcome physically is that the smallest amount of energy (other than zero) that a molecule (treated as an oscillator) inside a solid can have is $h\nu$. If the temperature of the solid is low, the energy content of the solid may be so low that not enough energy is present to excite the oscillations of frequency ν. The solid then behaves as though these degrees of freedom were not present at all—they take up no energy so that their contribution to the specific heat is zero.

The ratio of specific heats of gases cannot be deduced correctly from classical theory but can be explained by the quantum theory. If a molecule in a perfect gas has n degrees of freedom, then classical kinetic theory predicts that the ratio of specific heats γ should be $(2/n) + 1$. For high enough temperatures this is so but for low temperatures the ratio is what one would obtain if the molecules of the gas consisted of single atoms with no internal degrees of freedom. This is incomprehensible from the classical point of view, but is easily understood with the quantum hypothesis. Since each degree of freedom of a molecule associated with rotation or vibration represents action and action is quantized, a minimum amount of energy is required to excite any rotational or vibrational degrees of freedom of the molecule. At low temperatures, therefore, the energy in the gas is too low to excite these internal degrees of the molecule so that the molecule at these temperatures behaves as though it had only the three translational degrees of freedom. The quantum theory thus makes itself felt in the theory of gases.

We complete this chapter by emphasizing the revolutionary character of the quantum theory and the profound effect it has had on science and on our concepts of the structure of matter in general. Planck himself was aware as early as 1900 of the profound effects his theory was to have on science when he said to his son: "Today I have made a discovery as important as that of Newton." Later, in his retrospective writings, he states that he knew the significance of his discovery very early when, try as he might, he could not fit the quantum of action h into the frame of classical physics: "The failure of all attempts to bridge this gulf soon removed all doubt that the quantum of action plays a fundamental part of atomic physics and that with its appearance a new epoch of physical science has begun. For it bodes of something unheard of destined to reform thoroughly our physical thinking which, since the invention of infinitesimal calculus through Leibnitz and Newton, was based on the assumption of the continuity of all causal relations."

All the variegated structures of matter that we see today are possible because of the finiteness of h. If this constant were infinitesimal or zero, electrons could not circulate in stable orbits around the nuclei of atoms and chemical reactions could not occur. The periodic properties of the chemical elements exist only because of h, and molecules owe their existence to h. The importance of the quantum of action h for life processes is indicated by the continuity and the remarkable stability of the gene. If action were not quantized, even small environmental changes would change the genetic structure, but the existence of a quantum of action means that genes retain their structures unless enough high energy (such as a high-energy photon) is absorbed to disrupt this structure. In short, genes can only change their structures discontinuously and not gradually.

CHAPTER 15

The Electromagnetic Theory

We may therefore regard matter as being constituted by the regions of space in which the field is extremely intense. . . . There is no place in this new kind of physics for both the field and matter, for the field is the only reality.

—ALBERT EINSTEIN

In our discussion of the properties of radiation we noted that we can learn something about the structure of matter by carefully investigating how matter and radiation interact. Max Planck arrived at his quantum theory by picturing matter as consisting of harmonic oscillators which are set vibrating by radiation. As these oscillators vibrate, they are also pictured as emitting radiation having a frequency equal to the frequency of their vibrations. According to this picture, matter is constituted of electric charges which vibrate about some force center in response to the oscillating electric fields of the radiation striking the matter. Although this is a very rough picture, it tells us that matter and radiation do interact electrically and that we must have a correct theory of this interaction if we are to deduce something about the structure of matter from the analysis of the radiation emitted or absorbed by matter. To arrive at such a theory we first study the basic properties of electric charges.

Since matter in bulk is electrically neutral, we know that equal quantities of two opposite kinds of charge which we may call positive and negative exist in nature. For each unit of negative charge in an electrically neutral bit of matter a unit of positive charge is present and these two opposite charges exert an attractive force on each other to give the cohesive properties of matter.

That electric charges exist was already known to the ancient Greeks who had discovered that one can electrify a body by rubbing it with various kinds of substances. Thus, if a rod of ebony is rubbed with a piece of fur, the ebony and the fur both become electrically charged but with opposite charges. This fact is evident from the attraction between the fur and the ebony. If two pieces of ebony are rubbed with two separate pieces of fur, the two pieces of ebony repel each other as do the two pieces of fur, but each piece of ebony attracts each piece of fur. This fact shows us that like charges repel each other, whereas unlike charges attract.

The ancient Greeks investigated only slightly the nature of electric charges and the nature of the forces between them. The basic properties of charges and their interactions were not discovered until the 18th century when people like Cavendish in England and Coulomb in France measured the pull or push of one electric charge on another. Even without careful, detailed experiments we can easily see that the forces exerted by electric charges on each other are many times greater than the gravitational pull of small masses on each other. To see this we need only place some bits of paper on a tabletop and then bring an electrically charged ebony rod close to them. The bits of paper are quickly drawn to the charged rod, showing that the small electric charge on the rod exerts a greater electric pull on the bits of paper than the gravitational pull exerted by the entire mass of the earth.

Although the exact form of the law of force between electric charges was discovered independently by Cavendish and Coulomb, the law is known as "Coulomb's law" because Cavendish did not publish his discovery. Coulomb's law is expressed in the form of an algebraic formula that gives the force between two like electric charges q_1 and q_2 in a vacuum separated by a given distance r. Coulomb found that charge q_1 repels charge q_2 along the line connecting the two particles and q_2 repels q_1 with exactly the same force along the same line.

Before discussing Coulomb's law of force and giving its exact algebraic form, we consider the nature of electric charge and confront the basic question: "What is electric charge?" We are no more able to answer this question today than was Cavendish or Coulomb. We do not know what electric charge is any more than we know the precise nature of mass; we can only describe how electric charges behave and how the quantity or magnitude of a charge can be measured. We know that electric charges (on particles thus far observed) occur as integer multiples of a basic elementary charge [the charge on the electron (if negative) or on the proton (if positive)] and that electric charge does not exist by itself but always in conjunction with mass. Moreover, all elementary particles that have nonzero rest mass (can be brought to rest) have electric charge so that rest mass implies charge and charge implies rest mass. The only thing we can say about the nature of electric charge at this point is that it attracts an unlike charge (charge of the opposite sign) and repels a like charge (charge of the same sign). We shall see that electric charges can give rise to magnetic effects and electromagnetic waves (radiation), but right now we limit our discussion to the purely electrostatic properties of charge. Recent discoveries of the composite nature of the proton and neutron have changed our concept of the elementary charge because the constituent particles within the proton and neutron (the quarks) must have fractional multiples ($+\frac{1}{3}$ and $-\frac{2}{3}$) of the electron charge. If the quarks are the basic constituents of matter, then the question as to which charge is elementary is ambiguous. Although we cannot describe the nature of charge precisely, we can say how to determine the magnitude of a given charge. Before we do this, however, we must introduce labels for the two different kinds of charges. The most convenient convention is to call one kind of charge negative and label it with a minus sign (−) and call the other kind of charge positive and label it with a plus sign (+). This kind of labeling is useful because when we take the same quantities of the two kinds of charges and place them together, the effect is as though no charge were present at all. They cancel each other just the way a negative integer cancels its positive counterpart when they are added ($-3 + 3 = 0$). With this convention for labeling the two kinds of charges, we can carry out all the arithmetic operations with charges (e.g., addition, multiplication) that we do with other physical quantities such as mass, speed, energy, and so on. Electric charge, like mass, is a scalar quantity and not a vector. It does not matter which of the two kinds of charge we label negative and which we label positive, but once we have labeled the charges in a particular way we must stick to those labels. Physicists have adopted the following sign convention: The charge left on an ebony rod when it is rubbed with cat's fur is called negative and the charge on the cat's fur is called positive. With this convention we find that a glass rod acquires a positive charge when rubbed with silk and the silk acquires a negative charge. Thus, the charged ebony rod attracts the charged glass rod and the silk attracts the fur.

Coulomb's law of force for electric charges has the same algebraic form as Newton's law of gravity, which is not at all surprising since a point of charge produces a spherically symmetric electric field just the way a mass-point produces a spherically symmetric gravitational field. We thus have the inverse square law in both cases. Using the same arguments as we did in setting up Newton's law of gravity we see that the electric force depends on the product of the two electric charges involved. We have the following formula for the force between charges q_1 and q_2 separated by the distance r (as discovered by Coulomb): $F = q_1q_2/r^2$. This formula gives the strength of the electric force between two point charges of magnitudes q_1 and q_2. The direction of this force is along the line connecting the two charges. From Coulomb's law we see that electric charge has the dimensions of dynes$^{1/2}$ cm or gm$^{1/2}$ cm$^{3/2}$ sec^{-1}. The unit of electric charge can be introduced by applying Coulomb's law of force to two like charges which are separated by one centimeter and repel each other with a force of one dyne. If each of these charges has the magnitude q, Coulomb's law gives us (placing $F = 1$ and $r = 1$) $1 = q^2/1$ so that $q = +1$ or -1 (unit of positive or negative charge). The unit electric charge (positive or negative) is the quantity of charge which repels an equal charge with a force of one dyne when the two charges are one centimeter apart. This quantity of charge is called one electrostatic unit (e.s.u.). The charge on the electron is 4.77×10^{-10} e.s.u. It is interesting to calculate the electrostatic charge that would have to be distributed uniformly over the surface of the earth to support electrostatically a body against its own weight (the electrostatic force to equal the weight of a body) if there were one unit of charge on the body. We can do this calculation by noting than an electric charge distributed uniformly over the surface of a sphere behaves as though all the charge were concentrated at the center of the sphere. Since the weight of a body at the surface of the earth equals GMm/R^2 where M is the mass of the earth, R is its radius, and m is the mass of the body, then if q is the charge that must be placed on the surface of the earth to support the body against its weight, we must have (the q for the body is 1) $q/R^2 = GMm/R^2$ so that $q = GMm$. If the body has a mass of one gram, the surface of the earth must have a charge equal in e.s.u. to 6.7×10^{-8} times the mass of the earth to support the body against its own weight. The electrostatic effect of a unit electric charge is much greater than the gravitational effect of a unit of mass (a gram). The electrostatic force is thus a much larger force than the gravitational force. We may therefore expect to find that the structures that stem from the electric force are much smaller than those that stem from the gravitational force under ordinary circumstances.

In our discussion of the gravitational force we introduced the concept of the gravitational field to avoid the conceptual difficulty associated with action-at-a-distance. When Newton introduced his law of force, he found the idea of action-at-a-distance repugnant but he had no reason to reject it on that account. There was, however, a practical reason for introducing the gravitational field concept: it simplified enormously the mathematical techniques required to solve gravitational problems. Since the gravitational force is a vector quantity, working directly with the mutual gravitational forces of a group of interacting masses (the instantaneous forces at distances) means working with groups of vectors, which leads to considerable mathematical complexities. The gravitational field, on the other hand, can be defined in terms of the gravitational potential which is a scalar so that the mathematics of fields is much simpler than that of forces.

Though the choice of working directly with forces or with fields was a matter of taste or convenience before the 20th century, the choice was eliminated from physics with the development of the theory of relativity. We are now forced to work with fields because action-at-a-distance violates the relativistic concept that no physical effects can be propagated at speeds greater than the speed of light. Action-at-a-distance means the instantaneous propagation of the action of one body on another, whereas the field concept pictures the propagation of field effects from point to point or locality to locality in the field.

Although the theory of relativity now mandates that we describe the electric force of one charge on another in terms of an electric field, the electric field was, from its very inception, much more of a physical entity to scientists than the gravitational field. Much of this acceptance of the reality of the electric field was due, first of all, to Michael Faraday, who laid down the experimental foundation of electricity and magnetism, and, finally, to James Clerk Maxwell, who discovered the mathematical properties of the electromagnetic field.

To introduce the concept of the electric field, we consider the points in the space surrounding a point electric charge q and measure the force exerted on a unit charge when it is placed at any one of these points. Instead of considering the charge q as exerting a force on the unit charge, we remove our attention from q and focus it on the various points surrounding q. We now picture a vector physical entity (the electric field) which varies in magnitude and direction from point to point and which exerts a force on any electric charge placed at any point. The magnitude of the electric field at a point equals the magnitude of the force exerted on a unit charge placed at that point and the direction of the field is given by the direction of the acceleration of the charge. Using the unit charge we can determine the direction and magnitude of the field at each point and represent the field at any given point by an arrow having the appropriate length and direction. The totality of these arrows defines the electric field **E**.

A point charge, q, exerts the force q/r^2 on a unit charge placed at a distance r from it. Hence, the magnitude of the electric field stemming from a point charge q is q/r^2 at any point whose distance from q is r. The field in this case is radial; it is always directed along the line from the point to the charge. If **E** is the electric field at a point, then the force of the field on a charge Q placed at that point is $Q\mathbf{E}$. If we have a number of charges $q_1, q_2, \ldots, q_n$, the total electric field **E** arising from all of these charges is the vector sum of the individual fields so at any point $\mathbf{E} = \mathbf{E}_1 + \mathbf{E}_2 + \mathbf{E}_3 + \cdots + \mathbf{E}_n$. Electric fields are additive. If we wish to find the force exerted on a charge Q placed at any point in a region of space in which there is a distribution of electric charges, we first calculate the field of each charge at that point and then add all of these fields vectorially. If this vectorial sum is **E**, the force of the field on Q is $Q\mathbf{E}$. Although the above procedure is a precise and feasible way for determining the field stemming from a group of discrete charges, it is cumbersome mathematically because it involves vector additions. One can avoid this problem by defining the electric field in terms of the electric potential concept.

In our discussion of the gravitational field we saw that we can describe it in terms of a scalar function at each point called the gravitational potential at the point. The magnitude and direction of the field at each point can then be deduced from the potential by definite mathematical procedue. We proceed in the same way with the electrostatic field and introduce the electrostatic potential at a point in the electrostatic field as the work required to move a unit positive electric charge from infinity to the point in question. Thus, potential is work per unit charge or ergs per unit charge. If the potential at a point is ϕ, the work required to conduct a charge Q from the point to infinity is $Q\phi$.

To obtain the potential at a point in the electric field surrounding a point charge $+q$ or $-q$, we use Coulomb's law, noting that the magnitude of the field at a distance r from the charge is $+q/r^2$ or $-q/r^2$. Since this quantity is just the force exerted on a unit charge at a distance r from q, we can calculate the work done when we move the unit positive charge closer to q by a small amount Δr (the distance of the unit charge from q is reduced by the amount Δr so that it changes from r to $r - \Delta r$) by multiplying q/r^2 by Δr. If we sum all such contributions to the total work we do in transporting the unit charge from infinity to r, we obtain $+q/r$ or $-q/r$. This expression gives the electrostatic potential at the point of the electrostatic field stemming from the charge $+q$ or $-q$. This quantity, which is a scalar, is positive if q is positive and negative if q is negative. It changes from point to point, and decreases as one recedes from the charge q.

If the electric field arises from a collection of charges $q_1, q_2, q_3, \ldots, q_n$ at n different points, the potential at any point in the field is the sum of the potentials of all the charges. Thus, if q_1 is at the distance r_1 from the point, q_2 is at the distance r_2, etc., then $\phi = q_1/r_1 + q_2/r_2 + \cdots + q_n/r_n$. If the potential at the point A is ϕ_A and the potential at the point B is ϕ_B, a positive charge left to itself moves from A to B if ϕ_A is larger than ϕ_B, and the work required to take a unit charge from B to A is just the potential difference $\phi_A - \phi_B$. To take a charge Q from B to A we must do an amount of work $Q(\phi_A - \phi_B)$. We see that the potential concept is useful in computing work done in moving charges about or in computing the work (or energy) we can get from electric charges when they move about in an electric field.

The concept of the potential at a point is closely related to the potential energy of a charge at a point in an electric field. We must be careful here not to confuse potential with potential energy. The potential is a property of the field alone and is energy per unit charge whereas potential energy is the property of a charge placed at some point in a field. Thus, the potential energy of a charge Q at a point where the potential is ϕ is $Q\phi$. This quantity is energy and equals the work required to transport the charge Q from infinity to the point in question.

Consider now all the points in an electric field which are at the same potential so that no work is required to shift a charge from one of these points to any other one of them. This continuum of points lies on a surface which is called an equipotential surface. A charge placed on such a surface has no tendency to move along the surface, but only at right angles to the surface, since its potential energy at each point of the surface is the same. Since a charge at any point on an equipotential surface tends to move along a line that is perpendicular to the surface, we see that the electric field at that point must be perpendicular to the equipotential surface since an electric field along the surface would pull the charge. If we know the equipotential surfaces in an electric field, we know the direction of the electric field. The lines of force of the electric field must be perpendicular to each equipotential surface. The equipotential surfaces surrounding a point charge are concentric spheres whose common center is the point charge.

If we know the value of the potential ϕ at a given point P in an electric field as well as the values ϕ at points in the neighborhood of the given point (close to the given point), we can then calculate the magnitude of the electric field at the given point and its direction. In other words, we can find E at the given point if we consider any point P_1 close to P and consider the values of ϕ at these two points. If $\phi(P)$ is the value of ϕ at P and $\phi(P_1)$ is its value at P_1, we take the difference between these two values, namely $\phi(P_1) - \phi(P)$; if we divide this amount by the tiny distance ΔX between P and P_1, we obtain $[\phi(P_1) - \phi(P)]/\Delta X$. This is the component of the electric field at P in the direction from P to P_1. If we do the same thing with the two other points P_2 and P_3, choosing these points so that their directions from P are perpendicular to each other as well as to the direction of P_1 from P, we obtain three mutually perpendicular components of **E** at P. If we square these three components and add the squares, and then take the square root of the sum of the squares, we obtain the magnitude of the electric field at P. To obtain the direction of the electric field at P we consider the difference $\phi(P_n) - \phi(P)$, where P_n may be any point very close to P. As we shift from one point P_n to another, the difference $\phi(P_n) - \phi(P)$ changes, but there is always one point P_n for which this difference is a maximum. The direction from P to this point P_n is the direction of the electric field **E** at P. In other words, the direction of the electric field at any point is the direction along which ϕ decreases most rapidly as we move away from P. One and only one such direction exists for the field at any point P and this direction (the direction of the field) is perpendicular to the equipotential surface that passes through P. Note that one and only one equipotential surface passes through any point. Two different equipotential surfaces never intersect.

A convenient way to picture the electric field is by lines of force which we imagine as emanating from electric charges like quills from a rolled-up porcupine. The direction of the field at any point is then the direction of the tangent to the line of force at that point, and the strength of the field is proportional to the number of lines piercing a unit area of the equipotential surface passing through the point. Lines of force are always perpendicular to equipotential surfaces.

Along with their discovery of static electricity the Greeks had also discovered the magnetic properties of the lodestone, which finally led to the concept of the magnetic pole. The Greeks had learned that certain pieces of iron ore (the lodestone) align themselves in a north–south direction when suspended vertically by a string. Solids like the lodestone are called magnetic substances, of which iron is the best example. The ability to become magnetized is not a property common to all substances but limited to a few metals like iron and nickel and to certain alloys that can be made from these metals. A bar of metal like iron that is magnetized is called a bar magnet. When it is placed in a horizontal position so that it can rotate freely about a vertical axis, it aligns itself approximately along the north–south direction. The end of the magnet that points northward is called the north pole (N) of the magnet, and the other end of the magnet is called the south pole (S). When two bar magnets are brought together, they repel each other if they are arranged with like poles next to each other, but they attract each other if they are aligned with opposite poles next to each other. This simple experiment shows us that like magnetic poles repel each other whereas unlike magnetic poles attract each other. Just as a law of force governs the interaction between electric charges, so does a similar law of force govern the interaction between magnetic poles, which was also discovered by Coulomb. The law of force between magnetic poles is of the same form as the law of force between electric charges.

If M_1 is the strength (quantity of magnetism) of a given magnetic pole and M_2 is the strength of another pole, and if these two poles are separated by the distance r, then the force between the two poles is given by the formula $F = M_1M_2/r^2$. This expression is called Coulomb's law of force for magnetic poles. The direction of the force between two magnetic poles is along the line r connecting the two poles. A unit magnetic pole is one which, when placed one centimeter from an equal and like pole, repels it with a force of one dyne.

Although magnetic poles are the sources of magnetic fields just as electric charges are the sources of electric fields, an important distinction between poles and charges exists which introduces an asymmetry in their behavior. Whereas electric charges exist independently of other charges, this is not true of magnetic poles. A negative electric charge can exist all by itself as can a positive electric charge. In other words, negative and positive electric charges have their separate existences, but magnetic poles do not. A north magnetic pole cannot be separated from a south magnetic pole; north and south magnetic poles always come together at the opposite ends of the same magnet, and the two poles are of exactly the same strength. If a bar magnet is cut in half or if the north end of a magnet is cut off, we obtain two magnets, each with its own north and south poles. Moreover, the poles of the two magnets are equally strong.

Since the two poles of a magnet cannot be severed from each other, we always deal with pairs of magnetic poles. Such a pair is called a dipole, so that a bar magnet is called a dipole magnet. An important property of a dipole magnet is a quantity called its magnetic moment. If the length of a bar magnet is L and if M is the strength of its pole, then the magnetic moment of this magnet is ML.

Just as we can describe the interaction between electric charges via the electric field, we can describe the interaction between magnetic poles via the magnetic field. The magnetic field $\mathbf{H}$ (a vector field) surrounding a magnetic pole can be described at each point as the force exerted on a unit magnetic pole placed there, just as we defined the electric field $\mathbf{E}$. To describe the magnetic field we introduce magnetic lines of force, which we picture as emanating from a magnetic pole just the way electric lines of force emanate from electric charges. Note that whereas electric lines of force can extend to infinity and need not end on some other electric charge (electric lines of force need not form closed loops), magnetic lines of force cannot go to infinity; they must loop around and end on an opposite pole because poles come in opposite pairs. The magnetic field strength at a point in a magnetic field is defined as the number of magnetic lines of force piercing a unit area at right angles to the lines of force at that point. The lines of force of a magnet are much more crowded together at the north and south poles than elsewhere, showing that the field strengths near the poles are greater than at points more distant from the poles. One can actually demonstrate the lines of force in a magnetic field by spreading iron filings around a magnet. These filings align themselves along the lines of force which constitute the magnetic field between a north pole and a south pole.

How does a bar magnet that is free to rotate behave in a constant magnetic field? If H is the strength of the magnetic field and M is the pole strength of the magnet, then the field exerts the force MH on the north pole of the magnet and this force has the same direction as the field. The south pole experiences the force $-MH$ in the opposite direction. These two equal but opposite forces acting on the magnet constitute a torque and hence exert a turning action on the magnet, forcing the magnet to align itself parallel to the magnetic field. This turning is exactly what happens to a compass needle which aligns itself parallel to the earth's magnetic field. A magnet can also align itself in exactly the opposite direction but that alignment is unstable. If the length of the magnet is L and its axis (the line from the N pole to the S pole) makes an angle ϕ with the magnetic field, then the torque (force on each pole times the perpendicular distance between the two polar forces) acting on the magnet equals $MHL \sin \phi$. But ML is just the magnetic moment μ of the magnet so the torque $= \mu H \sin \phi$. If the magnet is free to rotate, this torque causes the magnet to turn so that its axis is parallel to the magnetic field and it remains in that position. But what happens when the magnet is spinning around its own axis so that it has an intrinsic angular momentum? In that case, owing to the principle of conservation of angular momentum, the axis of the magnet does not align itself along the magnetic field, but instead the magnet precesses with a definite angular velocity ω about the magnetic field. If a torque $\mathbf{T}$ is exerted on a rotating body and the angular momentum $\mathbf{J}$ of the body makes an angle ϕ with the force field that causes the torque, then the body precesses (like a spinning top) with an angular velocity $\omega = T/J \sin \phi$. The angle ϕ does not change, but the torque $T = \mu H \sin \phi$ so that $\omega = \mu H/J$. This formula is useful when we analyze the behavior of a spinning charge (electric) placed in a magnetic field.

Although current electricity (the flow of electric charges) was discovered by Benjamin Franklin in his famous kite experiment, where he showed that a lightning bolt causes a surge of electricity to move along a string, the serious study of this subject began when the Italian professor of anatomy Luigi Galvani, in 1791, accidentally discovered that the muscle of a wet frog's leg, lying on a metal plate, contracted when he touched the nerve of the leg with a metal scalpel, creating the first crude electrolytic cell. Galvani wrongly concluded that his contraction was due to the "animal electricity" in the frog's leg but the Italian physicist Alessandro Volta correctly explained the phenomenon. Volta demonstrated that the flow of electricity in the frog's leg was caused by the contact of the chemical solution in the frog's leg with the two different metals. Volta showed that he could obtain a constant flow of electricity by placing two different kinds of metal (copper and zinc) into a weak acid. In this way Volta constructed the first electric cell (the voltaic cell or pile) and launched the subject of current electricity on its remarkable development.

The study of current electricity by itself could not have led to modern electromagnetic technology, which is based on the relationship between electricity and magnetism; without the knowledge of this relationship we would not now have the remarkable variety of electric and electronic devices that characterize our age. Again we owe the discovery of electromagnetism to a happy accident. In 1819 Christian Ørsted, a Swedish scientist, discovered that when a compass needle is placed near a wire carrying an electric current, the needle always aligns itself perpendicular to the wire. Ørsted correctly concluded that this rotation was due to the torque exerted on the compass needle by the magnetic field generated by the current in the wire. Since the compass needle (a small bar magnet) sets itself at right angles to the wire, we conclude that in any plane perpendicular to the wire the magnetic lines of force of the magnetic field created by the current are concentric circles whose centers coincide with the point where the wire pierces the plane. The direction of field is given by the right-hand rule which states that if the thumb of one's right hand is pointed in the direction of the current, the fingers of this hand, when curled, point in the direction of the field. With this discovery many new avenues of research and technology were opened up to scientists. The electromagnet is one example of what can be done with electricity; if a current is sent through a coil, one obtains the equivalent of a bar magnet whose polarity is again determined by the right-hand rule. If one arranges his right hand around the coil so that his fingers point in the direction of the current in the coil, the thumb points in the direction of the north pole of the equivalent bar magnet. One greatly enhances the strength of a coil magnet by placing a bar of soft iron in the coil, thus obtaining an electromagnet, various kinds of which play such an important role in electromagnetic technology. If a piece of steel is placed in a coil magnet, the steel becomes permanently magnetized.

Since coils of wire carrying electric currents behave like bar magnets, such coils repel or attract each other depending on their relative orientations. Moreover, when such coils are placed in an electromagnetic field, the fields exert torques on the coils, causing them to rotate. This phenomenon is the basis of the electromagnetic motor. Indeed, an electric motor is nothing more than a coil of wire wound around some soft iron placed between the poles of a magnet. When electric current is sent through the coil, the magnetic field of the magnet turns the coil. Most electrical measuring instruments such as ammeters and voltmeters are based on this principle.

Electric charges can be made to flow along metallic wires if these wires are attached to metal plates immersed in an acid like sulfuric acid. Such a flow of charges is called an electric current. Although only the negative charges move in a current (the positive charges remain fixed and form the metallic lattice), the direction of the current is defined as opposite to that of the motion of the negative charges. This definition is entirely a matter of convenience.

The electrostatic unit of current is defined as a flow of 1 e.s.u. of charge per second across any plane at right angles to the wire. The current in any part of the same wire is the same since electric charge cannot pile up at any point along a wire. If n is the number per unit volume of freely moving units of charge in a wire of unit cross section and v is the velocity of each charge, the current is vn. If each freely moving charge carries the charge e, then the current is evn. If the cross-sectional area of the wire is A, the current is $Aevn$.

Not all substances conduct charges equally well. Those substances, like metals, which allow charges to flow through them readily, are called conductors. The other substances, like glass, are called nonconductors or dielectrics. There is no sharp division between conductors and nonconductors. Under certain conditions (e.g., glass heated to high temperatures), nonconductors become conductors.

Since a conductor is a substance in which the negative electric charges (the electrons) are free to move about and are free to respond to any forces acting on them, these charges arrange themselves on the surface of a conductor in such a way that there is no net force on any one of them. The surface of a conductor is always an equipotential surface since charges on such a surface have no tendency to move along the surface. If we consider a region completely surrounded by a metal surface and place an electric charge on this surface, the charge immediately distributes itself over the outer surface of the conductor so that the surface becomes equipotential. Outside this surface there is an electric field arising from the charges, but there is no electric field inside the metal conductor itself or within the hollow region surrounded by the conductor. All points within the conductor and on the inner surface of the conductor are also equipotential. They are at the same potential as the points on the outer surface of the conductor. The points in the hollow region surrounded by the conductor are all at the same potential (at the potential of the conductor itself). Hence, there is no electric field inside a region surrounded by a conductor because if a field were present, any line of force in this field would have to be perpendicular to the inner surface of the conductor and it would have to point in the direction of diminishing potential. Such a line of force would have to be perpendicular to some point B on the inner surface (where we may picture it as originating) and also to some point A (where we may picture it as terminating). But A would have to be at a lower potential than B, which contradicts the fact that the inner surface of the conductor is equipotential. There can be no charge on the inner surface of a conductor no matter how much charge is piled up on the outside surface.

If we have a region in space in which there is an electric field, a charge placed at any point in this field experiences a force. If we now surround the charge by a conducting surface, the charge experiences no force. A conducting surface is thus a shield against electric fields on the outside. If you are inside a hollow conductor, no electrical phenomena outside the conductor can affect you. Hence, you are safe from lightning inside all metal automobiles, airplanes, and office buildings.

The surface of a conductor is an equipotential surface because electric charges on such a surface are free to move about and to respond to any forces. We can have two metal surfaces at different potentials by placing different quantities of charge on the two surfaces (we charge one surface negatively and the other one positively, for example) or by placing the surfaces in different regions of a given electric field. If we connect the two surfaces by a conductor (a metal wire), charges flow from one surface to the other (from the surface that is at the higher potential to the one at the lower potential) until the two surfaces are at the same potential. The difference in the electric potential between two points on a wire drives electric current from one of these points to the other. If we want current to flow through a wire (a circuit) we must attach one end of the wire to a metallic surface that is at a high potential and the other end to a surface that is at a lower potential; this phenomenon occurs when you plug the prongs of an electric wire into one of the electric outlets in your house or when you press an electric button or throw a switch.

When two different metal electrodes such as copper and zinc are placed in sulfuric acid, the chemical interactions of these metals with the acid raise them to different electric potentials. Hence, current flows when these electrodes are connected by a conductor. As long as a difference of potential is maintained between the two endpoints of the conductor, current flows along the conductor. A potential difference must exist between any two neighboring points of a conductor; the closer the two points are to each other, the smaller is the potential difference between them. If we now mark off a series of points on the conductor and add up all the potential differences between successive points as we move from one end of the conductor to the other, the sum is the total potential difference between the endpoints of the conductor.

The study of the flow of steady (unchanging in time) currents through circuits is based on a very simple law known as Ohm's law and on some simple conservation principles. To state Ohm's law we introduce first the unit of potential difference and then the concept of the electrical resistance of a conductor. One electrostatic unit (e.s.u.) of potential difference exists between two points if 1 erg of work is done by the electric field in moving 1 e.s.u. of charge from the point of higher potential to that of lower potential. If the field moves Q e.s.u. of charge between 1 e.s.u. of potential difference, the work done is Q. In practical applications such as in household electrical appliances a much smaller unit of potential difference—called the volt—is used. The volt is 300 times smaller than 1 e.s.u. of potential difference so that 1 e.s.u. of potential difference = 300 volts.

It is also convenient in practical work to replace the e.s.u. of charge by a much larger unit called the coulomb. We define one coulomb as equal to 3×10^9 (3 billion) e.s.u. of charge. Using the coulomb as our basic unit of charge we now define the practical unit of current in a wire as the flow of one coulomb per second through any cross section of the wire. We call this unit of current the ampere (the amp).

We now consider a conductor of a given length across the ends of which there is a potential difference of voltage of 1 volt, and we measure the flow of current, using a current measuring device called an ammeter. Let this current in amps be I. If we now double the potential difference or voltage between the two ends of the wire (we use 2 volts), the current is $2I$. The reason is clear if we recall that the potential difference measures the work done on the charges in the wire by the electric field. As the potential difference or voltage is increased, more work is done on the charges so that they acquire more kinetic energy and move faster. But this means that the current is larger because current is proportional to the speed of the charges in the wire. If we double this speed, which happens when the voltage is doubled, we double the current. We see that current (I) is proportional to voltage (V). But this relationship is only part of the story because I also depends on the chemical nature and on the dimensions of the wire since all substances (more or less) offer some resistance (which depends on the nature and dimensions of the substance) to the flow of an electric current. If we lump all of these various characteristics which influence the strength of the current for a given potential difference into a single physical quantity which we call the resistance R of the wire, we find that the larger R is, the smaller is the current flowing through the wire. The magnitude of the current I depends directly on V (the potential difference between the ends of the wire) and inversely on R (the total resistance of the wire from one end to the other). The dependence of I on V and R is as follows: $I = V/R$ or $V = IR$. This expression is known as Ohm's law and it is the basis of electrical circuit theory. The physical dimensions of V, I, and R are as follows: V is expressed in ergs per charge, I is expressed in charge per second, and R is expressed in ergs seconds per charge squared. The practical unit of R is the ohm. A circuit has a resistance of 1 ohm if a potential difference of 1 volt produces a 1-amp current. The voltage between any two points of a circuit in which a current I is flowing is also called the IR drop between the points, where R is the resistance between the two points.

When an electric current flows, a simple algebraic relationship exists between the potential difference (voltage), the current (amperage), and the resistance R (ohms). Since a current is a flow of charge between points at different potentials, such a current is a source of available energy just the way water is that flows downstream or over a precipice. If I is the current in our circuit and V is the voltage, then I is the quantity of charge transported every second along the wire between a potential difference V. But since charges times potential difference is energy, we see that this current supplies us with an amount of energy IV every second. This quantity is called the electric power of the circuit, where power is defined as energy per unit time. The unit of power, which is called the watt, is defined as energy supplied per second by a current of 1 amp flowing across a potential difference of 1 volt. A 1-watt current supplies 10^7 ergs/sec, but since 10^7 is too large a number to work with conveniently in practical applications, we introduce this number of ergs as a new unit of energy, called the joule: A calorie is about 4.2 joules. Thus, a 1-watt current supplies 1 joule of energy every second. The energy supplied by an electric current in a time t is IVt (expressed in joules or watt seconds); this quantity is what you pay for when you pay your electric bill. Since very many watt seconds are supplied to a typical electrical appliance in your home when you operate it even for a short time, your electric company expresses the total energy supplied to you per month in kilowatt hours, where one kilowatt hour is 3,600,000 watt seconds (1 amp flowing across a potential difference of 1 volt for 3,600,000 seconds gives this amount of energy).

The final form in which the electrical energy is supplied to you depends on the kind of electrical appliance you are using. If it is a heater or a toaster, you want the final energy in the form of heat; if it is a motor, you want mechanical energy; if it is a lighting fixture or a TV set, you want light; if it is a magnetic device, you want magnetic energy; and so on. If an electric circuit consists only of wires (no appliances) so that all the energy supplied by the current appears in the wire itself and in the region surrounding the wire, the energy that appears in the wire is simply the kinetic energy of the moving charges, and if there were no resistance in the wire to the motion of these charges (electrons), these charges would move forever once they were set moving (as in the case of a superconductor; a superconductor acquires superconductivity when its temperature is reduced to absolute zero or near it). But the resistance in the wire interferes with the free flow of the electrons, slowing them down. A current in an ordinary conductor therefore quickly dies down when the conductor is disconnected from the electrical source because the resistance (actually friction) within the wire robs the moving electrons of their kinetic energy and transforms this energy into heat. The energy supplied by the source of the electricity in the wire itself appears as heat; the wire gets warmer as the current flows. In a superconductor, a current, once established, continues flowing even after the initial source of the current is removed because it generates no heat.

The amount of heat that appears in the wire every second is IV less any other kind of energy that is associated with the electric current. As we have seen, one other kind of energy is associated with the current in the wire—the magnetic energy of the magnetic field that surrounds the wire. In general, this amount is a small quantity for an ordinary circuit so that we may say that IV is the total heat developed per second in the circuit if the current is I and V is the voltage. From Ohm's law we conclude that the heat generated per second $= IV = I^2R$ since $V = IR$. The heat is proportional to the square of the current and to the first power of the resistance. To obtain a large amount of heat per second for a given voltage we must use a circuit that permits a large current to flow, which we achieve by reducing the resistance of the circuit. The formula $IV = V^2/R$ shows us clearly that the smaller the resistance is for a given voltage, the greater is the heat that is developed.

With Ørsted's discovery that an electric current is accompanied by a magnetic field a number of scientists became convinced that it should be possible to obtain an electric current from a magnetic field.

Most prominent among these scientists were Joseph Henry in America and Michael Faraday in England. Both Henry and Faraday attacked the problem of electromagnetic induction (using a magnetic field to induce current in a conductor) in about the same way. But Faraday succeeded before Henry did so that Faraday is credited with this amazing discovery.

Faraday, who came from a poor working-class family and was almost entirely self-taught, achieved the distinction of becoming the greatest experimental physicist of all time. His researches into electricity and magnetism were so extensive that one is amazed to learn that they were all the fruits of a single man's labor. Faraday appears to have been an experimentalist from the very moment that he began to think about the physical world.

Faraday had taught himself reading and writing and some arithmetic by the time he was thirteen, and he had already begun to probe into all kinds of phenomena that attracted his attention. He later described himself in a revealing passage: "I was a lively and imaginative person and facts were important to me. . . . I would trust a fact but always cross-examined an assertion." He preferred scientific books to all others, and, being apprenticed to a bookseller at that time, he had many books to choose from. He was particularly influenced and impressed by Mrs. Marcet's *Conversations in Chemistry,* but he refused to accept the assertions in that book unless he could check them through some experiments of his own, however crude. His need for the experimental verification of an assertion was best expressed in his own words: "So when I questioned Mrs. Marcet's book by such little experiments as I could perform, I felt I had got hold of an anchor in chemical knowledge and clung fast to it."

Although Faraday's experimental work dealt with many different aspects of electricity and magnetism, he is best remembered for his discovery of electromagnetic induction and for his experiments related to this phenomenon. This discovery was of fundamental importance to the development of modern electromagnetic technology and gave impetus to Maxwell's theoretical investigations of the electromagnetic field. As soon as Faraday heard about Ørsted's discovery, he was convinced that one should be able to obtain an electric current from a magnetic field, but it took some time before he discovered just how a magnetic field induces a current. He started out on the wrong track because he first thought that a current would appear in a conductor simply by virtue of the presence of the conductor in a magnetic field. To test this idea he placed the north pole of a magnet close to a coil of wire connected to an ammeter. Faraday thought that he would find a current flowing through the wire as indicated by the swinging of the needle in the meter, but this did not occur. He tried all kinds of static arrangements of magnet and coil, but nothing happened.

Before we describe how Faraday finally discovered induction, let us see if we can understand why a static magnetic field does not cause a current to flow in the coil. We know that the wire of the coil is a conductor and that it contains countless free electric charges (electrons) ready to move at the slightest electrical prompting (any slight difference of potential—that is, an electric field). But stationary electric charges do not respond to any magnetic field. If there were free magnetic poles in the wire (which there are not) these poles would move in response to the force exerted on them by the magnetic lines of force from the magnet, and we would then have a magnetic current in the coil. But stationary electric charges respond only to electric fields which is why Faraday found no current in the coil as long as the magnet was kept fixed relative to the coil. At this very point in his work Faraday was but a hair's breadth away from his final discovery because all he had to do was move either the magnet or the coil for a current to appear. Whether he did this quite by accident, noting by chance that a momentary surge of current appeared in the coil while he was moving the magnet from one arrangement to another, or as the result of some kind of reasoning does not really matter. He finally discovered that relative motion of the magnet with respect to the coil is a necessary ingredient for the induction of a current in a coil. If we accept the idea (as propounded by Faraday) that electrical and magnetic phenomena are symmetrical, then we see at once why we must have motion of the magnet with respect to the coil if we want electric charges to move in the coil (if we want to obtain an electric field in the wire from the magnetic field). We recall that Ørsted obtained a magnetic field from moving electric charges (a current in a wire) and not from fixed charges. In other words, if we want electric charges to produce a magnetic field, we must set these charges moving because otherwise nothing happens to a magnet placed near charges, no matter how many charges we have. Since moving electric charges generate moving electric fields, we conclude that a moving electric field is accompanied by a magnetic field.

If we want electric charges to move in a coil, we must have an electric field, which we obtain by moving a magnetic field around the charges; Faraday did this when he moved the magnet with respect to the coil. The only important thing here is the relative motion of the magnetic field and the coil. It does not matter whether we hold the coil fixed and move the magnet or hold the magnet fixed and move the coil. The final effect (moving charges in the wire) is exactly the same in both cases: Magnetism plus motion gives electricity.

If the charges in the wire are set into motion by the moving magnetic field, it must be because such a magnetic field gives rise to a force on the electric charges in the wire. We can discuss this force from two different points of view. On the one hand, when a charge is moving through a magnetic field (or when the field is moving past the charge), the magnetic field exerts a force on the charge that depends on the strength of the field and on the velocity of the charge. On the other hand, when a magnetic field moves relative to an observer, the moving magnetic field is accompanied by an electric field (as far as this observer is concerned) and this electric field accelerates electric charges relative to which the electric field is moving. Just as a moving electric field gives rise to a magnetic field, a moving magnetic field gives rise to an electric field; this is the essence of Faraday's discovery.

Two questions arise in connection with electromagnetic induction. The first of these relates to the direction of the flow of current in the conductor when conductor and field are moved with respect to each other, and the second relates to the strength of the potential difference (of the electromotive force) in the conductor. If a wire of length L moves through a magnetic field of strength H, how does the strength of the induced current depend on the direction of the velocity $\mathbf{v}$ of the wire relative to the direction of the field $\mathbf{H}$ and on the magnitude of $\mathbf{v}$?

Faraday discovered that if the wire is held perpendicular to the magnetic field (horizontally if the magnetic lines of force are vertical so that the length of the wire makes an angle of 90° with these lines of force H), the strength of the current induced in the wire depends on the angle that the direction of the wire's velocity $\mathbf{v}$ makes with $\mathbf{H}$. If the wire moves parallel to **H** (in the present case, either up or down), no current is induced. For currents to be induced the wire must move in such a way as to cut the magnetic lines of force. The greater the number of lines of force cut by the conductor in a given time, the stronger is the induced current. This means that we obtain maximum induction of current if the wire is moved at right angles to its own length and at right angles to the magnetic field **H**. Faraday found that when this is done, the direction of the induced current in the wire, the direction of the magnetic field **H**, and the direction of the velocity **v** of the wire form a right-handed orthogonal coordinate system. In other words, if the direction from the end B to the end A of the wire is taken as the X direction of a coordinate system the induced current flows in this direction if **v** is in the Y direction and the magnetic field **H** is in the Z direction of the coordinate system. The actual flow of charge (the electrons) is from A to B along the wire.

If we consider this from the point of view of a free charge in the conductor, we can express this part of Faraday's discovery as follows: If a positive electric charge is moved along the Y direction in a fixed three-dimensional coordinate frame in which a static magnetic field points in the Z direction, the charge experiences a force along the X direction. This settles the question concerning the direction of the induced current.

To relate the magnitude of the induced current to the strength of the magnetic field **H** and to the magnitude of the velocity of the conductor, we use the following experimental result that can be deduced from Faraday's experiments or from direct measurements: If 1 e.s.u. of charge is moved through a unit magnetic field (a field of 1 gauss) at a speed of 1 cm/sec, it experiences a force of exactly $1/c$ dyne ($\frac{1}{3} \times 10^{-10}$ dyne), which is at right angles to both the magnetic field and the velocity of the charge. This force (known as the Lorentz force, in honor of its discoverer H. A. Lorentz) causes a current to flow in a wire that is moving in such a way as to cut the magnetic lines of force. Here c is exactly the speed of light. The appearance of c here is quite remarkable and indicated to Maxwell, to whom we owe this discovery, that there is a deep relationship between light and electricity and magnetism. If we now consider a charge moving with speed v at right angles to a magnetic field of strength H, the charge experiences the force $(e/c)vH$ at right angles to **v** and **H**. Thus, an observer moving with speed v at right angles to a static magnetic field **H** detects an electric field of magnitude $(v/c)H$ at right angles to the magnetic field. Only the relative velocity of observer and magnetic field is important here. It does not matter whether we say that the observer is moving and the magnetic field is fixed or vice versa. Thus, whether or not a magnetic field accompanies an electric field and whether or not an electric field accompanies a magnetic field is determined by the state of motion of the observer relative to these fields. To an observer moving with respect to a particular field, both electric and magnetic fields are present whereas to an observer fixed with respect to the particular field only that field is present. That the presence of a field depends on the state of motion of the observer is an indication that electromagnetic induction has something to do with special relativity. When Einstein proposed the special theory in his famous 1905 paper, he showed that electromagnetic induction and, indeed, Maxwell's electromagnetic theory of light can be deduced by applying the Einstein–Lorentz transformation to electric and magnetic fields. This argument favors Einstein's theory since Faraday's and Maxwell's theories are universally accepted.

Knowing that the induction of a current in a conductor cutting magnetic lines of force is due to the electromagnetic forces acting on the free charges in the conductor, we can calculate the relationship between the induced potential difference in the wire (electromotive force), the strength of the magnetic field, and the length and the velocity v of the wire in which the current is induced. We note that if the force exerted by a magnetic field on a moving charge is $e(v/c)H$, as stated above, then the force exerted by a magnetic field H on a wire of length L and of unit cross-sectional area (1 cm^2) carrying current I at right angles to the field is $(I/c)LH$, and the direction of this force is perpendicular to both the direction of the field **H** and the wire. Each charge moving in the wire, and thus contributing to the current, experiences a force $(ev/c)H$ at right angles to the wire. If there are n electric charges per unit volume (per cm^3) in the wire moving at speed v and thus giving rise to the current nev, the force acting on just these electrons is $(nev/c)H$. If we take the entire wire of length L into account, the number of electrons moving along the wire with speed v is just nL (since the volume of the wire is L and there are n moving charges per unit volume), and the force on these electrons, and hence on the entire wire, is $(n/c)evLH$. But nev is just the current I. Hence, the magnetic field exerts a force of magnitude $(I/c)LH$ on a conductor of length L carrying current I at right angles to the field. This is the basis of the electric motor.

We consider a wire of length L moving at speed v at right angles to the magnetic field H and let I be the current that is induced. Since a current represents electrical energy, and energy is conserved, this energy must come from the work done on the wire to keep it moving. This means that the current set up in the wire must move in such a direction in the wire as to produce a magnetic field surrounding the wire that opposes the force pushing the wire. This phenomenon is known as Lenz's law which, as we see, is no more than a statement of conservation of energy. The current moves through the wire in a direction such that the magnetic field produced by the current is stronger ahead of the wire than behind it. In other words, the magnetic lines of force pile up in front of the wire so that the motion of the wire is resisted. Hence, work must be done to keep the wire moving and the energy expended in doing this mechanical work appears as the energy of the electric current induced in the wire.

Suppose now that the force exerted by the magnetic field $\mathbf{H}$ on the wire in which the current is induced is $\mathbf{F}$. If v is the speed with which the wire is moved through the field, the work done in a time t (force times distance) is just Fvt. But $F = HLI$ (from above) if I is the strength of the induced current. This quantity must be the energy delivered by the induced current I in the time t. Applying the principle of conservation of energy, we have then (on equating the work done on the wire to the energy delivered by the current) $IHLvt = IVt$ or $V = HLv$. Thus, the induced potential difference between the two ends of the wire equals the magnetic field strength (number of magnetic lines per unit area) multiplied by the product of the length of the wire and its speed. Since this last product Lv is just the area described by the wire in a unit time, the quantity HLv is just the number of magnetic lines of force cut by the wire in a unit time. This phenomenon is Faraday's law of induction. Since the number of lines of force threading an area is called the magnetic flux, we can also state Faraday's law of induction as follows: The induced potential difference (electromotive force) is proportional to the time rate of change of the magnetic flux through a circuit.

In the above example the moving wire may be the movable part of a circuit immersed in the magnetic field. A good example is a wire in contact with and moving along two parallel metal rails that are connected at one end. As the wire moves away from the connected ends of the rails, the area defined by the circuit increases so that the magnetic flux increases. If the moving wire does not cut the lines of force exactly at right angles but at some angle ϕ, then the induced potential difference in volts is $V = HLv \sin \phi/10^8$. The factor 10^8 is introduced to change from standard units to volts.

Lenz's law, which can be deduced from the principle of conservation of energy, has some interesting consequences, which at first appear quite amazing, but which can be easily understood on the basis of the law. We recall that the law states that whenever a current is induced, the current must flow in such a direction as to set up magnetic fields that oppose the motion that gives rise to the induced current. An excellent example is the induction of eddy currents which dampen the very motion which causes these eddy currents. If a copper slab is set swinging like a pendulum between the north and south poles of an electromagnet when no current flows through the electromagnet (no magnetic field between the poles), the slab of copper swings freely. But as soon as the electromagnet is activated (by current through its coils), the swing of the copper slab is quickly dampened. With current in the electromagnet a strong magnetic field is set up between the north and south poles and currents (eddy currents) are induced in the slab because of its motion. These currents set up a magnetic field in the slab whose direction is such as to dampen the motion of the slab, and it quickly comes to rest. In fact, the region between the poles of the magnet behaves as though it were filled with a very viscous substance (like honey). This phenomenon is used in many electrical devices.

Until Faraday's discovery of electromagnetic induction, electricity was used only sparingly in industry although much was being done in electrochemistry. The reason is that electric power for heavy industry depends on large currents, which, in those days, could not be obtained in any practical way from voltaic cells. The number of such cells that were required to produce even a few kilowatts of power was so large that it was completely impractical to do much with voltaic-cell electricity. But Faraday's discovery changed all that by opening up a technology for generating large electric currents from magnetic fields, which in turn can be produced, in practically unlimited magnitudes, from the induced electric currents themselves.

To see how to obtain as large potential differences (and hence large currents) as may be desired by using Faraday's laws of induction, we consider a circuit in the form of a single rectangle of wire between the poles of a magnet. As the rectangle is rotated between the poles the sides of the rectangle cut the magnetic lines and current is induced. As the rectangle is rotated, its area is alternately parallel and perpendicular to the magnetic lines of force. When the area is perpendicular to the magnetic field, the sides cut no lines of force so no current is induced at that moment. But as the area is rotated from perpendicular to parallel positions, lines of force are cut and current is induced. We thus obtain an alternating current as the rectangle is rotated. If it is rotated sixty times per second, we have a 60-cycle AC current. Since the induced potential difference or electromotive force depends on the product of the strength of the magnetic field, the length of the wires, and the speed of these wires, we can obtain very large potential differences by taking many loops of wire (a coil) wound around iron cores, rotating very rapidly in strong magnetic fields, thus obtaining the modern generators.

Since a generator is a device that transforms mechanical energy or work (the energy that is used to turn the loop or the coil) into electrical energy, it reverses what an electrical motor does. Hence, an electrical generator and an electrical motor are two different facets of the same device. In fact, if we send an alternating current through the rectangle, we make the loop rotate and we retrieve the mechanical energy we put into it. Thus, a coil of wire in a magnetic field can be used either as a generator or as a motor.

We have discussed various facets of electromagnetic fields associated with currents in conductors and we saw that when such fields are changing in the neighborhood of an observer these fields are accompanied by changing magnetic fields as seen by this observer. An observer in whose neighborhood an electric field is changing perceives a changing magnetic field surrounding a changing electric field, and if a magnetic field in his neighborhood is changing, he perceives a changing electric field. The question that should now be considered is how far away can an observer be from a changing field (magnetic or electric) and still detect the new fields that such changes generate? To be specific we suppose that an observer is standing near the north pole of a magnet and that the magnet is moved suddenly. The observer notes that while this motion is occurring electric lines of force are created that are at right angles to the magnetic lines (they form concentric circles around the magnetic lines). But while the magnetic field is changing, the created electric field is also changing, which means that this induced field itself is surrounded by magnetic lines of force at right angles to it, and so on. In other words, when a magnetic field (or an electric field) changes in the neighborhood of an observer a whole chain of intertwining electric and magnetic fields is created and this chain of fields extends out in all directions from the observer. We now ask what a distant observer detects when the electromagnetic field in the neighborhood of a nearby observer is changing. To be specific we suppose that an electric charge e is at rest with respect to two observers, one close to the charge and one far away. Each observer detects the electric field of this charge, but the distant observer finds a much weaker field (owing to the inverse square law) than the nearby observer. If the charge is now set moving at constant speed, relative to both observers, each of them detects a moving electric field accompanied by a magnetic field, which remains steady as long as the velocity of the moving charge is constant. The distant observer detects a much weaker magnetic field than the nearby observer does because the magnetic field decreases with increasing distance from the moving charge.

Suppose now that the velocity of the moving charge is changed. What does each observer detect? From Faraday's law of induction it follows that each observer will detect changes in the electric and magnetic fields in his neighborhood, but not at the same moment. The nearby observer will detect these changes (in the electromagnetic field) sooner than the distant observer does because such changes cannot be transmitted instantaneously. The reason is that an electromagnetic field carries energy, and any change in the field in a given region means that energy is either entering that region or leaving it. But according to the theory of relativity, energy, being equivalent to mass, cannot travel faster than light; indeed, in a vacuum, as was shown by Maxwell, electromagnetic energy travels exactly at the speed of light. Thus, an observer at any distance from a distribution of charges or magnets sees an electromagnetic pulse coming toward him at the speed of light when these electric charges and magnets are accelerated in any manner. In particular, if electric charges are set oscillating with a certain frequency, electromagnetic pulses of that frequency are propagated away from these charges at the speed of light in a vacuum. It can be shown that if an electric charge e is accelerated, the charge emits electromagnetic energy at a rate equal to $\frac{2}{3}(e^2/c^3)(a^2)$ where a is the acceleration of the charge and c is the speed of light.

To illustrate the emission of electromagnetic radiation by oscillating charges we consider two parallel plates (conductors) connected by a coiled wire conductor. Two such parallel plates, without the connecting coil, are called a condenser or a capacitor; it is a device that is used to store electric charge. If a given quantity of positive charge $+q$ is placed on one plate of the condenser and an equal quantity of negative charge $-q$ is placed on the other plate, we say that the condenser is charged, and the charge on the condenser is q. When a given quantity of charge is placed on a condenser, it produces an electric field between the plates and a difference of potential V between them, which is determined by the amount of charge q placed on the plates and a certain geometrical quantity called the capacity C of the condenser. The capacity C of a condenser depends only on its geometry (it is proportional to the area of the plates divided by their separation). For a given charge q the potential difference V between the two plates of a condenser varies inversely as the capacity of the condenser and for a condenser of a given capacity C, the potential difference depends directly on (varies directly with) the charge q so that $V = q/C$.

When we charge a condenser we must do work against the repulsive force of the like charges already on the plate. We have to push each new charge that we put on the plate of the condenser against the force of repulsion exerted by the like charges already on the plate. A charged condenser thus possesses an amount of energy which is proportional to the product of the charge on the condenser and the potential difference between its plates. In fact, we have E = energy = $\frac{1}{2}qV = \frac{1}{2}CV^2$. Since this quantity is stored electrical energy, we may compare it to the mechanical potential energy stored in the bob of a pendulum that is at the maximum height of its swing and call the energy in a condenser potential energy.

If we now connect the two plates by a conducting coil of wire, charges flow from the negative to the positive condenser plate, creating a current in the coil and establishing a magnetic field through the coil. At the same time the electric field in the region between the two condenser plates is decaying (decreasing). When all the negative charge has flowed to the positive plate, canceling the charges there, the two plates are electrically neutral so that there is no potential difference between the two plates nor an electric field between them. Thus, the condenser has lost its energy, and we must now ask where it has gone. If the connecting coil of wire has very little resistance, the energy of the condenser cannot have been dissipated in the form of heat within the circuit because the heat developed there depends on the product of the resistance and the square of the current. If the resistance is very small, the heat loss is negligible. Since the energy has not been dissipated as heat, it must still be present in the circuit in some form, and one does, indeed, find it in the form of the magnetic field surrounding the coil. Thus, the potential energy of the electric field between the condenser plates has been transformed into the magnetic energy of the magnetic field surrounding the coil.

Since there is no longer any charge on the condenser plates, one might suppose that no electric current is present in the coil but this is not quite true because the magnetic field surrounding the coil has reached its maximum value and begins to decay. The coil is thus surrounded by a changing magnetic field which induces a current in the coil in the same direction as the initial current. In other words, the energy in the magnetic field collapses back onto the coil, manifesting itself as an additional electric current moving in the same direction as the original current. This phenomenon is called self-induction because the magnetic field, whose changing magnitude causes this additional flow of current, arises from the original current flowing in the same coil. Associated with this self-induction is a quantity L called the coefficient of self-induction. It is essentially the total magnetic flux (total number of magnetic lines of force) through the coil when a unit current is flowing in the coil. Another definition of the self-induction L of a coil is the magnitude of the potential difference induced in the coil when the rate of change of the current in the coil is one unit of current per second.

Since the magnetic field surrounding the coil collapses back onto the coil, it causes the current to continue flowing in the same direction as it began flowing until the plates of the condenser are charged again but with opposite charges on them. The plate that was initially positively charged is now negatively charged and vice versa. Now the whole thing reverses itself and the current flows in the opposite direction and so on. In other words, we have here an oscillating electromagnetic system which, in many respects, is similar to a pendulum. The energy in a pendulum oscillates between a potential form (the electrostatic energy of the condenser) and a kinetic form (the energy in the magnetic field). A pendulum that is swinging freely ultimately comes to rest owing to friction, and this is also true of our oscillating electromagnetic system. The friction R in the circuit generates heat and the electrical energy is finally dissipated in accordance with Ohm's law.

An important difference between the oscillating pendulum and the oscillating electrical circuit which was first discovered and developed into a magnificent theory by James Clerk Maxwell is that the pendulum can come to rest only if there is mechanical friction present in the system, but this fact is not true of the electrical system. Maxwell pointed out that even if the electrical system were completely free of friction (superconductors), it would lose energy and finally cease oscillating. He noted that as the current oscillates in the wires, an equivalent oscillation of the electric field in the region between the condenser plates occurs. The field then waxes and wanes, reversing its direction periodically as the current in the coil changes its direction. He also noted that when the electric field between the plates is a maximum, the magnetic field surrounding the coil is a minimum and vice versa.

As Maxwell first pointed out, the alternating electric field between the plates of the condenser behaves just as though electric charges were moving back and forth; in other words, this oscillating electric field behaves like an oscillating current. Maxwell called it a "displacement current" and reasoned that this electric field itself, while changing, is surrounded by concentric lines of force (circles) of an induced magnetic field. In turn, this magnetic field is surrounded or linked by the concentric lines of force of still another electric field and so on out into space; thus, a whole chain of interlinking electric and magnetic fields is generated in the region between the plates of the condenser. Since these fields, which spread out in all directions, have energy, this energy must come from the initial energy of the condenser, which means that the oscillating current of the condenser loses energy steadily and finally comes to rest; thus, the condenser loses all its energy by radiating it away.

The electromagnetic vibrations produced by the oscillator described above move off into space at a finite speed, and if L is the self-inductance of the coil and C is the capacity of the condenser, the number of electrical oscillations that occur per second equals $1/2\pi\sqrt{LC}$. This quantity is called the frequency of the electric oscillator and it is also the frequency of the electromagnetic vibrations that leave the electrical system and move out into space.

Maxwell incorporated these various phenomena into a single all-encompassing mathematical theory which shows that the electromagnetic pulses that leave the oscillating electrical system move out in the form of a wave, which consists of an oscillating electric field and an oscillating magnetic field, which are oscillating at right angles to each other and at right angles to the direction of propagation of the wave. If an electromagnetic wave moves along the Y axis of a coordinate system, it consists of an electric field oscillating in the YZ plane of the coordinate system and a magnetic field oscillating (with the same frequency) in the YX plane.

Maxwell showed mathematically that two such oscillating fields in a vacuum must move at a speed c at right angles to the two oscillating fields. Since Maxwell's theory tells us that these oscillations travel at the speed c (the speed of light), Maxwell made the bold assumption that light itself is an electromagnetic wave and that the only difference between light and the wave emitted by an electric oscillator consisting of a condenser and a coil of wire is in their wavelengths. We saw that the frequency of the electromagnetic pulses emitted by the condenser–coil combination is $1/2\pi\sqrt{LC}$. Hence, the wavelength of these pulses is just $2\pi\sqrt{LC}$ times the speed of light; wavelength $= (2\pi\sqrt{LC})\ 3 \times 10^{10}$. If we want the wavelengths of these pulses to be as small as those of visible light, we must make L and C very small. In other words, if the condenser and coil were tiny enough, the electromagnetic waves emitted would have the wavelengths of visible light; atoms are just of the right size to produce such wavelengths when they vibrate.

With Maxwell's discovery of the electromagnetic properties of light a whole new realm of the physical world was opened up to scientific investigation. It was now possible to understand why and in what manner light interacts with matter, and this understanding could in turn be used to study the structure of matter. Light and matter interact (emission or absorption of radiation by matter) via electromagnetic fields. Light is emitted by matter because charged particles (of which all matter is composed) are set oscillating very rapidly, and light is absorbed by matter because the electromagnetic field carried by a beam of light moving through the matter sets the charged particles in the matter oscillating. Such stimulated oscillations remove energy from the beam.

Although Maxwell's electromagnetic theory of light was extremely appealing and beautifully structured, it was accepted only slowly because of the lack of experimental evidence. This experimental evidence was finally obtained, however, through the work of Heinrich Hertz in the last decade of the 19th century. Hertz was the first of a remarkable group of physicists who laid the foundations of the modern experimental era in physics. In 1887 he not only generated Maxwellian electromagnetic waves in a simple manner but also detected and studied them. His generator or transmitter was a condenser consisting of two small metal spheres which he could discharge across a small spark gap by activating a high-voltage induction coil whose ends were connected to the two spheres. When no spark is produced in the gap, the spheres remain charged at a definite potential and we have a static condenser. When the induction coil is operating, it causes a spark to jump across the gap, which ionizes the air gap making it a conductor. The spheres therefore discharge inducing electromagnetic oscillations that spread out from the gap.

To receive or detect these oscillations, Hertz used a small loop of wire with the ends not touching. By changing the distance between the spheres, Hertz tuned the generator (or transmitter) until its frequency exactly equaled the natural frequency of the receiving loop. When he did this, he found that a spark jumped across the ends of his receiving loop whenever his transmitting oscillator was operating.

With these simple devices, which were the precursors of our modern radio transmitters and radio receivers, Hertz demonstrated that the electromagnetic waves he generated were qualitatively the same as light. One of his most serious concerns was to show that his electromagnetic waves travel through a vacuum at the same speed as light, and when he did this, he was completely convinced that Maxwell's electromagnetic waves and light are identical. When he reproduced such phenomena as reflection, refraction, diffraction, and interference, his work was taken as the definitive experimental proof of Maxwell's theory.

Index